# Experimental Methods in the Physical Sciences

VOLUME 38

ADVANCES IN SURFACE SCIENCE

# EXPERIMENTAL METHODS IN THE PHYSICAL SCIENCES

Robert Celotta and Thomas Lucatorto, *Editors in Chief*

*Founding Editors*

L. MARTON
C. MARTON

Volume 38

# Advances in Surface Science

*Edited by*

Hari Singh Nalwa, M.Sc., Ph.D.
*Stanford Scientific Corporation*
*Los Angeles, California, USA*

*Formerly at*
*Hitachi Research Laboratory*
*Hitachi Ltd., Ibaraki, Japan*

## ACADEMIC PRESS

A Division of Harcourt, Inc.

San Diego   San Francisco   New York   Boston   London   Sydney   Tokyo

This book is printed on acid-free paper. ⊚

The articles in this book are selected from the Academic Press multi-volume work titled *Handbook of Surfaces and Interfaces*, edited by Hari S. Nalwa, and are uniquely arranged to focus on current advances in surface science.

Academic Press
A division of Harcourt, Inc.
525 B Street, Suite 1900, San Diego, California 92101-4495, USA
http://www.academicpress.com

Academic Press
Harcourt Place, 32 Jamestown Road, London NW1 7BY, UK
http://www.academicpress.com

International Standard Book Number: 0-12-475985-8

International Standard Serial Number: 1079-4042/01

PRINTED IN THE UNITED STATES OF AMERICA
01 02 03 04 05 EB 9 8 7 6 5 4 3 2 1

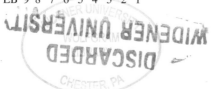

## CONTENTS

3. Secondary Electron Fine Structure—a Method of Local Atomic Structure Characterization

by YU. V. RUTS, D. E. GUY, D. V. SURNIN, AND V. I. GREBENNIKOV

4. Photonic and Electronic Spectroscopies for the Characterization of Organic Surfaces and Organic Molecules Adsorbed on Surfaces

by ANA MARIA BOTELHO DO REGO AND LUIS FILIPE VIEIRA FERREIRA

## 5. High-Pressure Surface Science
by VLADISLAV DOMNICH AND YURY GOGOTSI

# CONTRIBUTORS

Numbers in parentheses indicate the pages on which authors' contributions begin.

TERY L. BARR (111), *Materials Engineering and Laboratory for Surface Studies (LSS) University of Wisconsin, Milwaukee, Wisconsin, USA*

ANA MARIA BOTELHO DO REGO (269), *Centro de Química-Física Molecular, Complexo Interdisciplinar, Instituto Superior Técnico, 1049-001 Lisboa, Portugal*

VLADISLAV DOMNICH (355), *Department of Mechanical Engineering, University of Illinois at Chicago, Chicago, Illinois, USA*

YURY GOGOTSI (355), *Department of Materials Engineering, Drexel University, Philadelphia, Pennsylvania, USA*

V. I. GREBENNIKOV (191), *Institute of Metal Physics, Ural Branch, Russian Academy of Sciences, 620219 Ekaterinburg, Russia*

D. E. GUY (191), *Physical Technical Institute, Ural Branch, Russian Academy of Sciences, 426000 Izhevsk, Russia*

W. HEILAND (1), *University of Osnabrück, 49069 Osnabrück, Germany*

YU. V. RUTS (191), *Physical Technical Institute, Ural Branch, Russian Academy of Sciences, 426000 Izhevsk, Russia*

M. SCHLEBERGER (1), *University of Osnabrück, 49069 Osnabrück, Germany*

SUDIPTA SEAL (111), *Advanced Materials Processing and Analysis Center (AMPAC) & Mechanical, Materials and Aerospace Engineering (MMAE) University of Central Florida, Orlando, Florida, USA*

S. SPELLER (1), *University of Osnabrück, 49069 Osnabrück, Germany*

D. V. SURNIN (191), *Physical Technical Institute, Ural Branch, Russian Academy of Sciences, 426000 Izhevsk, Russia*

LUIS FILIPE VIEIRA FERREIRA (269), *Centro de Química-Física Molecular, Complexo Interdisciplinar, Instituto Superior Técnico, 1049-001 Lisboa, Portugal*

# VOLUMES IN SERIES
# EXPERIMENTAL METHODS IN THE PHYSICAL SCIENCES

(formerly Methods of Experimental Physics)

*Editors-in-Chief*
*Robert Celotta and Thomas Lucatorto*

# PREFACE

Although the first surface science experiments were performed early in the 20th century, it was midcentury before this area of research experienced explosive growth. The confluence of newly available technologies (e.g., ultra-high-vacuum methods) and a wide variety of important scientific and technological challenges (e.g., a better understanding of catalysis, the fabrication of new materials, and the control of corrosion) led to a rapidly expanding research effort. The expansion took place both in depth, as new experimental and theoretical methods evolved, and in breadth, as multidisciplinary teams of physicists, chemists, and materials scientists coalesced.

In 1985, Volume 22 of this treatise, *Solid State Physics: Surfaces,* edited by M. G. Lagally and R. L. Park, appeared. It covered most of the important tools required by the surface scientists of that time. In 1993, when it became clear what a profound impact the development of scanned probe microscopy would have, the first surface science volume was supplemented by Volume 27, *Scanning Tunneling Microscopy,* by J. A. Stroscio and W. Kaiser.

Here, in Volume 38, *Advances in Surface Science,* we again supplement our treatment of surface science. This volume broadly reviews the current methods of surface characterization and the use of photoelectron spectroscopy. It adds secondary electron fine structure to the arsenal of experimental tools available. Finally, it extends our experimental methods to the domain of organic materials and molecules and to high-pressure systems.

Will surface science soon reach a point of maturity where the next update will report only incremental advances? Surely not. The new area of nanoscale science, generally referred to as nanotechnology, is exploding on the scientific landscape with the promise of enormous advances in both technology and understanding. Many of its techniques will be unique. But many of the ways we will come to understand nanostructures will depend on meeting the challenge of applying currently mature surface science measurement methods to very small structures. Nanoscale analysis and characterization will depend greatly on our ability to creatively extend our current capabilities.

Surface science is, and will surely remain, an exciting and important area of research. We hope that as the great promise held by this discipline is fulfilled by the research community, the volumes of this treatise we have devoted to the experimental methods of surface science will have played a valuable supporting role.

ROBERT CELOTTA AND THOMAS LUCATORTO
*Gaithersburg, Maryland*

# 1. SURFACE CHARACTERIZATION: COMPOSITION, STRUCTURE AND TOPOGRAPHY

S. Speller, W. Heiland, M. Schleberger

University of Osnabrück, 49069 Osnabrück, Germany

## 1 INTRODUCTION

A major task in the field of surface physics and chemistry and of the field of interfaces of materials is the determination of the chemical composition. In comparison to the determination of the composition of bulk materials, only small amounts of matter are available for analysis. A surface layer of a solid contains on the order of $10^{15}$ atoms/cm$^2$. The usual probes, for example, ion beams, electron beams, X-ray beams, have diameters that are small when compared to 1 cm$^2$. Hence the actual number of atoms "seen" by the probe is small. Another fact encountered is the instability of surfaces and interfaces. The surface is a major defect of a solid, and its composition depends to a large extent on time, the atmosphere it is exposed to, and temperature. For example, bulk impurities tend to segregate on the surface, gases present above the surface tend to adsorb and, of course, these effects depend on time and temperature. Furthermore, the effects mentioned will depend on the surface structure and will also alter it. Trivial to say, an oxide covered surface has a completely different structure than the underlying material. Due to the two-dimensional nature of a surface or of an interface, the probing of its properties is bound to damage the analysis subject. For the same reason as for the two-dimensional character, the probing depth of any analytical technique is a major concern in this field. Hence the different techniques are to a different extent affected by the intricate coupling of composition and structure of surfaces and interfaces. Therefore, in many instances a combination of different techniques is needed to obtain sufficient information. Today's "ultimate" methods for surface topology determinations, scanning tunneling microscopy (STM, see Part III) and atomic force microscopy (AFM, see Part IV) are essentially insensitive to surface chemical composition. In comparison, the well-established secondary ion mass spectrometry provides a very high chemical sensitivity but no information about surface topology or structure.

Further aspects of surface analysis are the accuracy of determining composition or structural parameters. For practical purposes the possibility of combining different tools is very important. As an example, the STM is difficult to combine *in situ* with other techniques. Most surface analytical tools require ultrahigh-vacuum (UHV) for their operation, hence solid–liquid interfaces are hardly accessible. Another practical aspect concerns the analysis of surface processes, for

1

EXPERIMENTAL METHODS IN THE PHYSICAL SCIENCES
Vol. 38
ISBN 0-12-475985-8

example, thin-film growth. For that purpose "remote" probes are better suited in comparison to probes such as the STM or AFM. Therefore, for successful application of surface analytical tools, or for the financial investments in such systems, it is necessary to analyze which problems may be encountered and how to solve them.

Table I [1] lists surface analysis methods by the main physical phenomena used, that is, elastic scattering, ionization, bond breaking, diffraction, and charge exchange. As can be seen, the elastic scattering of electrons (EELS, HREELS) yields quite different information than ion scattering methods (LEIS 2.2, MEIS or RBS 2.1), that is, electron methods are sensitive to surface vibrations of adsorbate molecules and the surface phonons proper, whereas ion methods probe mostly surface chemical composition. However, a listing by the effects studied will cause a rather confusing table because HREELS provides quite good insight into the molecular structure of an adsorbed molecule, and MEIS and RBS combined with channeling will yield information about composition and the structure of the near surface region of a solid. In other words, there are only a few techniques that would show up in such a listing under one heading only, for example, STM under "topography." However, here it can be argued that by means of scanning tunneling spectroscopy (STS), that is, measuring the derivative of tunneling current to the bias voltage, dI/dU, the local electronic structure is also probed, thereby providing chemical composition information of great detail.

Still another point of view would be to consider the "imaging" qualities of the different techniques. Many techniques using electrons as a probing beam can be turned into imaging techniques by scanning the primary beam,

TABLE I.    Surface Analysis Methods for UHV Thin Films

| Physical phenomenon | Technique | Acronyms | Probe | Det particle |
|---|---|---|---|---|
| Diffraction | Low energy electron diffraction | LEED | Electron | Electron |
| Electron tunneling | Scanning tunneling microscopy | STM | Tip | Electron |
| Mech. Forces | | AFM | Tip | |
| Elastic scattering | Ion scattering spectroscopy | ISS, MEIS, RBS | Ion | Ion |
| | Electron energy loss spectroscopy | EELS, HREELS | Electron | Electron |
| | Scanning transmission electron microscopy | SEM TEM | Electron | Electron |
| Sputtering | Secondary ion mass spectrometry | SIMS, SNMS | Ion | Ion |
| Ionization | Photoelectron spectroscopy | XPS, UPS | Photon | Electron |

for example scanning AES (SAES), the resolution limited by the diameter of the beam. Similarly, SIMS or RBS can be extended to imaging. These techniques will not be discussed here in detail; instead, the emphasis will be on those techniques with inherent atomic resolution. However, the "depth-analysis" capabilities and their importance for interface analysis will be taken into consideration. A tentative comparison of the different techniques will be given.

## 2 ION SCATTERING METHODS

Ion scattering methods are distinguished by the energy ranges of the primary beams used. At low energies from $\approx300$ eV to 5 keV the method is named LEIS (low energy ion spectrometry) or ISS (ion scattering spectrometry), at about 100-keV beam energy MEIS (medium EIS) and at around 1 MeV the term RBS (Rutherford backscattering) is the common acronym. The physical principle of all techniques is the same—the primary ions are "elastically" scattered from the target atoms. The energy loss of the scattered ions depends on the mass of the target atoms and the scattering angle. In an experiment that uses angular resolved ion energy spectra, chemical surface analysis is based on the conservation of energy and momentum in binary collisions:

$$E_1/E_0 = \left\{\cos\theta_1 \pm (A^2 - \sin^2\theta)^{1/2}\right\}^2/(1 + A)^2 \tag{1}$$

where $E_1$ and $E_0$ are the energy of the scattered ion and the primary ion respectively, $\theta_1$ is the laboratory scattering angle of the scattered ion and $A = M_2/M_1$ is the mass ratio of the target atom and the projectile. For $A > 1$ only the plus sign is valid while for $1 > A > \sin\theta_1$ both plus and minus signs are valid. For scattering angles of $\theta = 90°$ and $180°$ the Eq. (1) for single binary collisions becomes particularly simple:

$$E_1/E_0 = (A - 1)/(A + 1), \quad \text{for} \quad \theta = 90° \tag{1a}$$
$$E_1/E_0 = (A - 1)^2/(A + 1)^2, \quad \text{for} \quad \theta = 180° \tag{1b}$$

For the recoiling target atom final energy $E_2$ is found from the binary collision calculation:

$$E_2/E_0 = (4A \cos^2\theta_2)/(1 + A)^2, \quad \text{for} \quad \theta_2 \leq 90° \tag{2}$$

Equation (2) is the basis for the elastic recoil detection analysis ( ERDA ) methods using heavy ions in the MeV region.

The mass resolution is obtained by differentiation of Eq. (1), which yields a dependence on the mass ratio and the scattering angle [2]. The mass resolution is better at larger scattering angles. The dependence on the mass ratio is obvious from the formula for $\Theta_1 = 90°$:

$$M_2/\Delta M_2 = (E_1/\Delta E_1)(2A/A^2 - 1) \tag{3}$$

This shows that the mass resolution decreases with increasing target mass for a given projectile mass. For light target masses H and He can be used and for heavier target masses heavier projectiles, Ne and Ar, become necessary. Consequently, heavy atoms on light substrates, for example, heavier metals on Si, are better resolved than heavy metals on heavy metals.

The classical scattering behavior of particles in the energy range of 100 eV up to a few MeV produces a phenomenon known as "shadow cone" or Rutherford's classical shadow [3]. The particles scattered from a single atom are deflected respective to their impact parameter by the repulsive potential (Fig. 1). The trajectories form a rotational paraboloid behind this atom, which is itself void of trajectories. Atoms lining up behind the first atom will not be hit by any ions. This shadowing effect can be used for structural analyses. As is obvious from Figure 1, the ion flux peaks at the edge of the shadow cone. For a Coulomb interaction potential the radius of the shadow cone and the ion intensity distribution can be estimated analytically [3]. For small scattering angles the momentum of the scattered ion but not the energy is altered. Under these conditions, the so-called momentum approximation of the scattering angle $\theta$ is inversely proportional to impact parameter $s$:

$$\theta = Z_1 Z_2 e^2 / E_0 s \tag{4}$$

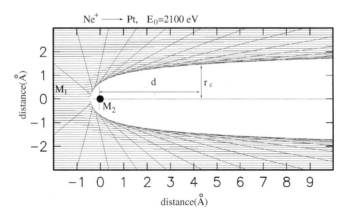

FIG. 1. Shadow cone for the scattering of Ne, mass $M_1$, ions of 1500 eV off a Cu atom, mass $M_2$ [8]. At a distance $d$ the shadow cone radius is $r_c$ (Eq. (6)). Reprinted with permission from E. Taglauer and John Wiley & Sons Ltd., © 1997.

If a second atom is located a distance $d$ from the first atom the distance between that atom and the trajectory with the scattering angle $\theta$ is given by

$$r_s = s + \theta d \tag{5}$$

For a Coulomb potential the minimum of $r_s(s)$ is then the edge of the shadow cone

$$r_c = 2\left(Z_1 Z_2 e^2 d / E_0\right)^{1/2} \tag{6}$$

The ion flux distribution as a function of $r_s$ can be estimated from

$$f(r_s)2\pi r_s \, dr_s = 2\pi f(s)s \, ds \tag{7}$$

This function peaks at the edge of the shadow cone. In the case of a chain of atoms on a surface, the position of the second atom is changed by the thermal vibrations, which cause a smoothing of the sharp edge of the intensity distribution and handle the mathematical singularity at the edge. At grazing angles of incidence of an ion beam on a chain or on a plane of atoms the intensity distributions of the individual shadow cones overlap and channeling is observed. The critical angle for channeling is approximated by Lindhard [3]:

$$\psi = e\sqrt{2Z_1 Z_2} / E d_{[hkl]} \tag{8}$$

where $hkl$ are the Miller indices of the atomic row. The main ingredients for the channeling model are the guided motion of the ions and the conservation of the perpendicular momentum. From the latter follows a simple way to estimate the critical angle, that is, if the "perpendicular" energy $E_\perp = E_0 \sin^2 \psi$ is smaller than $\approx 10$ to 20 eV channeling will be observed. At larger perpendicular energies the ions break through the potential barrier of the atomic chain or plane [4]. At low energies this effect is used for surface structure studies. At high energies channeling is very useful for determining the lattice locations of defects, lattice distortions at solid–solid interfaces, and the formation of compounds at both surfaces and interfaces.

The intensity of the backscattered ions in a doubly differential scattering experiment is given by the differential scattering cross section $ds$, which can be estimated for a given interaction potential. At high energies the pure Coulomb potential can be used to yield the well-known Rutherford scattering cross section [5], given here in the center of a mass system:

$$d\sigma/d\Omega = \left(Z_1 Z_2 e^2 / \left[4E_r \sin^2 \Theta/2\right]\right)^2 \tag{9}$$

where $E_r$ is the center of mass energy and $\Theta$ is the center of mass scattering angle. Here, $\Theta$ is related to the laboratory scattering angle $\theta_1$ by:

$$\tan \theta_1 = \sin \Theta / \left[(M_1/M_2) + \cos \Theta\right] \tag{10}$$

With decreasing primary energy the screening of the nuclear charge by the electrons becomes important and hence screened Coulomb potentials are used with a screening function $\Phi(r/a)$ where $r$ is the internuclear distance and $a$ is the screening parameter. Widely used screening functions are given by Moliere [6] as an approximation to the Thomas-Fermi atom model (TFM) and by Ziegler et al. (ZBL) [7] (which consists of four exponentials with $r/a = x$):

$$
\begin{aligned}
\Phi(x) = 0.0281 \exp(-0.2016x) &+ 0.2802 \exp(-04029x) \\
&+ 0.5099 \exp(-0.9423x) \\
&+ 0.1818 \exp(-3.2x)
\end{aligned} \tag{11}
$$

The screening length for the ZBL case is:

$$
a_{ZBL} = 0.4685 / \left( Z_1^{0.23} + Z_2^{0.23} \right) [A] \tag{12}
$$

Figure 2 shows a comparison of the Coulomb potential with the screened TFM potential for He scattering Ni [8]. Note the differences in the impact parameter range encountered in LEIS (ISS) and RBS, respectively. A consequence of the

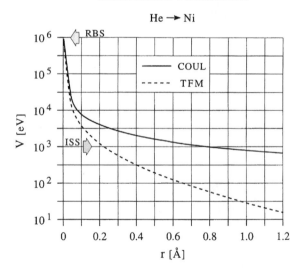

FIG. 2. Comparison of the Coulomb potential and the screened Thomas-Fermi-Moliere (TFM) potential for the scattering of He ions off Ni. The typical impact parameter range of Rutherford back scattering (RBS) and low energy ion scattering (ISS or LEIS) are indicated by arrows [8]. Reprinted with permission from E. Taglauer and John Wiley & Sons Ltd., © 1997.

screening is a smoother dependence of the scattering cross section on $Z_2$ in the LEIS regime compared to the $Z_2^2$ dependence of the Rutherford cross section.

For trajectory calculations the interaction potential is also needed. It enters into the scattering integral in the center of mass system via:

$$\Theta = \pi - 2 \tag{13}$$

where $r_{min}$ is the distance of closest approach during the collision, $s$ is the impact parameter, and $E_\perp$ is the center of mass energy [9]. The classical scattering integral is valid for conservative central forces in a system in which the classical laws of energy and momentum conservation apply. In general, Eq. (13) can only be solved numerically.

Due to to the densities of solids single scattering conditions are never realized but multiple scattering prevails. Numerical codes are necessary to calculate trajectories in detail. A code available for the interaction of ions with single crystals is the MARLOWE program [10, 11]. It is based on the binary collision approximation. The code allows the use of different potentials and inelastic energy loss functions. Thermal vibrations are included. Backscattering spectra, energy and angle resolved, can be calculated as well as average quantities, for example, particle and energy reflection coefficients, and sputter yields. For noncrystalline targets the code used most is TRIM [12, 13]. For a detailed discussion and comparison of different codes see Reference [14] for example.

Experimental details may be found in the literature [2, 15, 16]. Common features are ion source, mass separation, acceleration, beam collimator, goniometer for the target, energy analysis, and particle counting. Acceleration makes the most obvious difference between the different energy regimes. The physical size and the investment increase with increasing beam energy. The mass separation is done at low beam energies, right after the ion source, so the size and price for the separating magnets are the same. At low energies $E \times B$ filters (Wien filters) also can be used for mass separation and care must be taken to avoid the direct line of sight between ion source and target. Another difference is the type of energy analysis. In the RBS regime the solid state detectors most often used have an energy resolution of typically 15 keV, that is, of the order of 1% for 1 MeV particles. In the MEIS and LEIS regime electrostatic analyzers are favored that have energy resolution of the order of 0.5 to 1.0%. Time of flight methods (TOF) are also applied that afford lower primary beam current densities and hence less target damage. In all cases standard particle-counting techniques are in use.

## 2.1 High Energy Ion Scattering

In the RBS regime the binary collision model is extended to take into account that those primary ions that penetrate to greater depth are backscattered from atoms inside the bulk of the solid [15, 16]. On the path to this target atom and back to the surface the primary ion loses energy due to excitations of electron gas and ionization by target atoms. These losses are called electronic or inelastic losses. The losses are widely known and can be found in many tables [17]. As a function of depth $t$ and using the "kinematic factor" $K = E_1/E_0$ (Eq. (1)) the energy of the backscattered ion is

$$E_1(t) = K\{E_0 - t(dE/dx)_{in}\} - t(dE/dx)_{out}/|\cos\theta_1| \qquad (14)$$

This expression can be simplified for thin layers or short path length assuming $dE/dx = $ const. The depth resolution in the "surface approximation" is then:

$$\Delta t = \Delta E_1/\{K(dE/dx)|_{E0} + (dE/dx)|_{E1}/|\cos\theta_1|\} \qquad (15)$$

Computer programs are available for estimating RBS spectra and evaluating depth profiles [18]. A typical RBS spectrum is shown in Figure 3 using 1.9 MeV $He^+$ [19]. It is a PtSi layer system on alumina, $Al_2O_3$, with a thin W layer deposited first, followed by a Pt layer, which is covered with Si. During annealing of the sample the Pt and Si react and form a silicide. The low energy part of the spectra is in the intensity steps and is due to the Al and the O at the $Al_2O_3$-W in-

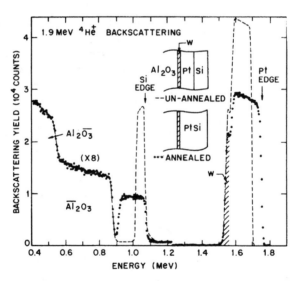

FIG. 3. The RBS spectrum of a Pt/Si layer system on a $Al_2O_3$ substrate. The dashed line is the spectra as deposited; the dotted line is after annealing and the formation of a PtSi compound. The W is used as a thin layer between the $Al_2O_3$ and the original Pt layer. The $Al_2O_3$ backing is seen as a step of Al (lower $Z$) and of O (lower $Z$). Note that the Si and Pt areas, that is, the number density of atoms/cm$^2$, is not changed in the annealing process [19]. Reprinted with permission from Z. L. Liau et al., *J. Appl. Phys.* 49, 5295 (1978), © 1978 the American Institute of Physics.

terface. The position of these steps agrees with binary collision Eq. (1) for Al and O shifted by the additional energy loss occurring when the $He^+$ ions pass through the overlayers. The extended continuous spectrum towards lower energies is due to the backscattered particles. The shape of this part of the spectra would be the same for a pure alumina sample. After annealing the thin layers of W, Pt, Si, and PtSi give rise to narrow features in the spectra. The energetic position is given by Eq. (1), modified where applied, by the additional energy losses of overlayers. The width of the layer spectra is related to layer thickness and the inelastic loss as expressed in Eq. (16). Yields $Y$ in the different parts of the spectra are estimated from the general formula applicable in scattering experiments from:

$$Y = I_0 N \Delta\Omega d\sigma/d\Omega \qquad (16)$$

where $I_0$ is the primary beam intensity [number of ions], $N$ is the number of target atoms [atoms/cm$^2$], and $\Delta\Omega$ is the solid angle of detection [sr]. The differential scattering cross section is given as [cm$^2$/sr]. Furthermore, 100% particle detection efficiency is assumed. For layers of thickness $Dt$ at depth $t$ the yield will assume a Rutherford cross section and a scattering angle of 180°:

$$Y(t) = I_0 \{Z_1 Z_2 e^2/4E(t)\}^2 N \Delta t \Delta\Omega \qquad (17)$$

A spectrum $E_1(t)$ is experimentally measured (Fig. 3) so Eq. (18) has to be converted to a spectrum $Y_1(t)dE_1$. With the kinematic factor $K$ and the notion that in the range of ion energies used the energy loss is slowly varying with energy, the ratio between the energy loss on the way in, $\Delta E_{in}$, and on the way out, $\Delta E_{out}$, will be approximately constant, that is,

$$A = \Delta E_{out}/\Delta E_{in} + K\left(E(t) - E_1\right)/\left(E_0 - E(t)\right)$$
$$\approx (dE/dx)|_{out}/(dE/dx)|_{in} \qquad (18)$$

By using the proper $dE/dx$ values, $A$ can be estimated explicitly. However, for light ions like $He^+$ and medium to high $Z_2$ targets $K \approx 1$ and an energy loss ratio out/in of approximately 1, Eq. (19) can be approximated by

$$Y(E_1) \propto (E_0 + E_1)^{-2} \qquad (19)$$

which is in good qualitative agreement with the shape of the experimental "bulk" spectra.

Peak heights $H$ and peak width $\Delta E$ of the Si and Pt peak of the annealed PtSi layer can be used to estimate the approximate composition of the layer. The ratio of the number of target atoms $N_a$ to $N_b$ is proportional to the peak height ratio, the width, and the inverse ratio of the cross sections. The latter ratio reduces to the square of the ratio of the respective atomic numbers $Z$:

$$N_a/N_b = (H_a/H_b)(\Delta E_a/\Delta E_b)(Z_b/Z_a)^2 \qquad (20)$$

The peak area of the spectrum of a thin layer is naturally proportional to the number of target atoms. For structural analysis using RBS or MEIS the shadow cone and the related effect of channeling are the basic tools. When a parallel beam of ions is directed onto a single crystal surface aiming at low index crystal direction most ions will penetrate into the crystal channels. These spectra are called aligned and show the surface peak. The scheme of an MEIS experiment and experimental spectra are shown in Figure 4 [20]. It shows the backscattering of 100.5 keV protons from a Pb(110) surface as a function of surface temperature. It is clearly seen that the surface peak intensity increases with increasing temperature and also broadens. There is no intensity increase below the surface peak due to the aligned geometry. Only at the highest temperatures close to the melting point of Pb is the typical bulk–scattering observed. Equation (7) is where we start if we are to understand the development of a surface peak. If we take two atoms, the flux distribution at the second atom $f(s_2)$ is related to the flux distribution of the first atom $f(s_1)$ by:

$$f(s_2)2\pi s_2 ds_2 = f(s_1)2\pi s_1 ds_1 \tag{21}$$

where the $s$ are respective impact parameters. If the flux distribution at the edge of the shadow cone is small compared to the thermal vibrational amplitude $f(s_2)$ it

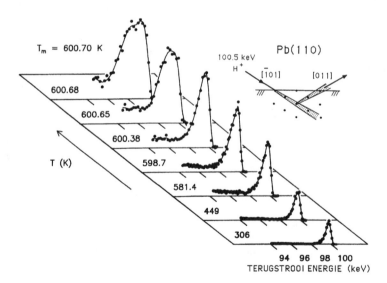

FIG. 4. Energy spectra of backscattered protons of 100.5 keV from Pb(110) with increasing target temperature from room temperature to close to the bulk melting point of Pb (600.8 K). The inset shows the shadowing/blocking geometry of the experiment [20]. Reprinted with permission from J. F. van der Veen, B. Pluis, and W. van der Gon, © 1990, Plenum Press.

can be approximated to a delta ($\delta$) function. Hence the flux at atom 2 in the region at the edge of the shadow cone and outside, that is, for $s_2 \geq r_c$, will be

$$f(s_2) = 1 + \left(r_c^2/2s_2\right)\delta(s_2 - r_c) \tag{22}$$

The index $c$ indicates that here a Coulomb potential is considered for the scattering. This flux distribution is then to be folded with the Gaussian position distribution $\Delta(s_2)$ to obtain the probability $P_2$ of hitting the second atom with a small impact parameter and thereby contributing to the backscattering yield. As large angle backscattering is considered here, the impact parameters will be small compared to the thermal vibrational amplitudes. Within the Debye model of thermal vibrations with amplitude $r$ the position distribution is

$$\Delta(s) = 1/\pi\rho^2 \exp\left(-r_c^2/\rho^2\right) \tag{23}$$

The probality for a close impact collision $P_2$ is then

$$P_2 = \left(1 + r_c^2/\rho^2\right)\exp\left(-r_c^2/\rho^2\right) \tag{24}$$

The total intensity of the surface peak is the sum of the scattering from the first atom and the second atom. More detailed studies show that depending on the ratio $r_c/r$, more atoms than two may contribute to the surface peak [15, 16]. Considering now the change of the surface peak in Figure 4 as a function of temperature, the initial increase is due to the increase of the thermal vibrational amplitude. The evaluation in terms of the number of atoms contributing to the surface peak shows this effect quite clearly (Fig. 5) [21]. At approximately 580 K, that is, 20 K below the bulk melting point, surface melting is observed which leads to the enhanced increase and broadening of the surface peak.

From the example given here, it is obvious how channeling can be used for structural analysis. It adds to the RBS thin-film analysis the capability to locate impurities or defects with respect to a host lattice. With a single crystal target using an aligned scattering geometry the variation of the scattering angle using a rotable energy spectrometer or solid state detector yields intensity distributions with pronounced minima for all major low index crystallographic orientations. Impurities in the host lattice will alter these minima. An interstitial in the middle of a channel causes a peak at the respective total energy loss and so mass, position, and depth can be measured. These channeling techniques can be combined with the detection of recoils (ERDA) which is one of the few techniques that affords quantitative analysis of light ion impurities in solids (H, D, etc.). Another line of analysis is opened when measuring the X-rays emitted due to the inner shell ionization by fast protons—proton induced X-ray emission (PIXE). Compared to electron-induced X-ray analysis, for example, with electron microprobes or SEM, the proton beams allow not only structural analysis using the channeling effect, but the remarkable feature is the possibility of bringing the protons

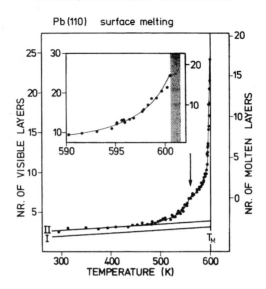

FIG. 5. Calibrated surface-peak area (Nr. of visible atoms) and the number of molten layers as a function of the temperature. The vertical line indicates the bulk melting point $T_M$. The arrow indicates the surface melting point. The inset is an expanded view of the highest 10-K interval. Reprinted with permission from J. W. M. Frenken and J. F. van der Veen, *Phys. Rev. Lett.* 54, 134 (1985), © 1985, The American Physical Society.

through a proper thin window into air. Degradation of beam quality is negligible and the PIXE technique is therefore applicable to archeological problems, art works, sculptures, paintings, and prints—essentially without damage [22].

## 2.2 Low Energy Ion Scattering

In the low energy regime (300 to 5000 eV) of ion scattering analysis the measured ion spectra essentially show only a surface peak. This is due to two reasons: i) the larger scattering cross sections lower the penetrations depth of the projectiles such that backscattering is reduced to a few layers of the target; and ii) there is a high probability for neutralization of the projectiles to cause a further decrease of the backscattering yield of deeper layers of the target. Hence, when measuring only scattered ions there is ample proof that these ions are scattered from only the top layer atoms [1, 2, 8, 24].

As an example Figure 6 shows low energy ion backscattering spectra from an $Au_3Pd$ surface using $He^+$ and $Ne^+$ as primary ions [23]. In Figure 6a it is a $He^+$ spectrum, measured with an electrostatic analyzer. The Au and Pd peaks are not resolved at $\Theta = 90°$ with He. In Figure 6b $Ne^+$ is used and the Au and Pd peaks are well separated. Due to the logarithmic scale the Gaussian peaks appear parabolic. The He spectrum (Fig. 6a) shows some low energy tailing reminiscent of the bulk yield of RBS spectra.

Writing the ion yield $I_i^+$ scattered from a species $i$ and the experimental quantity of transmission $T$ of the detecting system, including the detector sensitivity and the solid angle of detection $\Delta\Omega$, ion survival probability $P_i^+$ for the scattering from species $i$ has to be taken into account:

$$I_i^+ = I_0 T P_i^+ \Delta\Omega d\sigma/d\Omega \qquad (25)$$

The ion survival probability factor is the most important difference between the RBS and MEIS regime and low energy spectroscopy (LEIS). It depends on the individual projectile–target combination used. The probabilities of ion survival are

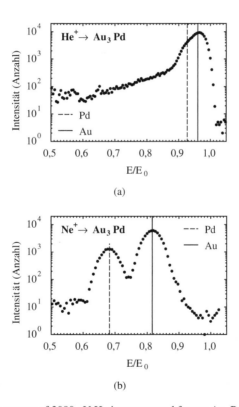

(a)

(b)

FIG. 6. (a) LEIS spectrum of 2080 eV He ions scattered from a $Au_3Pd$ alloy surface. The angle of incidence is $\Psi = 45°$ and the laboratory scattering angle is $\Theta = 90°$. Note the logarithmic scale and the lack of mass resolution [24]. (b) LEIS spectrum of 1560 eV Ne ions scattered from the same surface as in (a) under identical geometric conditions. Note the Gaussian shape (parabolic in the semi-log. plot) of the spectra and the mass resolution [24].

of the order of 5% for the noble gas ions He and Ne, which are used mainly for LEIS. Due to the large ionization energy of the noble gas ions, 25.4 eV for He and 15 eV for Ne, electron capture is very probable when the velocity of the projectile is below the Fermi velocity of the solid state electrons. Due to the much higher speed of the projectiles, mainly $H^+$ and $He^+$, in the RBS and MEIS range, the ion survival there is of the order of 100 and 90%, respectively. The first consistent model for ion neutralization at surfaces was developed by Hagstrum [25]. Figure 7 shows the essential features for the case of He in front of a metal with a conduction band of a width of approximately 5 eV. The possible charge exchange processes are indicated. Between excited states of the atom and the electrons at the Fermi edge of the solid electrons can be exchanged via resonant processes, RI = ioniza-

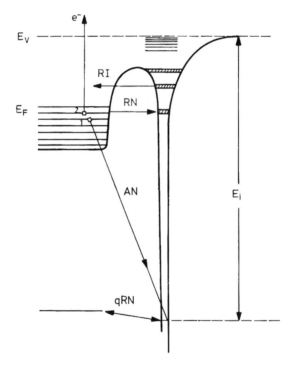

FIG. 7. Scheme of charge-exchange processes between an ion and a solid (metal) surface. Here $E_F$ is the Fermi energy of the solid and the difference to the vacuum energy $E_V$ is the work function. The ion has the ionization energy $E_i$. Excited states of the ion are indicated by the three hatched-line areas and Rydberg states merging into the continuum. Possible resonant exchange processes are indicated by RI (resonant ionization), RN (resonant neutralization) and qRN (quasi-RN). The AN is Auger neutralization with solid-state electrons involved [2]. Reprinted with permission from H. Niehus, W. Heiland, and E. Taglauer, *Surf. Sci. Rep.* 17, 213 (1943), © 1943, Elsevier Science.

tion and RN = neutralization. Some of these excited atoms survive the encounter with the surface and can be observed with classical optical methods. The yields of the excited atoms observable are of the order $10^{-4}$ [26]. The main use of the effect is, at this point, the study of surface magnetism [27]. In most cases either RN will end the life of the excited state or an internal relaxation process may occur, that is, Auger deexcitation, autoionization, or autodetachment [28]. By resonant capture negative ionic states also may be formed, a rare event with noble gas ions, but of importance when atoms or molecules with stable affinity levels scatter off surfaces [29, 30]. In many cases the dominant neutralization event is the Auger capture, which is due to the Coulomb excitation of conduction band electrons. By the approaching $He^+$ two electrons interact such that one electron is transferred to the He and the second electron is emitted to the vacuum carrying the released potential energy as kinetic energy. The energetic spectroscopy of these electrons is the basis of the ion neutralization spectroscopy (INS), which yields information about the surface density of states [25]. The deconvolution of the IN spectra needed to obtain the band structure information is complicated compared to the rather direct way by which ultraviolet photoelectron spectroscopy (UPS) gives band structure information. In the case of broadband metals, for example, Al, resonant capture of electrons into the 1s ground state is also possible [31]. The process is facilitated by the promotion of the He 1s level by interaction with the Al band electrons. The level promotions also play a role in the reionization of He or Ne in the close encounter with target atoms. The reionization occurs above certain threshold energies such that for the fate of an ion scattering of a surface it may be useful to distinguish three parts along the trajectory—the ingoing part, the close encounter, and the outgoing part [2, 24].

Another neutralization process, the quasiresonant neutralization qRN, is observed for target atoms with core levels that lay within $\pm 10$ eV of the He 1s level. The most striking example is Pb, where as a function of the He, primary beam energy oscillations in the ion yield are observed [32, 33]. The effect has theoretically been understood as Stueckelberg oscillation [34].

For practical purposes, especially chemical surface composition analysis, the neutralization effects lead to the fact that calibration is the sensible way to achieve quantitative analysis. On the other hand, neutralization is so effective that rare gas ions scattered from only the second layer of the target are neutralized. This is the major factor, beside the large cross sections, that makes LEIS sensitive only to top layer atoms. This statement has been corroborated by direct comparison of He and Li scattering [35] and supporting calculations of the ion trajectories [36]. For Li the ion survival probability is very high, up to 100% [29], such that the numerical trajectory calculations reproduce the experimental results with good quantitative agreement. The code used (ARGUS) is comparable to the MARLOWE code in the principle setup, that is, using the binary collision approximation, including the full 3D–crystal structure, inelastic losses, and thermal vibrations.

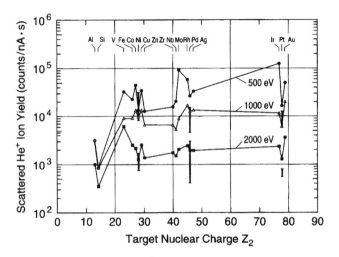

FIG. 8. Comparison of the yields of $^4He^+$ from different targets at 500, 1000, and 2000 eV. The vertical bars indicate the scatter of the data from a round robin from different laboratories [38]. Reprinted with permission from H. H. Brongersma et al., *Nucl. Instr. Meth.* B142, 377 (1998), © 1998, Elsevier Science.

Another experimental finding is the independence of the ion survival probability for a given projectile–target pair from the neighborhood of the target atom, that is, matrix effects are rare in LEIS. This finding affords a straightforward calibration using clean surfaces as standards. Where this is not possible (gases), adsorbed monolayers may serve the purpose. In Figure 8 experimental ion yields are collected [37, 38] and can be used to estimate relative surface concentrations. Included in the graph are the results of a round robin experiment [38] showing that different laboratories can measure these ion yields within ±20%. As can be seen, the dependence on the target atomic number is rather smooth and increases slowly with increasing $Z$. This behavior reflects the weak $Z$ dependence of the differential scattering cross section for the screened Coulomb potential. For the Rutherford regime the cross section increases with $Z^2$. The high ion yield, for example, for He–Al, is due to the reionization effect [37, 38].

As in the case of MEIS and RBS, low energy ion scattering (LEIS) is very useful for obtaining structural information. As in the cases of MEIS and RBS the shadow cone phenomenon is the main tool for structural analysis. However, the surface sensitivity of LEIS is unique due to the neutralization effect discussed in the previous section. This is especially the case when the scattered ions are measured with an electrostatic analyzer (ESA). Figure 9 shows spectra of a $Au_3Pd(110)$ surface after sputtering and annealing, respectively [39]. Typical for LEIS, the Au peak is in this case lower when Pd is on the surface. Quite generally

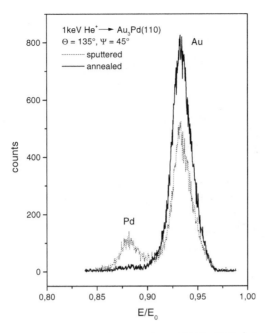

FIG. 9. The LEIS spectra (raw data, linear scale) of 1000 eV He$^+$ backscattered from Au$_3$Pd(110) after sputtering (dotted line) and after annealing (solid line). Due to the larger scattering angle $\Theta = 135°$, the Au and Pd peaks are separated (compare with Fig. 6a) [39]. Reprinted with permission from J. Kuntze et al., *Phys. Rev.* B60, 9010 (1999), © 1999, The American Physical Society.

there is a linear relation between the coverage with $N_i$ atoms/cm$^2$ of a species $i$ on a substrate and the intensity $I_s$ of ions backscattered from the substrate with atom density $N_s$,

$$I_s^+ = I_0 T P_s + \Delta\Omega (N_s - aN_i)ds/d\Omega \qquad (26)$$

The parameter $\alpha$ is an empirical shadowing parameter containing both the blocking effect due to the shadow cone and the neutralization effects. The other terms in Eq. (26) are identical to those of Eq. (25). There is ample experimental evidence for this linear behavior for many adsorbate systems, for thin-film systems, for example, metal on metal, metal on semiconductors, etc. [1, 2, 8, 24, 37]. Results of this kind can be obtained with very simple equipment, evaluation needs no computer codes, and calibration is possible by using clean surfaces of the species involved. Furthermore, the information as to whether a species is adsorbed on top, or is incorporated into the top layer, or diffuses into the bulk, can be obtained by LEIS for both single crystal and polycrystalline samples. Conversely, segregation processes also are accurately observable by LEIS. Figure 9 is an example of segregation, in this case for Au on $Au_3Pd$, forming a complete monolayer of Au.

The direct application of the shadow cone for surface structure evaluation is more involved. There are three possible experimental schemes. If the grazing angles of incidence used are lower than the critical angle for channeling (Eq. (8)), the crystal structure can be analyzed using the forward scattered beam. In that case the individual shadow cones overlap and surface channeling is observed. The spatial intensity distribution can be imaged using a position sensitive detector [40]. Figure 10 [41] shows the variation of such intensity distribution for Pb(110) at room temperature and at 580 K, that is, at the stage of surface melting (see Figs. 4 and 5 for the MEIS data). Structural changes are also under similar experimental conditions observable using only a small angle detector [42, 43]. In these applications thin-film growth is monitored. When we start with a flat surface, the detected

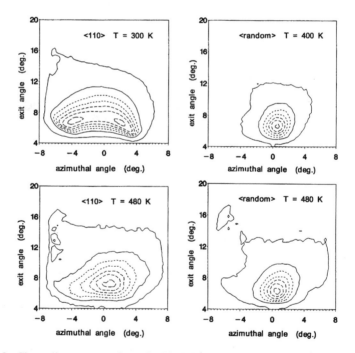

FIG. 10. Channeling patterns from the Pb(110) surface at different temperatures with 2000-eV Ne ions incident at $\Psi = 5°$. The exit angle is measured from the surface and the azimuthal angle is measured from the [110] surface crystallographic direction [41]. (Reprinted with permission from S. Speller et al., *Phys. Rev. Lett.* **68**, 3452 (1992), © 1992, The American Physical Society.

scattering intensity is high but decreases during film growth due to scattering off adatoms or island edges. When the first monolayer is completed a new intensity maximum is observed and so forth. A quantitative evaluation of this data needs computer codes, but the modeling of a rough surface is difficult. Thus, in most cases the method is used only qualitatively.

Surface structure information is also obtained when placing the detector for the backscattered ions at a large angle and then varying either the azimuthal angle or the angle of incidence. In the first case, that is, keeping the angle of incidence constant and below the critical angle of channeling for the low index direction of the surface under study, the variation of the angle of incidence shows the crystal structure in real space. Figure 11 is a result of such a backscattering experiment for Pb(110) [41]. The major low index surface directions, that is, the surface semichannels, are seen as pronounced minima that vanish at the temperature of surface melting. Naturally, the channeling pattern of Figure 10 changes from the half moon shape to an almost isotropic distribution in that temperature range. The channeling minima of Figure 11 are a measure of the quality of the surface, that is, steps, adatoms, and impurities will cause an increase in the minimal intensity. This observation is of the same quality as in the case of channeling in the RBS regime; however, a quantitative evaluation is less practical due to problems inherent in modeling disordered surfaces.

Additional and quantitative information of structural parameters is obtainable when using the combination of a fixed azimuthal angle, preferably for a low index direction, and a fixed, large scattering angle and hence the angle of incidence can

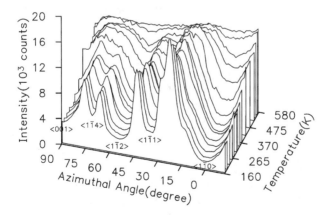

FIG. 11. Surface blocking patterns, intensity vs azimuthal angle $\Theta$, for 2000-eV Ne scattered from Pb(110) at temperatures between 160 and 580 K. The angle of incidence is $\Psi = 11°$ and the scattering angle is $\Theta = 165°$ [41]. Reprinted with permission from S. Speller et al., *Phys. Rev. Lett.* 68, 3452 (1992), © 1992, The American Physical Society.

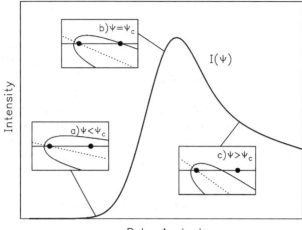

Polar Angle $\Psi$

FIG. 12. Schematic concept of the (neutral) impact collision ion scattering spectroscopy (N)ICISS for 180°. By the measurement of the critical glancing angles $\Psi_c$ for the onset of backscattering at a known surface geometry, that is, lattice constant $d$, the shadow cone size can be determined experimentally (compare with Fig. 1) [21]. Reprinted with permission from J. W. M. Frenken and J. F. van der Veen, *Phys. Rev. Lett.*, 54, 134 (1985), © 1985, The American Physical Society.

be varied. This experimental scheme was introduced as ICISS (impact collision scattering spectrometry) [44, 45]. Figure 12 shows the concept behind such an experiment. It is obvious that the experiment can serve a number of purposes. For a known surface lattice constant $d$ the shadow cone radius $r_c$ is measured. Hence, the shadow cone can be numerically calculated and fitted to the experiment. In essence, the shadow cone is calibrated from a known structure and can be used for further measurements for a given experimental situation, that is, target, projectile, energy, scattering angle, etc. In the procedure for estimating the shadow cone, information about the thermal vibrational amplitudes of the surface atoms is obtained too. Figure 13 shows experimental data for the $Au_3Pd(100)$ surface [46]. In this case a time-of-flight method (TOF) is used to avoid complications due to neutralization. The acronym NICISS is used for this scheme [2]. The experimental data are directly comparable with calculation based on a two-atom scattering model that takes into account scattering potential, thermal vibrations and lattice structure [23]. The agreement between calculation and experiment is of the order of $10^{-5}$ (least square). In the case of Au on $Au_3Pd(100)$, the lattice constant is determined to $4.00 \pm 0.02$ Å, which is smaller than the bulk value of bulk Au [46]. The spectra extended over a larger angle of incidence range for the [110] surface direction show a signal from Au and Pd, whereas the scattering along

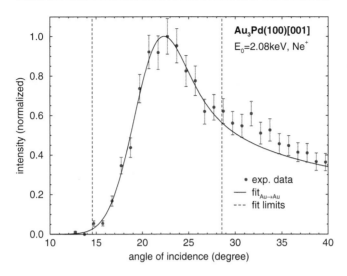

FIG. 13. The NICISS of the Au$_3$Pd(100) surface along the [001] surface direction with 2080-eV Ne at a laboratory scattering angle of 165°. Only Au-Au collisions have to be taken into account. The solid line is the calculation of a two-atom model fitted to the experimental data within the vertical dashed lines [46]. Reprinted with permission from M. Aschoff et al., *Surf. Sci.* 415, L1051 (1998), © 1998, Elsevier Science.

the [001] surface direction yields a Au signal only. This reflects the fact that Au forms the top layer and Pd is found only in the second layer. Furthermore, along [001] the second layer is shadowed by the first layer. This is not the case for the [110] direction for which the second layer is visible to the ion beam [23]. Further examples of applications of (N)ICISS are found in these References [1, 2, 8, 37]. The technique has been improved by using a position-sensitive detector in combination with a time-of-flight technique [47–50]. Here the detector is set at a large scattering angle as in NICISS and not in the forward direction as in the direct observation of surface channeling. This combination affords rather direct structural anlysis. Quantitative analysis is obtained by comparing the results of a computer code that simulates the structure in 3D with ion scattering. The authors labeled the technique with the acronym SARIS (scattering and recoiling imaging spectrometry).

Information on deeper layers can be obtained from the results of the combination of sputtering and LEIS. In the case of adsorbates or catalyst surfaces this has been proven to be successful due to the fact that when He$^+$ is used for analysis the sputtering is rather slow and gentle. An example is shown in Figure 14 [8, 51]. It is a sample of an alumina-supported Rh catalyst. At low He$^+$ fluences only Rh and O are detected. With increasing fluence the metal peak increases and, with

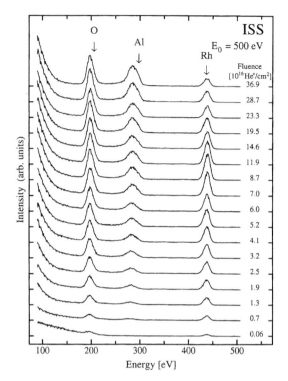

FIG. 14. The He$^+$ LEIS spectra from an Al$_2$O$_3$ surface covered with approximately one monolayer of Rh [8]. Reprinted with permission from E. Taglauer and John Wiley and Sons Ltd., © 1997.

some delay, also the O and Al peaks. After some sputtering the Rh peak decreases again, whereas the O and Al peak become stationary. The interpretation of such spectra is straight forward. Initially the surface including the Rh is covered with oxygen. While the O is sputtered the Rh becomes visible followed by the Al after further sputtering. As the Rh is present as a thin layer only it will be sputtered off, and O and Al will be the prevailing peaks. Note that Figure 14 represents a poly-crystalline sample. The LEIS analysis permits catalytic processes to be studied *in situ* [52]. The SARIS technique is also applicable for adsorbed layers, especially when making use of the recoils [48, 53]. For recoil analysis Eq. (2) is valid and for a TOF system it converts to with a flight path of length $L$ to

$$T = L(M_1 + M_2)(8M_1E_0)^{-1/2} \cos \Theta_2 \qquad (27)$$

where all terms are equal to the ones of Eq. (2). Recoil detection is important for the analysis of light adsorbed atoms, for example, H, D, which are not detectable by other surface analytical tools.

An interesting notion is the time-resolving power of SARIS, currently experimentally limited to a time resolution of approximately 1 s. This is, however, in a range where at least slow growth processes, adsorption, and desorption, or surface reactions can be followed *in situ* and in realtime.

### 2.2.1 Secondary Mass Spectrometry.

Secondary mass spectrometry (SIMS) is part of the large field of mass spectrometry and thus most of the instrumentation is shared with mass analytical techniques for gas phase analysis. However, special developments occurred for SIMS with respect to the ion sources used for sputtering, postionization of the sputtered material, and imaging of the samples. The basic process is the sputtering effect, which is closely related to the process of nuclear stopping [54]. Primary particles hitting a solid lose energy by colliding with target atoms (this is the nuclear loss part) and by excitation of bound target and projectile electrons as well as by exciting plasmons and electron-hole pairs, which is called the electronic loss (see Section 2.1). Depending on the primary energy and on the particular trajectory of the primary particle it will lose energy in a few or in many collisions with target nuclei. These target atoms, set in motion by the impact according to Eq. (2), will cause further target atoms to be removed from their lattice sites as long as there is enough energy available. At energies typically used for SIMS, that is, around 10 keV, and for heavy primary particles, typically Ar, several tenths of atoms are set in motion in a relatively small volume. This phenomenon is called collision cascade. If the collision cascade happens to intersect with the surface of the target some of the atoms in motion may leave the surface. To do so, the atoms have to overcome the surface binding energy. From this qualitative description it is obvious that there is an optimal energy regime for sputtering and hence SIMS, that is, at medium energy above the threshold for sputtering and for primary energies where the ions penetrate into the solid. At too low energies single binary collisions prevail and direct recoils governed by Eqs. (24) are observed mainly. At too high energies the primary ions penetrate deeply into the target and the collision cascades seldom intersect with the surface. The sputtering phenomena and its applications are discussed in great detail in several reviews (see, e.g., [55, 56]). An approximation for estimating sputter yields, that is the ratio of the mean number of emitted atoms to the number of incident ions, is given by [54]:

$$Y = \Lambda \alpha N S_n \tag{28}$$

where $\Lambda$ is a parameter containing the atom density $N$ of the target and the surface binding energy $U_s$, that is,

$$\Lambda \sim 0.042/N U_s \text{ (Å/eV)} \tag{29}$$

The binding energy is approximately equal to the heat of vaporization of 2–4 eV. The factor $\alpha N S_n$ is approximately the deposited energy, where $\alpha = 0.25$ for practical purposes, $S_n$ is the nuclear stopping power, and $dE/dx|_n = N S_n$

is the nuclear energy loss. A typical value, for example, for Ar incident on Cu at normal incidence at the stopping power maximum, is $NS_n = 124$ eV/Å.

With a target density of $N = 8.5 \times 10^{-2}$ atoms/Å$^3$ and a binding energy of 3 eV, the sputter yield is estimated from Eqs. (28) and (29) to be $Y = 5.1$ atoms/ion. The experimental value is $\approx 6$ atoms/ion [16].

The sputtered particles have a wide energy distribution with a maximum at approximately $1/2U_s$. As most mass spectrometers will lose mass resolution when exposed to such a wide range of energies, energy filters are used for SIMS. Direct line of sight also has to be avoided between target and detector, because fast neutral sputtered atoms and neutralized primary ions will cause a background signal. For different reasons though, the primary beam also ought to include a deflection because low energy neutral particle originating from the ion source may reflect from the sample surface and cause spurious results. For very low concentration analyses care has to be taken to avoid resputtering from surrounding structural elements, for example, apertures, lenses etc., by the fast neutrals onto the target. These considerations with respect to the design of the primary ion source, the effects of fast neutrals, light emission from the ion source, etc. may also disturb the depth analysis when combined with LEIS, XPS, or AES.

Although SIMS is in principle destructive, it is for that reason that it can provide depth analysis. The yield of sputtered particles of a multicomponent target has to be for reasons of particle number conservation proportional to the bulk concentration under steady state conditions. At the beginning of the sputter process preferential sputtering is found, that is, the lighter components are sputtered preferentially. For example, on oxide surfaces a depletion of oxygen by sputtering is a quite general observation. The preferential sputtering effects also play a role in LEIS, where the actual surface concentration is measured, and also in XPS and AES, where an average over some surface layers is measured as surface concentration. The LEIS cannot in general measure bulk concentrations but AES and XPS may measure bulk concentrations if preferential sputtering effects are minimized.

The yield of sputtered ions, which is the signal in most SIMS experiments, depends mostly on the sputter yield $Y(E_0, \alpha)$ and the ion formation probability $P^+(E_1, \beta)$, where $E_0$ and $E_1$ are the primary energy and the energy of the sputtered particle, respectively, $\alpha$ is the angle of incidence relative to the surface normal, and $\beta$ is the observation angle, that is, the ion optical axis of the mass analyzing system relative to the surface normal. The other parameters are equivalent to those of Eq. (22), $T$ is the transmission including the detector sensitivity of the system, $\Delta\Omega$ the solid angle of acceptance, $I_0$ the primary beam intensity, and $c_a$ the concentration of the species $a$, which yields:

$$I_a^+ = I_0 T c_a P_a^+(E_1, \beta) Y(E_o, \alpha) \Delta E_1 \Delta\Omega \qquad (30)$$

The ion formation probability $P^+$ of sputtered singly charged ions varies over several orders of magnitude. The probabilities depend strongly on the chemical environment, for example, oxygen causes large increases of ion yields. In contrast to LEIS, sputtered ion yields are thus connected with huge matrix effects. Due to these ion yield problems, SIMS systems always apply a scanning of the ion beam and an electronically gated detection system in order to prevent edge effects. The lower current densities at the rim of the primary beam leaves higher impurity concentrations, which in turn cause higher ion yields. For quantification of SIMS postionization methods of the sputtered material using plasma discharges, electrons [57] or photons [58]. The abbreviation SNMS (secondary neutral mass spectrometry) is used in this context. An example for a depth profile SNMS analysis shows Figure15 [57]. It is a W-Si multilayer system with an individual W-Si layer thickness of 3.6 Å. Because W-Si is insulating the sputtering was performed with Ar gas discharge. The plasma source provides a very low primary energy distribution, which in turn causes very uniform sputtering. In addition, the intermixing of the species in the sample is reduced at low primary energies, which the serves to keep the depth resolution high even into great depths. Another benefit of low energy is the finding that in many cases the sputtering craters remain flat, which also contributes to good depth resolution. In the case here (W-Si), the plasma discharge was used for post-ionization. Mass analysis is based on a quadrupole mass filter and an electrostatic energy analyzer.

An average sensitivity of up to $10^{-3}$ for all elements is achieved by using SNMS. Higher sensitivities are possible with the laser post-ionization schemes,

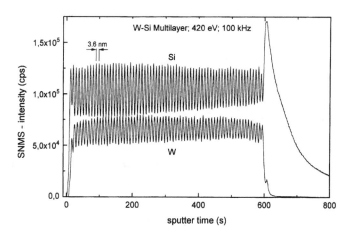

FIG. 15. The SNMS depth profile of Si–W layers [57]. Reprinted with permission from H. Oechsner, *J. Mass Spectr. Ion Proc.* 143, 271 (1995), © 1995, Elsevier Science.

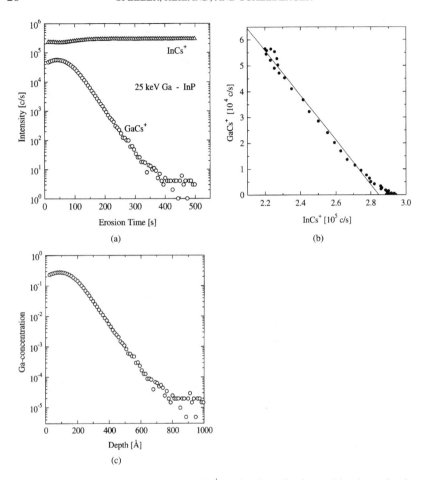

FIG. 16. (a) Depth profile using 3 keV $Cs^+$ projectiles of a focused ion beam implant of 25 keV $Ga^+$ in InP monitoring $GaCs^+$ and $InCs^+$ ions. (b) Correlation of the signals shown in (a); the slope yields the relative sensitivity factor $S_{GaCs}/S_{InCs}$. (c) The SIMS data of (a) converted into the Ga concentration distribution using the factors derived in (b) [59]. Reprinted with permission from H. Gnaser, *J. Vac. Sci. Tech.* A12, 452 (1994), © 1994, The American Institute of Physics.

where with two-laser resonant photoionization, concentration levels of as low as 2 ppb of $^{56}Fe$ in Si have been measured [58].

Another path for minimizing the ion yield effects is the application of $Cs^+$ as primary ions. It has been shown that with Cs and especially semiconducting materials, for example, Si-H, Si-C, Si-N, Ga-As and In-P, the yield of $MCs^+$ ions is widely independent of the status of the samples [59]. The respective target atom in question is M. Figure 16 shows the results of a depth analysis of 25 keV

Ga implanted into InP using 3 keV $Cs^+$ ions for sputtering. For these systems Eq. (26) can be simplified by introducing the so-called useful yield or practical sensitivity $S$:

$$I^+_{MCs} = I_0 Y c_M S_{MCs} \qquad (31)$$

The quantity $S_{MCs}$ is the number of detected $MCs^+$ ions per sputtered M atom. For fixed experimental conditions this term includes all the factors, such as, for example, $T$, $\Delta\Omega$, etc. For a binary system with components $A$ and $B$, for example. In and Ga as shown in Figure 16, and with $c_A + c_B = 1$, the yield of component $A$ is given by:

$$I^+_{ACs} = S_{ACs} I_0 Y - I^+_{BCs}(S_{ACs}/S_{BCs}) \qquad (32)$$

If the sensitivity factors are independent of the concentration, the $I^+_{ACs}$ is linearly related to the yield $I_B Cs$ as shown in Figure 16b. If $Y$ is known or is evaluated from the depth of the crater then the data can be converted in a straightforward fashion to concentration (Fig. 16c). This assumes then that the sputter yield does not change with erosion time, which can happen when there are major changes in concentration. For example, in a thin-film system (e.g., Fig. 15) the sputter yield $Y$ will change when going from W (low) to Si (higher). The extension of the concentration profile over several orders of magnitude as in Figure 16c is quite typical for SIMS depth analysis including the facility for 3D depth profiling [60–62].

# 3 DIFFRACTION METHODS

The diffraction of the electromagnetic radiation of both electrons and atoms provides one of the most widely used tools in solid state physics, material science, and surface physics. There are as many textbooks on the use of these tools as there are lectures delivered at universities worldwide on the underlying physics involved in their use. Thus, section will be restricted to some fundamental aspects and to the more recent developments of the applications of these probes.

Diffraction is described by either Bragg or von Laue equations. The Bragg equation relates as a condition for constructive interference the wavelength of the electromagnetic radiation $\lambda$ to the interplanar distance $a$ and the observation angle $\theta$

$$n\lambda = a \sin\theta \qquad (33)$$

For particles, the wavelength is estimated from the de Broglie equation. For electrons the notion

$$\lambda(\text{Å}) = \sqrt{150.4/E(\text{eV})} \qquad (34)$$

is useful. The acceleration energy of the electrons is denoted by $E$. The equation (29) gives the position of possible diffraction spots in real space only. For quantitative analysis, knowledge of the diffraction intensities is necessary. It depends on the type of experiment and radiation used how complicated the evaluation of the intensities will be. For X-rays, the calculation of intensities is rather straightforward. The diffraction of fast electrons (RHEED) can be treated in the so called kinematic approximation, which is still comparably simple, but LEED requires very elaborate programs due to the multiple scattering effects of the slow electrons used. In the atom diffraction experiments being used in basic research, problems arise from the interaction potentials between the slow atoms and the surface. All diffraction methods basically measure the long range order of either a crystal or a crystalline surface. There is hardly any short-range order information obtainable that makes diffraction methods completely complementary to a method like STM. There is, furthermore, a fundamental difference in the depth of information. The penetration depth of X-rays is large, typically of the order of 100 Å, and the penetration or escape depth of electrons are on the order of 10 Å at the minimum of the mean free path. Here STM and LEIS are the techniques that have monolayer sensitivity. Another general notion is the fact that RHEED and X-rays for surface analysis have to use glancing angles of incidence, which then causes large areas to be probed. Further, in the case of the X-ray standing waves method (XSW), large areas are probed [63]. Surface sensitivity of this technique is obtained only for monolayer adsorption. An advantage of X-ray techniques is that of applicability in air or liquids (electrolytes), that is, UHV is not a condition *sine qua non.*

### 3.1 X-Ray Diffraction

From the point of view of general knowledge on X-ray techniques for bulk structure analysis we will address very briefly the applications in surface science. The principal geometry for a glancing incidence X-ray experiment is shown in Figure 17 [64]. Information about surface structural parameters can be obtained from measuring the specular intensity with $k_{fz} - k_{iz} \equiv q_z$, that is, the conditions are such that there is a finite perpendicular momentum transfer and there is no transfer of parallel momentum $k_{f\parallel} - k_{i\parallel} \equiv q = 0$. The measured intensity is related to the lateral average of the electron density at the surface. In order to evaluate these experiments full dynamic calculations are necessary [65]. At very glancing angles total external reflection is obtained. The momentum transfer is more complicated. The parallel momentum remains unchanged as in the case of specular reflection. The perpendicular momentum transfer is given by

$$k_{fz} - k_{iz} = k(\sin^2 \alpha_f - 2\delta + 2i\beta)^{1/2} - k(\sin \alpha_i - 2\delta + 2i\beta)^{1/2} \qquad (35)$$

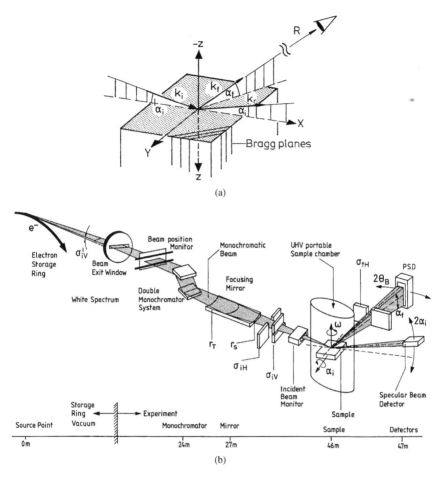

FIG. 17. (a) Schematic view of glancing angle diffraction: $k_i$ and $k_f$ are the wave vectors of the incident and the scattered beam, $k_r$ is the wave vector of the reflected beam [66]. Reprinted with permission from H. Dosch, *Springer Tracts in Modern Physics* (1995) and the Springer-Verlag, Heidelberg. (b) Schematic view (real space) of a glancing angle X-ray experiment for surface structure analysis. The beam line scheme corresponds to the RöWi 1 beam line of HASYLAB. PSD stands for position sensitive detector. Reprinted with permission from [66], © 1995, Springer-Verlag, Heidelberg.

The wave vectors are complex in the case of total external reflection. The angles $\alpha$ appearing in Eq. (31) are initial angle of incidence and outgoing angle, $\alpha_i$ and $\alpha_f$, respectively. Here $\beta = \mu\lambda/4\pi$, where $\mu$ is the linear absorption coefficient and $\lambda$ is the X-ray wavelength. Further, $\delta = \alpha_c^2$, where $\alpha_c$ is the critical angle for total external reflection, that is, $\alpha_c = \lambda(NZr_e/2\pi)^{1/2}$ and $N$ is the density of atoms with charge number $Z$ and $r_e$ is the classical electron radius. The intensity of the scattered X-rays is in part in Bragg peaks from the ordered structures and in part in a diffuse intensity due to local correlations [64, 66].

It is interesting to note that for subcritical angles of incidence and subcritical outgoing angles the scattering depth is typically reduced to 50 Å compared to several $10^4$ Å at steeper angles. Thus the information depth is large compared to those of LEIS, LEED, AES, and XPS. However, the benefit of this information depth is the possibility of analyzing thin films, so called $\delta$-layers, or thin buried layers, and effects such as surface melting and segregation [64, 67, 68].

## 3.2 LEED and RHEED

Low energy electron diffraction (LEED) is currently one of the most widely used surface analytical tools. Historically, LEED was the first experiment to show the wave nature of the electron [69] but it took several decades before it entered into surface physics [70]. The reason for the delay was the lack of controllable ultra-high-vacuum (UHV). The experimental setup is rather simple and did not change much over the years (Fig. 18). In addition to the system shown here, which is called front view, reverse view systems are quite common. In reverse view the luminescent screen is on a glass hemisphere and the diffracted intensity is measured by a vidicon sytem through the glass. More recently, micro-channelplates (MCP) replaced the luminescent screens. These can be combined with a fluorescent screen or with a position-sensitive anode. In the latter case, direct data processing is possible without a vidicon. This is of interest for low current LEED because when studying sensitive material, for example, organic molecular layers, the primary electron beam causes damage. Another development is high resolution systems with transfer length of the order of 1000 nm [71–75]. The basic principle of LEED is, of course, diffraction of the electrons at the 2D surface mesh of the solid. The main diffraction spots, called Bragg peaks, are located at the positions estimated from the Braggs equation and the electron wavelength is obtained from the de Broglie equation. The electron energies used are in the range of 20–200 eV.

Visual observation of the diffraction spots, whether on the screen or via vidicon or MCP on the screen of the computer, are already very useful for probing the symmetry of the crystal surface. It is of course necessary to observe the spots at different primary energies in order to avoid misjudgments due to a particular destructive interference at a given energy. Further, step structures can be elusive if the energy of the primary beam is not varied.

For quantitative crystallographic analysis the intensity of the diffraction spots has to be measured as a function of the primary beam energy respective of the voltage, that is, the so-called I–V curves. The theory for dynamic LEED analysis has been developed over many years [73, 76–79]. Kinematic theory fails in the

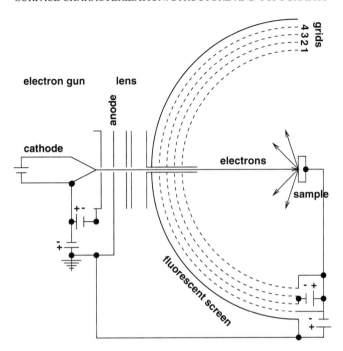

FIG. 18. Scheme of LEED experiment with a fluorescent screen, three-grid system, Faraday cup, and Vidicon system. The diffraction spots are observed visually directly or via the camera and a monitor. The beam intensities can be measured using the Faraday cup or a digitized Vidicon system [1]. Reprinted with permission from M. Schleberger, S. Speller, and W. Heiland, *Exp. Meth. in the Phys. Sci.* 30, 241 (1998), © 1998, Academic Press.

case of LEED due to multiple scattering effects, for example, interlayer diffraction. Results of many I–V curve analyses are sample in Reference [80].

As an example for such an analysis Figure 19 shows I–V curves of a $Au_3Pd(110)$ surface in comparison with calculated results [39]. The experimental data were obtained from the clean surface using a video LEED system. The calculated data are based on the Barbieri–Van Hove symmetrized automated tensor LEED package [81, 82]. For estimating the effect of the random enrichment of one of the components, either Au or Pd, the average $t$-matrix approximation was used [83]. The real part of the inner potential was set to 10 V and optimized using reliability factor (R-factor) analysis. The imaginary part of the inner potential was fixed at 5 V. The calculations were extended over an energy range of 80 to 380 eV.

An important point in the I(V) curve analysis is the evaluation of the quality of the comparison between experiments and LEED calculation. The so-called

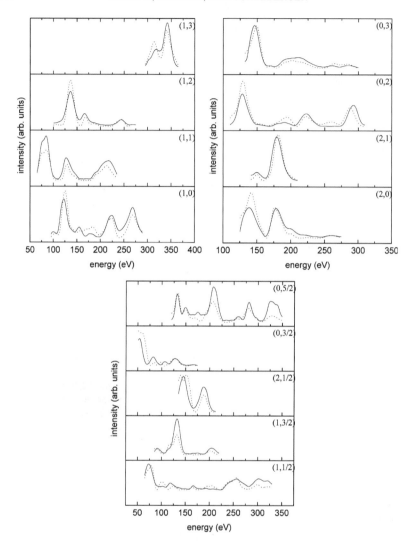

FIG. 19. Comparison between experimental (solid lines) and calculated (dotted lines) I-V curves of a Au$_3$Pd (110) surface. All data are acquired at normal incidence. The calculated data correspond to the minimum of R factor analysis [39]. Reprinted with permission from J. Kuntze et al., *Phys. Rev.* B60, 9010 (1999), © 1999, The American Physical Society.

R-factor analysis on the basis of Pendry, $R_P$, or modified Zanazzi and Jona, $R_{MZJ}$ is usually successfully applied [84, 85]. The results of the LEED experiments and calculations in comparison with a theoretical estimate are summarized in Table II. The structural parameters are defined in Figure 20 [39]. The

TABLE II.    Structure and Composition of $Au_3Pd(110)$ Determined by LEED and a Monte
Carlo-Embedded Atom Model Calculation (Sim.)

| Parameter | $Au_3Pd(110)$LEED | $Au_3Pd(110)$ sim. |
|---|---|---|
| $d_{12}$ | $1.23\pm0.04$ Å | $1.29$ Å |
| $d_{13}$ | $1.43\pm0.04$ Å | $1.43$ Å |
| $\Delta z_3$ | $0.14\pm0.04$ Å | $0.03$ Å |
| $d_{34}$ | $1.41\pm0.04$ Å | $1.41$ Å |
| $\Delta y_2$ | $0.01\pm0.07$ Å | not calc. |
| $c_1$ | $1.0\pm0.13$ | $0.96$ |
| $c_2$ | $1.0\pm0.13$ | $0.94$ |
| $c_{bulk}$ | $0.75$ fixed | $0.75$ fixed |

The parameters $d_{12}$ to $\Delta Y_2$ are explained by Figure 20; $c_1$, $c_2$, $c_3$ and $c_{bulk}$ are the Au concentrations in the first, second and third layer and in the bulk, respectively. The interlayer distances refer to center-of-mass planes. The error bars have been estimated using the variance of the Pendry R factor [39].

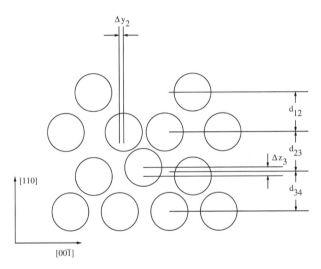

FIG. 20. Schematic missing row model of a fcc (110) surface (1 × 2 reconstruction). The structural parameters are, for example, evaluated for $Au_3Pd(110)$ using LEED and a combined Monte Carlo and embedded atom model calculation (Table II) [39]. Reprinted with permission from J. Kuntze et al., *Phys. Rev.* B60, 9010 (1999), © 1999, The American Physical Society.

$Au_3Pd(110)$ surface is reconstructed as the clean Au(110) surface in the so-called missing row (1 × 2) structure. The top surface layer is strongly enriched in Au as shown by the LEIS data as well [39]. Athough good agreement is obtained for the case shown here—and many other LEED analyses—we note that even the

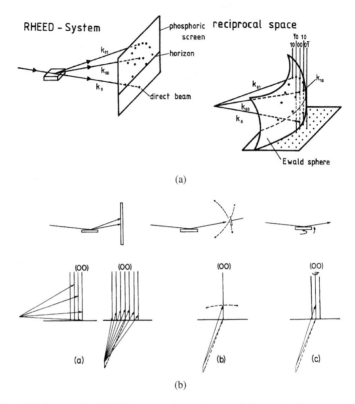

(a)

(b)

FIG. 21. (a) Scheme of a RHEED system in real space (left) and in $k$-space (right) with Ewald sphere construction and the Miller indices for different beams, which correspond to intersections of diffraction rods with the Ewald sphere. Different experimental schemes realized in RHEED in real space and in reciprocal space. (b) The registration of patterns as in (a). (b) Measurement of angular profiles. The dashed line shows the movement of the detector or the pass of the diffracted beam across a fixed detector or a screen. (c) Measurement of rotation plots or rocking curves. Rocking curves are obtained for a set of fixed crystallographic directions, usually low index directions, and by varying the glancing angle of incidence between 0° and 5° equivalent to 0 mrad to 100 mrad. (Reprinted with permission of the authors and Academic Press). (c) Schematic RHEED patterns for different types of defects on a single crystal surface. (Reprinted with permission from [1], © 1998, Academic Press.)

very best LEED analyses cannot *unambiguously* determine surface crystal structures [86].

This situation is continuously improving, for example, by taking the diffuse parts of the LEED intensity into account. As in the case of X-ray structure analysis, the intensity between the Bragg peaks carries additional information [73, 87].

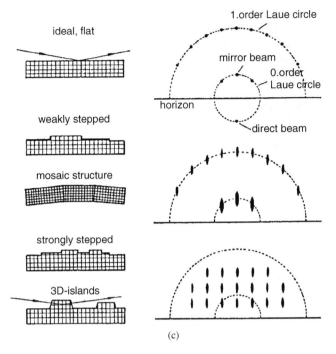

(c)

FIG. 21. Continued.

Reflection high energy electron diffraction (RHEED) takes the electron energy into the 10 keV range. This implies then grazing incidence (Fig. 21). Figure 21a shows the scattering geometry in both real space and $k$-space. The situation is of course equivalent to X-ray scattering. However, for the evaluation of RHEED patterns a kinematic approach is a good approximation. Several arrangements are being used experimentally (Fig. 21b). The observation of the pattern (a), for example, for the observation of thin-film growth, is very straightforward. Alternatively, angular profiles (b) or rocking curves (c) are measured. The rocking curves contain full crystallographic information [72, 88]. Another interesting aspect of the RHEED technique is the possibility of studying defect structures in a more direct way than by LEED. Figure 21c [5] shows the important defect structures in, for example, thin-film growth.

For the effect of 3D islands in RHEED, Figure 22 serves as our example, in comparison with an STM topographic image (Fig. 23) of the same surface [89]. The surface is $Pt_3Sn$ (001), which is prepared by sputtering and annealing. The sputtering causes depletion of Sn in the near surface region—preferential sputtering—which is the cause for pyramidal formation. These structures are metastable and they are formed at moderate temperatures of $\approx$600 K. Anneal-

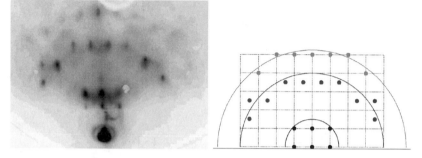

FIG. 22. Experimental (right) and schematic (left) RHEED patterns of the Pt$_3$Sn(001) surface after sputtering and low-temperature annealing. The main features are transmission spots lying on horizontal lines rather than on Laue circles. The electron energy is 12 keV, direction of incidence is along [100]. Compare with Figure 21c, Figure 23 (STM) and Figure 24 (LEED). The transmission spots are due to small pyramids on the surface [89]. Reprinted with permission from M. Hoheisel et al., *Phys. Rev.* B60, 2033 (1999), © 1999, The American Physical Society.

FIG. 23. An STM image of a Pt$_3$Sn(001) surface after sputtering and annealing low temperature annealing. The scan width is (3200 Å)$^2$, the bias voltage $U = 0.9$ V, the tunneling current $I = 1.0$ nA [89]. Reprinted with permission from M. Hoheisel et al., *Phys. Rev.* B60, 2033 (1999), © 1999, The American Physical Society.

ing at higher temperatures (1000 K) destroys the pyramids and large flat terraces are formed with residual atomic chains [89]. It is interesting to note that the metastable structures of the Pt$_3$Sn(001) surface have been found in an early LEED study [90]. The pyramids cause a streaky LEED pattern [90], which was accessi-

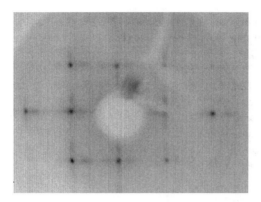

FIG. 24. The c(2 × 2)—LEED pattern of the Pt₃Sn surface under identical conditions as in Figs. refrheedpt3sn and 23. The streaks surrounding the main spots are facet spots due to the pyramidal structure seen in the STM topograph (Fig. 23) [89]. Reprinted with permission from M. Hoheisel et al., *Phys. Rev.* B60, 2033 (1999), © 1999, The American Physical Society.

ble for a quantitative interpretation at the time. Figure 24, the LEED pattern observed in Reference [89], agrees qualitatively with earlier observations. The main spots are due to the basic c(2 × 2) structure of the surface. The streaky intensity between the main spots is caused by the pyramids showing up in the STM topography. In the RHEED pattern (Fig. 22) the pyramids cause the diffraction spot to be on horizontal lines rather than on the Laue circles. This finding is an example of the schematic representation in Figure 21c. The metastable Pt₃Sn(001) serves here as an example of the need for different surface analytical tools to solve a given problem. The LEED pattern alone is certainly not sufficient. The RHEED pattern shows the presence of 3D structures. The convincing answer in the case here is given by the STM topograph. However, the diffraction methods and the STM have no answer to the chemical state of the surface, that is, the relative surface concentration. To answer that question, LEIS or AES is required. After sputtering the Sn concentration is lowered, partly restored to the stoichiometric value after 600-K annealing, and finally restored after 1000-K annealing [89]. Naturally, the question as to *why* the pyramids are formed is not answered by any of these analytical results. The other two low index surfaces of the Pt₃Sn alloy, (111) and (110), form a honeycomb network and facets, respectively, under comparable sputtering and annealing conditions [91, 92]. It is worth noting that when we have the RHEED electron gun in the system, AES can be done by simply adding an electrostatic energy analyzer (ESA). In a similar way, LEIS can be introduced, with some refinement of the sputter gun that is, to provide for a better angular beam definition. Then an ESA can be added for LEIS. In many cases the addition of an

ESA leads to an obstruction, for example, the ovens, Knudsen cells, etc. needed
for deposition of thin films prevent a direct view of the substrate. This is one reason why RHEED is in rather widespread use in thin-film deposition systems—the
electron gun and screen are remote and grazing incidence does not interfere with
film deposition. Therefore, in many cases RHEED is a real *in situ* technique. By
comparison, the conventional front view LEED system blocks essentially the full
space and the rear view system blocks half of the space available. In this respect
MEIS and RBS are also remote systems that do not take much space around the
substrate or target. In some systems transport is installed as a solution (it is the
only solution when adding STM to the tool box).

## 4 ELECTRON SPECTROSCOPY

Nanotechnology, nanostructured materials, nanotubes, etc.—during the last
several years, the Greek prefix $\nu\alpha\nu\omega$ (= dwarf), used to denote $10^{-9}$, has become
almost a synonym for modern material science. As the nanometer length scale has
become observable *and* controllable for scientists, the number of experiments investigating nanostructures have exploded and so have the number of applications
for projects addressing nanostructures. The recently founded Center of Compe-
tence for Nanotechnology has been provided 150 million German marks and the
market volume for nanomaterials in the year 2001 is an estimated 75 billion German mark—these figures underscore the scientific, economic and political importance of the subject. These new artificial designer solids have or at least promise
properties that may differ totally in terms of properties from conventional, natural
materials. The realization of nanoscale architecture yields macroscopic functions.
Thus, nanostructures are interesting from the scientific point of view and at the
same time have the potential for technological applications.

Much of the work in the nanostructure business has been done with metal layers
on metal substrates for various reasons. Technologically, however, nanostructured
metal/semiconductor systems are by far more important. Semiconductor substrates are nowadays the starting point for virtually any electronic device. They are
commonly available and inexpensive. Heterostructures consisting of a magnetic
metal and a semiconductor exhibit new physical properties with possible technological applications such as the spin polarized field effect transistor [93, 94]. In
trilayers of Fe/Ge and Fe/Si, an antiferromagnetic exchange coupling has been
found that can be increased by thermal heating [95]. Although it is not yet clear
by what mechanism the exchange coupling is mediated, it is safe to assume that
the nanostructure of the multilayer holds the key to the understanding of this intriguing new class of artificial solids.

When fabricating nanostructures one always faces the problem of how to design the solid in a way that the required properties are achieved. They depend of

course strongly on the actual details of the structure. Many systems are fabricated by means of molecular beam epitaxy (MBE). In general, on a certain substrate made of material $A$, a second material $B$ is deposited from a suitable source. In thermal equilibrium, the growth mode of any such system is governed by surface thermodynamics [97]: the energy balance of the different free energies of the substrate, the deposit, and the interface determine the growth mode and thus the morphology of the system. If the sum of the interfacial energy $\gamma_I$ and the deposit energy $\gamma_D$ is larger than the substrate energy $\gamma_S$ then 3D island growth will occur. This will leave the substrate partially exposed. Island growth is often referred to as Frank van-der-Merwe growth (FM). In the case of layer-by-layer or Vollmer–Weber (VW) growth (i.e., $\gamma_I + \gamma_D < \gamma_S$), the covering of the whole surface yields an optimum energy reduction. However, $\gamma_I$ often increases with increasing layer thickness and additionally the misfit strain has to be absorbed. Therefore, after a few layers the VW growth mode is terminated and switches to island formation. This morphology leaves the energetically favorable first layer partly exposed but at the same time allows for strain reduction. This is known as Stranski-Krastanov (SK) growth. In reality, however, these structures are often not realized because the prerequisite of thermal equilibrium is usually not given.

Transition metals generally have the highest surface free energy, whereas noble metals and semiconductors have substantially lower surface energies. This implies that even if growing of material $A$ on material $B$ works fine, the opposite will not work—a common problem in the fabrication of multilayers with their inherently symmetric structure. Another frequently encountered problem is that of the possibility of interdiffusion at the interface and/or chemical reactions between the deposit and the substrate. It is well known that, especially with the very reactive materials Si and Ge, both can happen, depending on temperature and growth conditions.

The ultimate goal for the scientist working in the field of nanostructures is the specifically controlled fabrication of a certain nanostructure and control of the corresponding interrelation *nanostructure—physical characteristic*. Naturally, suitable analytical tools are a prerequisite for that. This chapter reviews applications of the inelastic peak shape analysis to nanostructered semiconductor systems. We introduce the basic concepts for studying nanostructured materials by peak shape analysis. By analyzing spectra from a number of different semiconductor systems, we quantitatively characterize their nanostructure, addressing the structural aspects as well as the chemical ones. We show in detail that the analysis of the inelastic background in (X-ray photo)electron spectra is a reliable, nondestructive and fast technique to study such systems and discuss its advantages and weaknesses.

# PART I. QUANTITATIVE ANALYSIS OF ELECTRON SPECTRA

## 4.1 Formalism

Both X-ray photoelectron and Auger electron spectroscopy are ideally suited for (non-destructive) surface analytical purposes because of two main facts: the mean free path of electrons in solids in this energy range is short (on the order of 10 Å) and the binding energy of the excited core electrons is element specific. The energy resolved spectrum of emitted electrons thus displays peaks at kinetic energies corresponding to elastically scattered electrons and an associated background due to inelastically scattered electrons.

Because traditional compositional analysis requires the elastic peak area to be determined, a subtraction of the background of the spectrum is necessary. A common method in XPS involves simply subtracting a straight line (which is very unreliable because of the arbitrary determination of the starting points of the line). Another possibility is background subtraction according to Shirley [98]. There it is assumed that the background intensity at one point is proportional to the intensity of the total peak area above the background and the integration constant is chosen to match the background at lower kinetic energies. This purely mathematical method is widely used because of its simple implementation and it has the advantage that it is independent of the choice of starting points. However, it has been shown that errors of typically 10–50% in the quantitative result are possible [99].

Electrons Traveling Through Solids.    In order to treat the physical background correctly, one has to consider the scattering processes of the electrons traveling through the solid. Tougaard and Sigmund have studied the influence of elastic and inelastic scattering on the shape and intensity of energy spectra of electrons emitted from solids [100]. Subsequently, a reliable method to separate the peak from its associated background was developed [101, 102]. The measured energy spectrum $J(E, \vec{\Omega})$ arises from the multiple scattering events that the electrons undergo during their way through the solid and is given by the primary excitation spectrum of the photoelectron emitters $F(E_0, \vec{\Omega}_0, x)$ and the propagator function $P(E_x, x; E, \vec{\Omega})$ containing the effects of inelastic and elastic scattering events,

$$J(E, \vec{\Omega}) = \int dE_0 \int d^2\vec{\Omega}_0$$

$$\times \int dx \, F(E_0, \vec{\Omega}_0, x) \, P(E_0, \vec{\Omega}_0, x; E, \vec{\Omega}) \qquad (36)$$

where $x$ is the depth of excitation, $E_0$ is the initial energy, and $\vec{\Omega}_0$ is the direction of an excited electron.

For quantitative analysis, the primary excitation spectrum is needed. The authors assume that the inelastic scattering cross section $K(E, T)$ with $T$ being the energy loss in individual collisions, is identical for all electrons within the spectrum to be deconvoluted. They include multiple inelastic scattering but neglect angular scattering from elastic events. With $\lambda(E)$ the inelastic mean free path, $f(x)$ the in-depth profile of emitters, $E$ the kinetic energy, $\Omega_D$ the solid angle of the detector, and $\Theta$ the angle between surface normal and detector, the measured flux of electron, $J(E)$ is given by

$$J(E, \Omega_D) = \frac{1}{2\pi} \int dE_0 \, F(E_0, \Omega_D) \int ds \, e^{-is(E_0 - E)}$$
$$\times \int_0^\infty dx \, f(x) \, e^{(-x/\cos\Theta)} \sum(s) \tag{37}$$

where

$$\sum(s) = \frac{1}{\lambda} - \int_0^\infty K(T) e^{-isT} \, dT$$

Finally, the primary excitation spectrum $F(E)$ can be derived from

$$F(E, \Omega_D) = \frac{1}{2\pi} \int dE' \, J(E', \Omega_D)$$
$$\times \int ds \, e^{-is(E-E')} \frac{1}{P(s)} \sum(s) \tag{38}$$

where

$$P(s) = \int_0^\infty f(x) e^{(-x/\cos\Theta)} \sum(s) \, dT$$

**Inelastic Mean Free Path.**    The measured spectrum is a convolution of the unscattered atomic peaks with the loss spectrum. For deconvolution of a measured spectrum, both the inelastic mean free path (IMFP) and the inelastic scattering cross sections have to be known. We will first consider the problem of obtaining the inelastic mean free path, defined as the mean distance that electrons can travel inside a solid without suffering from inelastic scattering. Known to most surface scientists is the compilation of experimental data by Seah et al., the so-called *universal curve* [103]. It shows the dependence of attenuation length on the emitted electron energy for several elements. In the energy region of interest, that is, 100–3000 eV, the scatter is, however, large. In a semiempirical approach Tanuma et al. have calculated the IMFPs of 27 elements in the energy range 200 to 2000 eV [104]. The authors use optical data to take the dependence of the inelastic scattering probability on the energy loss into account and the theoretical Lindhard dielectric function to describe the dependence of the scattering probability on momentum transfer. The subsequent fit to the Bethe equation for inelastic

electron scattering in matter results in a general formula for the calculation of
IMFPs without the need of experimental data

$$\lambda(E) = \frac{E}{E_P^2 \, \beta \, \ln(\gamma E)} \, \text{Å} \tag{39}$$

where $E_P = 28.8(N_V \rho / A)^{1/2}$, $N_V$ the total number of valence electrons per
atom, $\rho$ the bulk density, and $A$ the atomic weight. The term $E_P$ represents the
free-electron plasmon energy, that is, for the oscillator strength for those electrons
that give the main contribution to the inelastic scattering, $\beta$ and $\gamma$ are element
specific values (see [104]).

**Inelastic Scattering Cross Section.**  The inelastic scattering cross section that
defines the probability that an electron loses energy $T$ per unit energy loss and per
unit path length traveled is one of the key parameters in quantitative peak shape
analysis. In the dielectric response formalism of solid electron interaction, the
cross section may be evaluated from the wave vector and frequency dependent
complex dielectric function $\epsilon(q, \omega)$ [105, 106]

$$K(E, \hbar\omega) = \frac{1}{\pi E a_0} \int_{q^-}^{q^+} dq \, \frac{1}{q} \, \Im\left(-\frac{1}{\epsilon(q, \omega)}\right) \tag{40}$$

where $q^{\pm} = (2m/\hbar^2)^{1/2}(E^{1/2} \pm (E - \hbar\omega)^{1/2})$ and $a_0$ is the Bohr radius. For zero
momentum transfer, $\epsilon(0, \omega)$ is related to the refraction index $n$ and the extinction
coefficient $k$

$$\epsilon(0, \omega) = (n + ik)^2 = \epsilon_1 + i\epsilon_2 \tag{41}$$
$$\epsilon_1 = (n^2 - k^2)$$
$$\epsilon_2 = 2nk$$

The term $\Im(-1/\epsilon(q, \omega))$ is referred to as the dielectric loss function. Structures
in this function can be correlated to bulk plasmon excitations. In the vicinity of
a surface the differential cross section for inelastic scattering has to be modified
to describe the excitation of surface plasmons. The surface energy loss function
is proportional to $\Im(-1/\epsilon(q, \omega) + 1)$. In general, the dielectric function is not
known with respect to energy and momentum transfer. Theoretical approaches to
determine the cross section therefore have to rely on model dielectric functions.
Experimentally, cross sections are determined by either optical absorption experiments or analysis of reflection energy loss spectra [107, 108] (see Section 4.3).

   In a systematic study applying the latter method to several materials, it was
found that the inelastic electron scattering losses of solid materials exhibit common characteristic features [109]. Therefore, the inelastic scattering cross section

can be replaced in most cases by a universal cross section

$$\lambda(E)K(E, T) = \frac{BT}{(C - T^2)^2} \tag{42}$$

where $B$ and $C$ are energy dependent parameters that have to be determined for each element. Some materials, for example, aluminum, exhibit a narrow plasmon loss, and are modeled significantly better by a three-parameter function [110]

$$\lambda(E)K(E, T) = \frac{BT}{(C - T^2)^2 + DT^2} \tag{43}$$

where $B$, $C$, and $D$ are energy dependent parameters. Finally, very few materials with a dominant narrow structure in the cross section are better described by an experimentally determined cross section. These can be obtained from electron energy loss spectra measured in reflection under normal incidence conditions, as was shown by Tougaard and Chorkendorff [107].

## 4.2 Data Analysis

It is important to recall that the relative intensities of the peak and its associated background depend strongly on the spatial distribution of the emitting atoms inside the solid. As pointed out earlier, electrons that initially constitute a distribution around the primary energy $E_0$ will lose energy due to inelastic scattering. This will distort the peak shape, the peak height, and the background of the measured spectrum. The degree of distortion depends on the in-depth distribution $f(x)$ of the emitters and the differential inelastic scattering cross section $K(E, T)$. That is, the shape and intensity of the background signal comprise the information on the spatial distribution of the emitters. This is shown in Figure 25, where different deposits of iron on silicon lead to a change of the elastic iron peak intensity as well as a change in the shape and intensity of the inelastic background.

Any XPS spectrum can be corrected for inelastic scattering if the in-depth profile $f(x)$, the inelastic mean free path $\lambda$, and the inelastic scattering cross section $K(T)$ are known, as shown in Figure 26. We begin with the XP spectrum of a pure homogeneous reference sample (in this case Cu). Using the tabulated IMFP ($\lambda_{Cu}(950 \text{ eV}) = 8 \text{ Å}$ [104]), and the universal inelastic scattering cross section [109], we can calculate the inelastic background of our copper spectrum, since we know the in-depth emitter profile to be $f(x) = 1 \ \forall \ x$. The parameter $B$ in the inelastic scattering cross section is adjusted so that the inelastic background becomes zero in a wide energy range (150 eV) below the elastic peak. Usually, $B$ is in the range of 2900 $\text{eV}^2$. After subtraction of the background (dashed line in Fig. 26), the intrinsic, atomic Cu peaks are obtained (solid line in Fig. 26) and can then be used as a reference for the quantitative analysis of nanostructures that involve Cu.

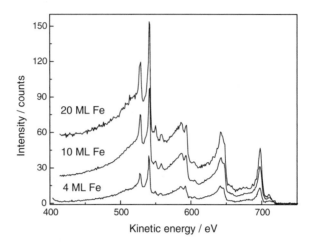

FIG. 25. The XPS spectra from the Fe *3p* core level excited by Mg $K_\alpha$ radiation. Shown are spectra for different amounts of iron deposited on an amorphous silicon substrate [96]. Reprinted with permission from M. Schleberger et al., *Phys. Rev. B* B60, 14360 (1999), © 1999, The American Physical Society.

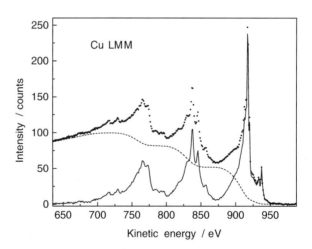

FIG. 26. Inelastic background of an X-ray-excited Cu spectrum. The dots represent the spectrum as recorded, corrected for the transmission function of the analyzer. The dashed line represents the inelastic background, calculated with appropriate $\lambda$, $B$, $K(E_0, T)$, and $f(x)$. The resulting atomic Cu reference spectrum is depicted by the solid line [113]. Reprinted with permission from M. Schleberger, D. Fujita, and S. Tougaard, *J. Electr. Spectr.* 82, 173 (1996), © 1996, Elsevier Science.

The analysis of a given nanostructure proceeds as follows: Let us assume that the sample to be analyzed consists of a substrate material $S$ and a deposited material $D$ that is distributed in some way, described by the unknown emitter profile $f_D^{Sam}(x)$. From a reference sample, consisting of the pure material $D$ extending homogeneously into infinite depth, a spectrum $J_D^{Ref}(E)$ is recorded. This spectrum is corrected for the known depth profile $f_D^{Ref}(x) = 1$. This yields the atomic spectrum $F_D(E)$ of our reference sample, such as the Cu-spectrum in Figure 26. Next, we guess a specific profile $f_D(x)$ for our sample with unknown nanostructure. The background according to this profile is calculated and subsequently subtracted from the measured spectrum $J_D^{Sam}(E)$. If the chosen profile $f(x)$ adequately models the real nanostructure, the corrected spectrum $F_D^{Sam}(E)$ will be identical the reference spectrum $F_S^{Ref}(E)$. The principle underlying the data analysis is shown in Figure 27 in the form of a flow diagram.

For the deposit/substrate system the total measured spectrum is simply the sum of the substrate spectrum and the deposit spectrum (noninterfering peaks are of course best suited for the analysis). As the two distribution functions are com-

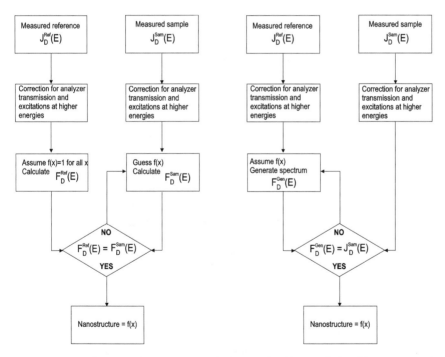

FIG. 27. Flow diagram of the data analysis when using Tougaard algorithms to correct measured spectra.

plementary, we know $f_S^{\text{Sam}}(x)$ for the substrate if $f_D^{\text{Sam}}(x)$ for the deposit has been determined. Accordingly, a cross check can easily be made if the iterative procedure is applied to the spectra of material $S$ as well. This is especially useful if the amount of material $D$ is small compared to material $S$. Then the signal from substrate material $S$ is only weakly distorted and analysis of this spectrum alone will not yield any results. Another possibility is the generation of a spectrum from a pure reference assuming a specific nanostructure. This generated spectrum $F^{\text{Gen}}(E)$ is then compared to the experimental one $J^{\text{Exp}}$ (see Fig. 27, right panel). Again, a trial-and-error procedure will yield the best structural model.

In the case of buried interfaces or compounds, the following approach can be taken—a reference spectrum of a sample with a given stoichiometry is recorded. After background subtraction ($f(x) = 1$, $\lambda$ see Section 4.2) and the usual corrections, we obtain the intrinsic spectrum for this compound. From this we generate a spectrum corresponding to a compound layer with a given thickness $d$ and starting depth $x$. From a reference spectrum of the pure material we generate a spectrum corresponding to a layer of the pure material, again with given thickness $d$ and starting depth $x$. By varying $d$ and $x$ for both layers and adding the two resulting spectra, we can build any possible structure consisting of a compound, for example, at the interface of the heterostructure and a pure layer on top. By using several references of different compounds, composition gradients can in principle also be modeled. Alternatively, if no compound references are available, appropriate island structures of the pure material can be used to model the corresponding stoichiometries. If this approach is used, all spectra should be taken in the normal exit geometry.

In-Depth Profiles. It is of course impossible to try all conceivable functions for $f(x)$. In order to investigate growth modes it is fully sufficient to choose between the three basic types of growth: Vollmer-Weber; Frank-van der Merwe; and Stranski-Krastanov. Additionally, an exponential profile can be helpful for the investigation of interdiffused materials. These structures can be represented by the following basic models depicted in Figure 28 (the (fit-)parameters needed to specify the structure are given in square brackets): A (buried) flat layer [depth, thickness], an island structure consisting of one or more island types [number of islands, height and coverage for each island type], a layer + island structure [thickness of flat layer, number of island types, height and coverage for each island type], and an exponential depth profile [surface concentration, bulk concentration, decay length]. More complex structures such as buried nonsharp interfaces can be modeled by adding the basic structures accordingly.

In practice, models consisting of more than three islands types turn out to be less useful because such structures involve too many parameters. Due to the limitation to three or fewer different island types, island distributions with broadly differing island types are difficult to resolve. One should always keep in mind that the result of the analysis is only a simple model of the true nanostructure. If, for

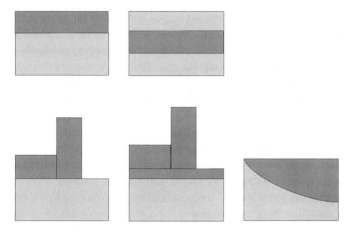

FIG. 28. Models of different nanostructures. From left to right: flat layer on top, buried layer, island, island + layer, exponential depth profile. If more complex structures are required, these basic types can be combined.

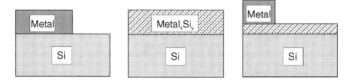

FIG. 29. Different nanostructures with the same in-depth profile of emitters if spectrum is taken in the normal exit geometry [116]. Reprinted with permission from M. Schleberger et al., *J. Vac. Sci. Technol.* B13, 949 (1995), © 1995, The American Vacuum Society.

example, a good fit is obtained with two island types, one 10 Å high, covering 50% of the surface area and the other 50 Å high covering 10% of the surface area, the interpretation should read something like this: The real nanostructure consists of many small islands of an average height of 10 Å covering half of the surface, a few approximately 50 Å high islands covering 10%, and 40% of the substrate surface area is not covered with the deposited material at all. If alloy or compound formation is present, another possible interpretation, as shown in Figure 29 should be taken into account. This problem is discussed in more detail in Part II.

The Inelastic Mean Free Path.     In the analysis the inelastic mean free path enters as a fixed parameter. As pointed out earlier, methods to calculate the IMFP for each material are available. However, when using Eq. (38), it is most likely not the elemental inelastic mean free path that should enter the deconvolution algorithm. Consider, for example, our substrate/deposit system again. Let us assume that the deposited material is distributed homogeneously in a thin layer covering the whole surface. An electron excited in the substrate will have to travel through the substrate and through the covering layer, encountering different inelastic mean free path lengths in the two different materials. While losing energy in scattering events, the inelastic mean free path also changes.

This situation becomes even more complicated if the nanostructure is not as simple, for example, if it consists of an alloy or a compound. In this case, the simplest approximation is to extrapolate the values for the pure materials according to the (known) composition of the compound (see for example, [111]). This may be a valid approximation, but it has to be tested. This can be done in the following way: The reference spectrum of the compound (homogeneous stoichiometry into infinite depth) is analyzed with varying $\lambda$ while $B$ is kept fixed. This nanostructure can be mimicked by an appropriate island model. With the IMFP for the pure material taken from the literature (e.g., [104]), the IMFP of the compound can be determined because it is now the only unknown parameter in the analysis: The agreement between the two spectra is good only if the right $\lambda$ for the analysis of the compound spectrum has been chosen.

However, even if $\lambda(E)$ and $K(E, T)$ could be evaluated exactly for the different materials one would have to know the exact path along which the electrons travel. Especially in crystalline materials, focusing effects may occur that affect the path length distribution rather strongly. According to Eq. (38) the atomic spectrum can be calculated if in-depth profile $f(x)$ is known, or as we pointed out $f(x)$ can be calculated. In both cases it is required, however, that the IMFP is constant with respect to $x$. At the surface another problem occurs. Here, due to the image charge potential, the IMFP might be totally different from the bulk value. These surface effects would have more influence as less material is deposited. The inelastic mean free path is thus never exactly known and the elemental value has to be considered as a rough estimate only—not more accurately than $\pm 10\%$. Incorporating a variable IMFP is, in principle, possible, but it has to be decided whether the additional work is worth the trouble.

**The Inelastic Scattering Cross Section.**   When analyzing data from nanostructures we need to choose an appropriate inelastic scattering cross section. Depending on the depth of excitation and the specific structure of the overlayer, the electron will scatter inelastically to a greater extent at substrate atoms or at deposit atoms. Hence, the cross section of the material that the electrons most have to travel through should be used. As pointed out earlier, the cross section can, in principle, be calculated. However, conventional methods rely on adaptation to experimental optical data, which might not be available for the material under investigation. Additionally, optical absorption measurements are bulk sensitive and neglect contributions from the surface. A much better way is the calculation of the cross section from experimental reflection electron energy loss spectra (REELS) as suggested by Tougaard et al. [107].

From the REELS spectrum, the quantity $\lambda L/(\lambda + L) \times K(E_0, T)$ can be extracted, where $\lambda$ is the IMFP and $L$ is the characteristic length for the distribution of path length experienced by the electron prior to emission (from the sample). This can be approximated to $\lambda K(E_0, T)$ because $L \gg \lambda$. Using the IMFP from Tanuma et al. [104], we arrive at $K(E_0, T)$. The energy losses are small compared to the primary energy and, therefore, $K(E_0, T) \simeq K(T)$ is a valid approximation. Figure 30 shows $\lambda K(T)$ of different materials as calculated from experimental REELS data. Note that the losses obtained by this method are not exactly identical to the losses to be deconvoluted due to different trajectories of the relevant electrons through the surface layer [107].

If neither optical data nor REELS data are available, the so-called universal cross section is a reliable alternative. For the analysis of Fe/Ge and Fe/Si samples, for example (see Part II), a cross section is needed that describes the loss features of all the materials, Fe, Si, Ge, $Fe_xSi_y$, and $Fe_xGe_y$, reasonably well. Investigating several metal/Si systems, we compared the quantitative results obtained with both the universal and the calculated cross section. Only minor deviations in the numbers proved that the universal cross section is indeed a good approximation to the real one. Thus, even materials with prominent plasmon losses, such as Si, can be analyzed using the universal cross section.

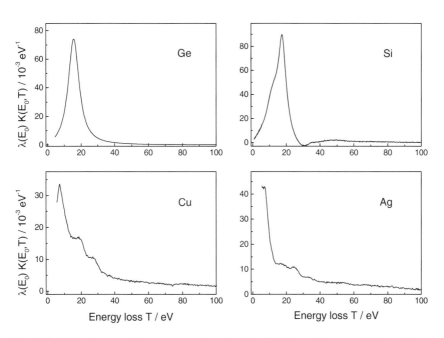

FIG. 30. Inelastic scattering cross sections for Ge, Si, Cu, and Ag as calculated from REELS data.

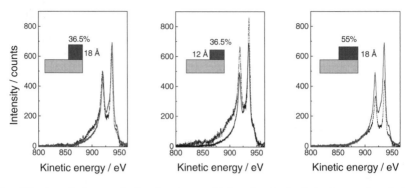

FIG. 31. A typical example showing the sensitivity of an inelastic background analysis. A total amount of approximately 5 ML Pt was deposited on a Si(111) substrate. Shown are the Pt4$d$ core levels excited by Mg $K_\alpha$ radiation. The solid line represents the reference spectrum from pure Pt, the dotted line depicts the fit of the measured spectrum from 5 ML Pt on Si(111), assuming the nanostructure shown in the insets [112]. Reprinted with permission from M. Schleberger et al., *Surf. Sci.* 331–333, 942 (1995), © 1995, Elsevier Science.

**Sensitivity.** In this subsection we wish to show how sensitive the analysis is with respect to the structural parameters that are determined. This should provide a measure to evaluate the limitations of the peak shape analysis. As a first example we present the analysis of a thin Pt film deposited on a Si(111) substrate. For experimental details we refer the reader to Reference [112]. The analysis of the Pt4$d$ spectrum was performed according to the procedure described in Section 4.2. The best result for this system was obtained with a single island type model, with the single island 18 Å high and covering 36.5% of the surface. The corresponding fit is shown in the left panel of Figure 31. If we now change the parameters slightly that is, either the coverage or the height of the islands, we obtain the fits shown in the right panels of Figure 31. Both are clearly worse.

The parameters are of course not independent of each other. Due to the fixed amount of material deposited on the substrate, the product of coverage times height is constant. Nevertheless, it is not sensible to vary the parameters individually within their possible limits. On the other hand, this means that even if an exact match cannot be obtained, it is still possible to extract information about the nanostructure in the sense that certain nanostructures can be completely ruled out. This is illustrated in Figure 32. Again, a single island was the best structure to model the spectra taken from a Ge deposit on Si(100), as illustrated in Figure 32a. Now we change the model from island to layer + island, assuming an additional 1.4-Å thick wetting layer of Ge corresponding to one monolayer (ML) of Ge. This gives an optimum fit with an island coverage of 41% (Fig. 32b). In the second try, we assume a wetting layer of two monolayers. The island coverage is

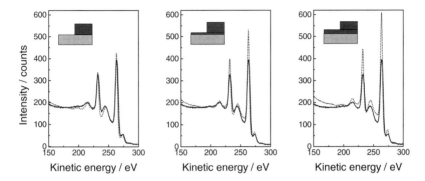

FIG. 32. The Ge$2p$ core level peaks excited by Al $K_\alpha$ radiation. The solid line represents the Ge$2p$ reference spectrum, the dotted line represents the spectrum from the Ge/Si(001) sample, assuming the nanostructure shown in the respective insets. The Stranski-Krastanov morphology can be ruled out for this system [127]. Reprinted with permission from M. Schleberger et al., *J. Vac. Sci. Techn.* A15, 3032 (1997), © 1997, The American Vacuum Society.

now reduced to 36% for the optimum fit (Fig. 32c). The fit with 1 ML is slightly worse than for pure island growth, while the fit with 2 ML is clearly worse. This behavior continues for thicknesses >2 ML. Consequently, pure island growth is the best model structure and we can exclude the existence of a pure wetting Ge layer of more than 1-ML thick.

### 4.3 Auger Electron Spectra

The theoretical considerations given in Section 4.1 are valid for electrons traveling through a solid material. The physical picture that leads to the deconvolution algorithms is independent of how these electrons were originally excited. In the case of X-ray excitation, it is safe to assume that the photoelectrons are excited uniformly to essentially infinite depths. Therefore, we can analyze our reference spectra with $f(x) = 1\ \forall\ x$. In the case of electron excitation, however, the assumption of a homogeneous excitation profile is no longer valid because the attenuation length of the electrons is much shorter than for X-rays (nm vs $\mu$m).

In analyzing electron-excited Auger electron spectra the deconvolution must take the depth excitation function into account. This function is usually not known and, hence, the inelastic peak shape analysis method is rarely used to analyze conventional Auger spectra.

In a number of experiments we investigated the distribution of emitted Auger electrons as a function of the primary energy of the incident electron beam. The basic idea was the following: We take X-ray-excited Auger spectra (XAES) from a pure homogeneous material. As we know the excitation function to be $f(x) = 1$,

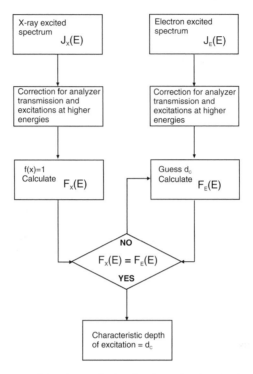

FIG. 33. Flow diagram of the data analysis when investigating the characteristic depth of excitation in electron-excited Auger electron spectra.

we can calculate the corresponding atomic spectrum. Next we record *electron-excited* Auger spectra (EAES) from the same material under the same conditions. Guessing a certain $f(x)$, we subsequently calculate the corresponding atomic spectrum from the EAE spectrum. If we now compare the atomic XAE spectrum with the atomic EAE spectrum we obtain agreement between the two, only if the excitation function is properly modeled by our choice of $f(x)$. In the analysis of nanostructures we use $f(x)$ as the function describing the in-depth profile of emitters. Here, we apply $f(x)$ to describe the excitation function. The principle is shown in Figure 33.

We took EAE spectra from several materials at different excitation energies, ranging from 500 to 4 keV. All spectra were corrected for the transmission function of the analyzer. For further experimental details see Reference [113]. The EAE spectra need to be corrected for the background due to backscattered primaries and due to the secondary electron cascade [114]

$$B_P(E) = AE^{-m} \tag{44}$$

$$B_S(E) = BE_p^{1+q}E^q \tag{45}$$

where $A$ and $B$ are material dependent constants and $p, q$ are fitting parameters. If the peaks under consideration are either very high or very low in energy, the correction can be limited to the subtraction of the primary background or to the subtraction of the secondary electrons, respectively.

For analysis, we used inelastic scattering cross sections calculated from REELS data. As already mentioned, this should, in principle, be superior to cross sections calculated from optical data because it also reproduces surface loss features such as surface plasmons. As the fine structure of $K(E_0, T)$ depends on the primary energy [108], we chose the energy of the primary electron beam to match the energy range of the Auger lines. In the subsequent analysis the depth profile in a first approximation was assumed to be a flat layer cut off at the characteristic depth of excitation $d_c$

$$f(x) = \begin{cases} 1 & \text{if } 0 \le x \le d_c \\ 0 & \text{if } x > d_c \end{cases} \tag{46}$$

Asssuming a given $d_c$ and, therefore, a given $f(x)$, the measured EAE spectra are corrected and compared to the XAE reference spectrum. The best agreement between the reference and the corrected spectra is obtained when the correct $f(x)$ has been assumed. That means that the characteristic depth of the excitation function has been determined. Note that all uncertainties in the inelastic mean free path or the parameter $B$, for example, do not come into play here. The analysis is based on only one free parameter, namely the characteristic depth $d_c$. As an example we show the fits of copper EAE spectra for three different energies (see Fig. 34). In this way we determined $d_c$ for a number of different materials. The result of our analysis is shown in Figure 35.

Electrons that travel in a solid lose a few hundred eV per 100 Å pathlength. In this energy range the ionization cross section remains roughly constant [115]. A large fraction of the primary electrons will be backscattered due to elastic scattering effects. This takes place over the length of one transport mean free path (TMFP) $\lambda_T$, which is defined as the length between two elastic scattering events. For example, for copper the TMFP is $\approx 260$ Å for an excitation energy of 4 keV [113]. The other relevant quantity in this context is the inelastic mean free path $\lambda$. A comparison of the characteristic depth obtained with our approach and TMFP or IMFP, respectively, shows that $d_c$ is almost proportional to $\lambda$. The characteristic depth is always found to be smaller than $\lambda_T$, which indicates that the attenuation of the zero-loss signal is determined mostly by the IMFP. The depth of averaged excitation regions is almost proportional to the elastic penetration depths $d_P = \lambda \cos \Theta$. For materials with a short inelastic mean free path and at sufficiently high excitation energies, quantitative peak shape analysis of electron-

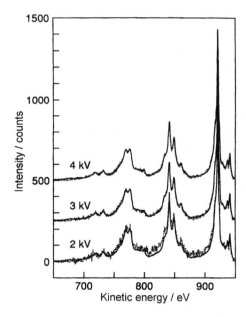

FIG. 34. Three of the corrected measured electron excited Cu LMM Auger spectra (dotted lines) for different primary energies. The background due to the secondary electron cascade and the backscattered primary electrons has been subtracted prior to correction. The reference from the X-ray excited Cu LMM Auger spectrum (see Fig. 26) is represented by the solid line [113]. Reprinted with permission from M. Schleberger, D. Fujita, and S. Tougaard, *J. Electr. Spectr.* 82, 173 (1996), © 1996, Elsevier Science.

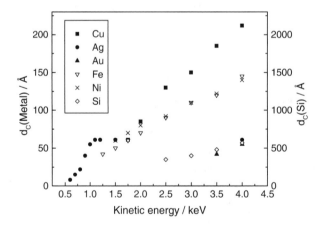

FIG. 35. The characteristic depth of excitation $d_c$ for several materials determined by the procedure shown in Figure 33. The right axis is valid only for the Si data. Data taken from [113, 244] with permission from all authors and Elsevier Science.

excited Auger spectra is thus possible without taking the depth excitation function explicitly into account.

## PART II. NANOSTRUCTURES OF SEMICONDUCTOR SYSTEMS

### 4.4 Metals on Si(111)

One of the major problems to be solved in the fabrication of technologically important semiconductor systems is that of finding the proper conditions for ideal growth. Many papers have addressed this issue but due to the many possible systems and modes of preparation, there is no attempt to provide a complete list of references here. The growth mode and possible interdiffusion of thin metallic layers into silicon is very difficult to determine by conventional surface analysis. Therefore, we have applied the inelastic background anlysis of X-ray excited electron spectra to a variety of metal/Si(111) systems [112, 116]. As a typical example for the analysis, Figure 36 shows the fit for a platinum deposit on Si(111) assuming a layer-by-layer-growth model (Fig. 36a) and a single island model (Fig. 36b). The better fit is evidently obtained with the second model. In this way, room-temperature (RT) growth mode as well as annealing behavior have been studied

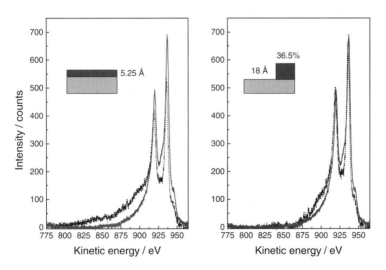

FIG. 36. Analysis of 5 ML of Pt deposited on Si(111). Shown are the Pt4$d$ core levels excited by Mg $K_\alpha$ radiation. The solid line represents the reference spectrum from pure Pt, the dotted line depicts the fit of the measured spectrum assuming the nanostructure shown in the insets. The flat layer is clearly not the right structure [112]. Reprinted with permission from M. Schleberger et al., *Surf. Sci.* 331–333, 942 (1995), © 1995, Elsevier Science.

in detail for several materials (Cu, Ag, Pt, Au). In general, the agreement between the fit and the experimental data is quite good for the metal-Si(111) systems and the reliability of the structural models is thus high. Experimental details can be found in the original papers.

Note that by inelastic peak shape analysis the in-depth distribution of atoms is determined. This is sometimes not sufficient to establish the complete structure when metal silicides are formed. For example, a pure island structure might as well be interpreted as a distribution corresponding to a metal/silicon alloy layer plus possibly islands of pure metal (a structure similar to SK), because the in-depth distributions of atoms are identical for these structures (see Section 4.2, Fig. 29). These two possibilities may be distinguished by observing chemical shifts of the metal peaks.

Cu/Si(111).    In the case of copper, we find that relatively high islands are formed at room-temperature deposition. Annealing leads to a decrease in height as well as in coverage and is evidence for diffusion of copper into the bulk, possibly in the form of a copper silicide. For copper at RT an island growth mode has been observed by both Auger electron spectroscopy (AES) [117] and scanning tunneling microscopy (STM) [118] and Cu is known to form metal-silicides at RT [119]. It has been reported that at elevated temperatures a monolayer of silicide forms, on top of which a 3D island of presumably copper silicides grows. This is in very good agreement with our analysis.

Pt/Si(111).    The deposition of platinum on Si(111) leads to the growth of high islands. These islands at first become even higher when annealing the sample and remain high thereafter. This behavior is unique and underlines the very strong tendency of platinum for island growth over a wide temperature range. Again, diffusion of platinum into the bulk occurs. For platinum it has been reported that film growth at RT involves the formation of 3D islands and that the deposited layer consists of unreacted metal grains covered with silicidelike material [120]. This is fully consistent with the results of the present work, which points to strong island formation at RT. This structure is reported be stable up to 200 °C; above 200 °C the silicide phase is growing. At temperatures higher than 400 °C all the Pt has reacted and Si segregates to the surface [120]. The latter explains the loss of material observed by us at elevated temperatures and is also consistent with the island type structure we found.

Ag/Si(111).    Silver is the only metal in the present work that possibly leads to a Stranski-Krastanov morphology upon deposition on Si(111). This type of nanostructure is preserved during slight annealing but changes to pure island growth at higher temperatures. As with platinum we find agglomeration upon annealing, that is, the islands grow higher, while the coverage decreases. For silver it is well known [121, 122] that at RT the growth mode is characterized by the formation of short islands at a very high coverage. At temperatures >250 °C tall 3D islands on top of a thin 2D layer are formed (SK growth mode), as was again confirmed by SEM [123] and ion collision scattering spectroscopy [124]. The findings at RT are in accordance with our data, although we are not able to verify whether it is a single island or an island plus thin layer structure. The change from short to tall islands at higher temperatures is confirmed by our analysis. At higher temperatures SK growth consisting of a thin layer of *pure silver* plus islands of pure silver, has been reported [122, 124]. This contradicts the present results. Our analysis clearly shows that silicon is present at the surface. Therefore, the structure must be either pure islands of Ag or a thin layer of *silver alloy* plus possibly islands of pure silver, that is, a kind of SK growth.

Au/Si(111).    Upon deposition of gold on Si(111) islands are formed. These islands grow higher if more gold is added. During annealing the height decreases steadily with increasing temperature. The coverage decreases fast at the beginning, but then remains constant (25–30%) up to $\simeq 750\,^{\circ}$C. Again, gold diffuses into the bulk during annealing. For gold the formation of an alloyed interfacial region takes place at RT. It has been suggested that for coverages exceeding 5 ML, a 5-ML-thick $Au_3Si$ alloy is formed. Between this silicidelike layer and the silicon substrate the metallic Au nucleates [125]. As we found a chemical shift of the gold peak, the structure cannot be pure island formation. Combining the results in [125] with ours, we arrive at the following conclusions: On top of the Si substrate we have a metallic gold layer covering roughly 60% of the surface, on top of which a 5-ML-thick $Au_3Si$ alloy is formed. At temperatures of up to 400 °C the migration of Si atoms is enhanced, which results in a progressive decrease of the average gold concentration [122]. This would explain the loss in material we observed. The SEM studies showed that at temperatures >400 °C, 3D islands form that cover only 20% of the surface [126]. Consistent with this, our analysis in that temperature range shows island formation and a constant coverage of $\approx25\%$. As with silver, the pure island formation might also be an SK-type growth consisting of a thin layer of *gold alloy* possibly plus islands of pure gold.

In conclusion we have shown that the growth mode for copper, platinum, and gold is island formation. Only for silver do we find evidence for a Stranski-Krastanov morphology. While all materials show island formation, the detailed structural parameters as well as the change in morphology during annealing are different. For copper and gold we find that both exhibit a loss of material at relatively low temperatures. Both elements are known to form silicides at low temperatures. We assume that because of the silicide formation some of the metal is diffusing deep into the bulk, out of range of our probing depth of $\approx\ \leq5\times\lambda$. For Ag and Pt a redistribution of material also takes place at rather low temperatures but the metal atoms remain at the surface. Figure 37 schematically shows the different nanostructures and the corresponding change during annealing for the different materials.

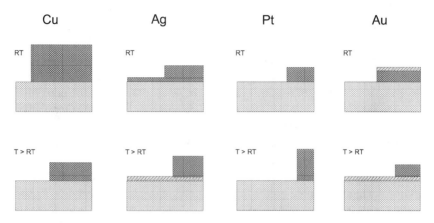

FIG. 37. Schematic nanostructures of Cu, Ag, Pt, and Au as determined by inelastic peak shape analysis. Shown are the respective structures for deposition at room temperature (RT) and the generic change after annealing the sample (T > RT). Dark gray, metal; light gray, silicon; hatched, silicide. Drawing is not to scale.

## 4.5 Ge on Si(001)

The functionality of integrated electronic circuits depends mainly on the details of the electronic band structure of the material. The bandwidth of possible architectures could be greatly enhanced if the bandstructure could be customized. One way to obtain this is by using heterostructures consisting of two different materials grown epitaxially on top of each other. The band offset between the two materials leads to a band bending due to the space charge at the interface. Silicon/germanium is not a lattice matched system and the misfit influences the electronic band structure significantly. This can be used for band engineering. Additionally, the chances that Si/Ge heterostructures can be easily adapted to conventional Si-based integrated circuits are high. Therefore, this material combination is interesting from the basic research and the technological points of view. Potential applications of such structures include integrated optical as well as electronic devices.

We ran a number of experiments with Ge deposited on Si(001) wafers and studied the morphology with inelastic peak shape analysis of X-ray excited electron spectra. Additional measurements by Rutherford backscattering (RBS) and atomic force microscopy (AFM, see Part IV) were performed to give complementary information on the nanostructure of this system. For details of the experiment and the data analysis the reader is referred to the original publications [127, 128].

We prepared five different samples with increasing amounts of germanium. The nominal amount that was deposited was equivalent to 4, 8, 12, 15, and 35 ML (sample no. 1-5) of germanium, respectively. Analyzing our data we found that for

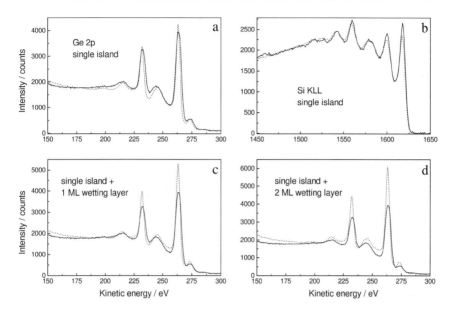

FIG. 38. Analysis of a germanium deposit on Si(001), sample No. 3. Shown are the Ge2$p$ data and the Si$KLL$data (upper panel). In the lower panel a Stranski-Krastanov growth mode with a 1 ML (left) and a 2 ML (right) wetting layer was assumed. The solid line is the generated spectrum from the pure references, the dotted line represents the measured spectrum from sample No. 3 [127]. Reprinted with permission from M. Schleberger et al., *J. Vac. Sci. Techn.* A15, 3032 (1997), © 1997, The American Vacuum Society.

all samples the growth mode of Ge on Si(001) is island formation. As an example Figure 38a shows the best fits within the single island model for sample No. 3. The islands are about 45±5 Å high and cover approximately 45±5% of the Si surface. The analysis of the corresponding Si$KLL$ spectrum is shown in Figure 38b. Again, a single island gives the best fit and the parameters are very close to those from the Ge2$p$ analysis. The model island is about 52±10 Å high and covers ca. 53±5% of the surface area. This result agrees well with the so-called hut clusters that have been observed for this system with STM [129].

By applying the same procedure as described above to all the samples we determined the nanostructure of all samples as shown in Figure 39. The results of the analysis of the Si$KLL$ and the Ge2$p$ lines are consistent. For the first two deposits, the coverage increases strongly, whereas the height varies only slightly. For the last three deposits, the coverage is constant and only the height increases.

Since SK-growth has previously been reported for this system [130, 131], we tried also a model structure that consisted of a thin layer with a single island type on top. The best fits we obtained for this Stranski-Krastanov structure from sample

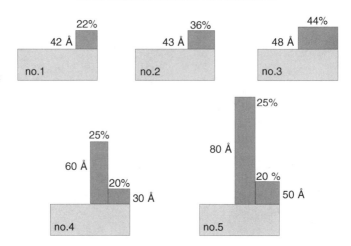

FIG. 39. Structure models for the different samples (Nos. 1–5). Light gray, silicon; dark gray, germanium. Numbers as determined by inelastic peak shape analysis of Ge2$p$ as well as Si$KLL$ spectra. Even at the highest deposit no complete coverage of the surface area is obtained. The schematic drawings are not to scale.

no. 2 are shown in the lower part of Figure 38. A wetting layer of 1 ML (Fig. 38c) and of 2 ML (Fig. 37), respectively are shown. In comparison with the pure island growth (Fig. 38a and b upper panel), the Stranski-Krastanov model results in optimum fits which are worse. Consequently, pure island growth is the model structure that gives the best fit to our data, and we must exclude the existence of a pure wetting Ge-layer more than 1 monolayer thick.

We compared our results with structural parameters obtained from AFM measurements (see [127, 128] for experimental details). The AFM results are represented in Figure 40a,b,c,d corresponding to sample Nos. 2, 3, 4, and 5. Sample No. 1 is not included because the AFM picture is very similar to that of sample No. 2. The morphology of the first two samples differs from the nanostructure found by inelastic peak shape analysis. This might be due to the geometry of the tip, as the curvature and quality of the AFM tip will affect the structures observed in AFM images [132, 133]. The oscillations in Figure 40a may be the traces of a poorly resolved island structure with an average island spacing of $\simeq$60–100 Å. If we assume that the islands have the height and coverage determined from our XPS analysis ($\simeq$40 Å, 36%) a normal tip curvature radius of 20–600 nm [132] will be far too big to resolve such a structure. For the very high deposits (sample No. 5), the AFM is better suited to determine the nanostructure. Due to extremely high islands, the XPS signal is saturated and no longer yields no reliable height information.

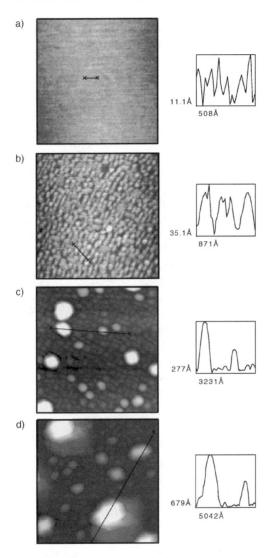

FIG. 40. The AFM images and corresponding line scans from sample Nos. 2–5. The area covered in all frames is 5000 × 5000 Å, the scale for the line scans as indicated [128]. Reprinted with permission from A. Cohen Simonsen et al., *Thin Solid Films* 338, 165 (1999), © 1999, Elsevier Science.

In conclusion, the deposition of Ge on Si(001) leads to the formation of a variety of islands, ranging from small, regular islands, covering roughly half the surface area of the substrate, to the well-known hut clusters of medium size and, finally, to extremely high single islands. No wetting layer is observed for this system. Without further tricks, say, for example, the use of surfactants, the growth of flat, epitaxial layers of Ge on Si(001) is thus highly unlikely.

## 4.6 Amorphous Fe/Si and Fe/Ge

Multilayers consisting of iron and silicon, and germanium, respectively, have been studied by several groups because of their magnetic properties. The field of spin-electronics with ferromagnetic metal/semiconductor heterostructures is rapidly growing and, therefore, information about compound formation, alloying and the structure of metal/semiconductor interfaces is much needed. Conventional magnetic multilayers consist of a magnetic material (Fe, Co, Ni, or alloys thereof) and a metallic but nonmagnetic spacer. The magnetization of the magnetic layers is coupled via the exchange coupling. As the exchange coupling itself is oscillatory, the actual coupling behavior of the multilayer depends strongly on the thickness of the spacer [134]. Many different material combinations have been experimented with and as a consequence the giant magnetoresistance (GMR) was discovered [135], that is, the dependence of the ohmic resistance of such a multilayer on the magnetization. If the magnetization of the magnetic layers is parallel (ferromagnetic coupling), a current applied perpendicular to the layer is higher than in the case of antiparallel alignment of the magnetization (antiferromagnetic coupling). This can be explained in terms of the spin-dependent scattering of the charge carriers in the multilayer [136] and will not be discussed in detail here. The GMR effect is technologically important because it allows realization of simple and cheap detectors for magnetic fields. Today many sensors based on the GMR effect (hard disks) and magnetic memory chips (MRAM) are currently being developed.

In 1994 Toscano et al. [137] used a semiconductor consisting of amorphous silicon as a spacer material. The general idea behind this was to utilize the semiconducting properties of the spacer to influence the magnetic behavior of the multilayer. When the temperature of a semiconductor is increased, the number of charge carriers increases. If these charge carriers mediate the coupling, such a *semiconducting magnet* should exhibit a positive temperature coefficient—increasing temperature yields an increasing magnetic exchange interaction. Toscano et al. found antiferromagnetic coupling for their multilayers and could also show that the coupling strength between the two ferromagnetic layers increased with increasing temperature. Their results were very controversial and were discussed vigorously, with other groups who were claiming that they found only ferromagnetic coupling with no inverse temperature dependence. It was stated that the results of Toscano and his coworkers were due to the formation of a silicide in the spacer layer and therefore their results had not been correctly interpreted. This was the starting point for our work with iron/germanium multilayers and our XPS measurements of amorphous Fe/Si and Fe/Ge bilayers.

**Magnetic Measurements.**    Our method of choice to measure the coupling behavior was spin-polarized secondary electron emission (SPSEE). After bombardment with 1 keV electrons, spin-polarized low-energy electrons are emitted. The spin analysis is carried out by a Mott detector. The secondary electrons are emitted from the first few layers only and the method is thus surface sensitive.

To measure the exchange coupling across a semiconducting spacer the following sample structure is realized: On top of our substrate (A $Fe_5Co_{75}B_{20}$ metglass ribbon) we evaporate a layer of pure amorphous iron. Due to the direct coupling of the iron to the metglass, this constitutes our magnetic driver, yielding a high remanent magnetization and a low coercive field. Next, a thin amorphous layer of the semiconducting material is evaporated. This spacer layer serves as the medium for the indirect exchange coupling. The topmost layer consists of a thin layer of amorphous iron again. It acts as a magnetic detector layer. The top iron layer shows virtually no remanence and responds quasilinearly to low external fields, such as the coupling field from the source layer. Therefore, the remanent polarization of the top layer may be correlated with the strength of the exchange coupling.

As the Fe/$a$-Si/Fe system was the only system reported to show thermally induced effective exchange coupling, we tried another system, that is, trilayers consisting of Fe/$a$-Ge/Fe. Like Toscano et al., we prepared our samples at a temperature of $T = 30$ K in order to avoid diffusion during evaporation. For germanium we found that the preparation at $T = 30$ K as well as a subsequent annealing to 200 K is absolutely necessary to obtain antiferromagnetic coupling. Neither preparation at a higher temperature nor preparation at 30 K without subsequent annealing would result in antiferromagnetically coupled layers (for experimental details see Reference [95]). The trilayers show antiferromagnetic coupling at a spacer thickness of between 20–25 Å, which is slightly thicker than for silicon. As shown in Figure 41 the coupling is indeed temperature dependent with a positive coefficient: The coupling strength between the two coupled layers increases with increasing temperature. The external field $H_{Comp}$ that is necessary to compensate the exchange coupling is, with 12 Oe at $T = 230$ K, twice as large as the field necessary at $T = 30$ K. The effect is fully reversible in the temperature range between 40 and 230 K. From the compensation field we calculate the coupling strength to be $J_{Exc} = 1.5$ merg/cm$^2$. This is comparable to the value found for Fe/Si trilayers but by a few orders of magnitude smaller than for metallic multilayers.

**XPS Measurements.**    The most likely explanation for the exchange coupling in the case of semiconductors is resonant tunneling through localized states in the gap of the spacer. This would explain the observed temperature behavior and is supported by our findings that the preparation at $T = 40$ K and the subsequent annealing of the trilayer are absolutely necessary to obtain the antiferromagnetic coupling. As the interfaces in the trilayers seem to play a vital role for the magnetic properties, we used XPS and the inelastic peak shape analysis to

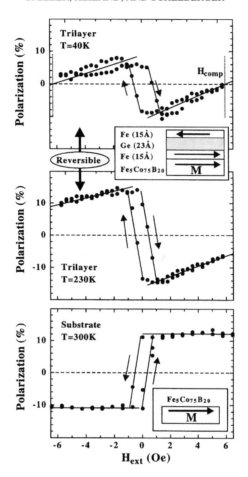

FIG. 41. Spin polarization of secondary electrons vs applied magnetic field for the Fe/Ge/Fe trilayer at different temperatures (upper panels). In the lower panel the magnetic response of the substrate is shown. The coupling behavior of the Fe/Ge/Fe trilayer is obviously antiferromagnetic [95]. Reprinted with permission from P. Walser et al., *Phys. Rev. Lett.* B80, 2217 (1998), © 1998, The American Physical Society.

study amorphous Fe/Ge and Fe/Si systems. To exclude the signal from the second interface, the samples were prepared as bilayers of iron and germanium, and silicon, respectively. The preparation conditions were the same as for the magnetic measurements, except for the Cu(111) substrate (see References [96, 111]).

As an example we show the analysis of 20 Å iron deposited on germanium. Figure 42 shows the optimum fits that could be obtained for different structural models. Obviously, the nanostructure consisting of a $Fe_{50}Ge_{50}$ compound and a

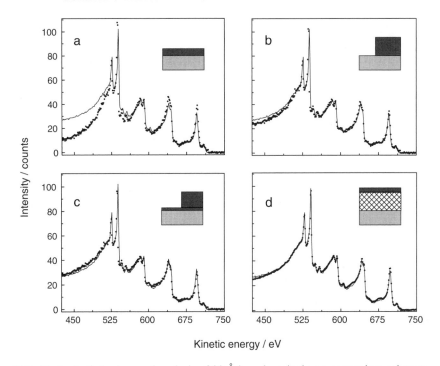

FIG. 42. Inelastic background analysis of 20 Å iron deposited on a germanium substrate. Dark gray, iron; light gray, silicon; hatched, compound. Shown are the optimum fits that could be achieved with the nanostructures shown in the insets. The morphology $d$ consisting of a compound layer + a pure iron layer on top clearly yields the best fit [111]. Reprinted with permission from M. Schleberger, *Surf. Sci.* 445, 71 (1995), © 1995, Elsevier Science.

flat layer of pure iron on top yields a superior fit. Our results indicate that the nanostructure as well as the chemical reaction of the Fe/Si interface are very similar to those for the Fe/Ge interface. From the chemical shift measurements (see Reference [96]) and the inelastic background analysis we conclude that the deposition of small amounts of Fe onto an amorphous Si or Ge substrate results in the formation of a layer of reacted compound material. The stoichiometry of the layer depends on the amount of deposited Fe but is homogeneous. The percentage of Fe increases with increasing layer thickness. For silicon as well as germanium our results indicate that above nominal Fe overlayer thicknesses of 10 Å the formation of a pure Fe layer on top of the Fe/Si compound layer takes place. These results are in good qualitative agreement with the interpretations of Kilper et al. [138]. Our result, that above a certain amount of evaporated Fe the total quantity of Si does not increase any more, confirms experimental findings from samples prepared at room temperature [139]. On the other hand, our measurements clearly exclude the formation of a sharp interface between the Fe film and the Si or Ge substrate. Also the idea of one alloy in the spacer independent of the overlayer thickness [140] or a composition gradient at the interfaces [141] must be ruled out.

In conclusion, we have shown that the spacer layers of the trilayers prepared at low temperatures by Toscano et al. [137] and Walser et al. [95] are indeed pure amorphous semiconductors. The thickness dependence, however, has to be recalibrated because we find evidence for interdiffusion at the Fe/Ge as well as the Fe/Si interface. Silicon (germanium) deposited on Fe, on the other hand, does not react within the resolution of our experiment. Therefore, in the case of Si the spacer layer thickness is reduced by $\approx \delta = 10$ Å compared to nominal spacer thickness. The $\delta$ is the amount of semiconductor material that interdiffuses into the Fe top layer. In addition, we have shown by investigating the chemical shifts of the core level peaks that the thickness of the interdiffused layer depends on the substrate temperature. This might be one reason for the quite distinct magnetic results obtained from Fe/Si/Fe layers that have been prepared at different temperatures. However, while the chemical shifts of the corresponding core level peaks for layers prepared at 40 and at 290 K, respectively, differ considerably, the difference in the amount of interdiffused material as obtained by inelastic background analysis is rather small. The question as to how the differences in magnetic behavior between the different systems can be explained still remains unanswered.

# 5 SCANNING PROBE MICROSCOPY

Deeper understanding of the relation between function and structure requires linking observed physical properties directly to the geometric atomic structures. Diffraction methods (see Section 3) can yield very precise information on the crystallography of surfaces and solids, but nonperiodic, very complex surface structures are in general not accessible by such methods. Scanning probe microscopy (SPM) techniques are the methods of choice here. A starting point was the scanning tunneling microscope, which allows measurement of the topography of surfaces with very good resolution up to subatomic features. Methodical spin-offs provide maps of various physical properties on a very local scale (up to Å).

# PART III. SCANNING TUNNELING MICROSCOPY

Scanning tunneling microscopy (STM) has become an extremely powerful method for surface topography and structure. This was the first technique capable of atomic resolution on flat surfaces. The first atomically resolved images of the $(7 \times 7)$ structure of Si(111) were published in 1982 [142]. In 1986 the Nobel price for physics was awarded to Binnig and Rohrer for the development of the STM method together with Ruska for the development of the electron microscopy method.

## 5.1 Method

For STM, both electrodes, the probe and the sample, must be conductive. A sharp tip is used as a probe, and it is usually produced by electrochemical etching of a tungsten wire in KOH or NaOH solution (see Section 5.1.1). The tip is brought close to the sample surface by piezoelements until a tunnel current in the nA-range flows through the vacuum or air gap.

A gap in between two conductive materials represents an energetic barrier for electron waves. In the media the wave can propagate quasifreely and in the barrier it is damped exponentially with the penetration depth,

$$I \propto e^{-2\kappa d} \quad \text{with} \quad \kappa = \sqrt{2m\bar{\Phi}}/\hbar \qquad (47)$$

$m$ is the electron mass, $d$ is the spatial gap width, and $\bar{\Phi}$ is the energetic barrier height, which is related to the mean work function of both electrode materials. Compared with free surfaces the work functions of proximate electrodes are lowered significantly due to the image charge.

The maximal possible current flowing through an atomically sharp tip is of the order $I_0 = U/R_K$ with gap voltage $U$ and Klitzing constant $R_K = h/e^2 \approx 25 \text{ k}\Omega$. Actual tunnel currents are of the order 1 nA. The saturation of preamplifiers used in STM is typically reached with 50 nA.

For steady electron tunneling conditions a small bias $U$ must be applied between sample and tip. A tunnel current of a few nA indicates that the distance between probe and sample is of the order of some 10 Å, that is, typical wave lengths of the valence and conduction electrons close to the Fermi level. With crystalline materials the wave lengths of the electrons contributing to the tunnel current depend on the effective mass $m^*$ and on the relative energy with respect to the bottom of the bulk or surface band $E_0$ from which the tunnel electrons originate:

$$\lambda_{\text{de Broglie}} = \frac{h}{\sqrt{2m^*(E - E_0)}} \qquad (48)$$

As the amplitude of the electron wave is exponentially damped in the gap the tunneling current is an extremely sensitive measure of the changes of distance [143]. This high sensitivity is vital in STM.

After the approach of the probe to the sample a regulation loop of the current is activated, that adjusts the $z$ fine piezoelement according to the feedback set-

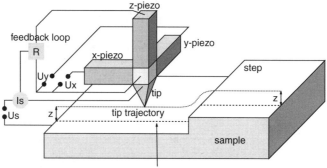

projection of the tip trajectory onto the sample surface

FIG. 43. Illustration of the working principle of the STM method.

point current. The tip scans a point lattice on the surface by moving the $x$ and $y$ piezoelements stepwise. The $z$ piezovoltages necessary to keep the tunnel current constant are taken in a data set and give a height map of the surface $z(x, y)$ (Fig. 43). When the tip is sharp and clean and the surface is flat and clean, atomic resolution is easily obtained because generally more (or less) electrons are accessible above the surface atoms than in between them. Thus, regulation retracts or lowers the tip according to the atomic structure. In this so-called constant current mode (CCM) the current (i.e., constant *microscopic* distance) is kept constant and changes in height of the tip are measured. This is the most commonly used mode. The measured height maps are actually a surface of constant surface density of states ($\rho_{sample}(r_{tip}, E)\rho_{tip}(E - eU) \overset{!}{=}$ constant). The $z$ signal is the result of both the geometrical and electronic properties of the sample surface. With higher gap voltages $U$, it is necessary to realize that the *integrated and weighted* charge density between the Fermi level $E_F$ and the energy of the applied gap voltage $eU$ enter in the tunnel current.

$$I \propto \int_{E_F}^{E_F+eU} dE \; \rho_{sample}(r_{tip}, E)T(E, U) \qquad (49)$$

here $T(E, U)$ is the transmission coefficient. This formula neglects the variation of the density of states of the tip. With metallic samples the measured map of $z$-piezo heights corresponds roughly to an envelope of the atoms in the surface. With semiconducting samples the electrons are less delocalized and the density of states (DOS) is more structured, hence STM "topography" is afflicted to a larger extent with electronic effects and is energy dependent ($eU$).

A different, much faster mode is with regulation switched off or slowed down. Then the height, that is, the $z$ piezolength, stays constant while scanning. The tunneling current, that is, also the local tip-surface distance, changes with $x$ and $y$ and is recorded. The disadvantage of this constant height mode (CHM) is that it only gives a qualitative image of the surface topography, which means no exact heights are obtained. Actually, this mode measures the regulation error that behaves qualitatively like the differentiation of the topography. The edges of geometrical and electronical features are emphasized in such images.

Today low-temperature STMs at liquid He temperature, variable-temperature STMs offering cooling and annealing, high-speed-STMs allowing movie shots, and liquid-phase STMs for the study of electrolytic surface reactions have been developed and are commercially available.

**5.1.1 Probes.** Most commonly used are tungsten probes that have been electrochemically etched in KOH or NaOH solution. For example, a 0.4-mm thick tungsten wire, initially annealed, can be used as an anode and a copper wire of equal thickness can be used as a cathode. The etching can be done in 2N KOH solution, with 16 V DC voltage and a decreasing current of $\approx 20 \ldots 3$ mA. With some conditions, for example, for tunneling on $H_2O$ films, problems arise with tungsten tips. The PtIr tips are an alternative. They can be either etched or just cut off by a pair of pliers.

With flat surfaces a mesoscopically blunt tip can yield atomic resolution because minitips form after treatment with voltage pulses up to 10 V during tunneling. The most protruding minitip is the active tunneling area. The rougher the sample the more probable are artifacts because the tip contacts larger objects sideways and the active tunnel region changes occasionally during the scan. Thereby higher objects can be imaged multiply in larger frames (ghost images) or steps on single crystal surfaces can be imaged with heights that correspond to fractions of the interlayer spacing.

## 5.2 Topographies

In view of the huge number of surface systems that have been successfully studied by STM we give in the following section examples for the different categories of systems. It must be mentioned that the interpretation of STM images can become difficult or even impossible unless other standard surface analytical tools, such as LEED/RHEED (Section 3.2), AES (Section 4.3), or LEIS (Section 2.2) are employed.

**5.2.1 Clean, Low-Indexed Surfaces.** The structures of clean surfaces are of fundamental interest because they are the basis for more complicated systems offering practical applications. On their own, clean surfaces have importance as quasi-2D systems, which can show special effects like relaxations, reconstructions, phase transitions, surface-specific defects, local mass fluctuations, and roughening transitions. In the following we concentrate on face-centered cubic (fcc) metals. The geometry of the three low-indexed fcc surfaces is shown in Figure 44.

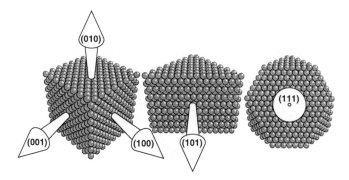

FIG. 44. Sphere models of the three low-indexed planes of the face-centered-cubic-crystal structure.

The equilibrium structure of many clean surfaces has been cleared up or corroborated by the help of STM. Early findings were the missing row reconstructions on Au(110) [144, 145] and Pt(110) [146, 147]. These reconstructions exist together with a mesoscopic step net structure called fishscale pattern [147, 148]. This structure could be detected only by STM and not by standard crystallographic methods such as LEED. The origin of the reconstruction is the higher energy of the {111} microfacets. The fishscale pattern serves to hide antiphase domain walls in order to keep the surface energy high. The surface energy is defined as the excess free energy per unit area [149]. On fcc(110) surfaces the fishscale pattern is never present with reconstructions other than $(1 \times 2)$, for example, with the $(1 \times 4)$ reconstruction of Pt(110) [150].

A remarkable exception was the Ir(110) surface. Its equilibrium structure is a mesoscopically corrugated hill-and-valley structure with {331} facets exposed, 13.3° inclined with respect to (110) (Fig. 45) [151]. Up-and-down sequences of these facets show a period of 10–100 Å (Fig. 45). Additionally, a mesoscopic waviness with periods of approximately 1000 Å is observed. With certain adsorbates, for example, oxygen, an unreconstructed surface with large terraces instead of the hill-and-valley structure can be prepared and atomically resolved [152]. At elevated temperatures a $(1 \times 2)$ reconstruction is obtained [153]. To date there is no calculation for the faceted structure of Ir(110) and its origin is unexplained. One may argue that such structures are surface stress induced, as was evoked with the corrugated iron structure of Pt(110), which is observed on a very large scale, exhibiting periods of ≈1500 Å [154].

5.2.2 Stepped Surfaces. Steps on vicinal surfaces are interesting because they represent a set of one-dimensional (1D) nanostructures. A regular step array is most often observed [155–158] though at low temperatures the energetic minimum can be a faceted surface [159]. The origin of this order is the step-step

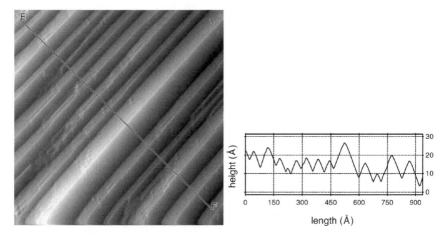

FIG. 45. The STM image of Ir(110) after standard preparation (a) and line section between points E and F. (b) 770 Å, −0.9 V, 1.0 nA. The structure consists of up- and-down sequences of {331} facets inclined by 13.3° [152]. Reprinted with permission J. Kuntze (doctoral dissertation).

repulsion. In a simple approach the step-step interaction can be divided into three contributions: entropic; electronic; and elastic interaction.

Elastic interaction occurs when the displacement fields from steps substantially superpose. Atoms located in the vicinity of steps tend to relax stronger compared to those farther away. The resulting displacements or lattice distortions decay with increasing distance perpendicular to the steps. Atoms situated in between two steps experience two opposite forces and cannot fully relax to an energetically more favorable position as would be the case with quasiisolated steps. The line dipoles at steps are due to Smoluchowski smoothing [160] and interact electronically. Only dipole components perpendicular to the vicinal surface lead to repulsion whereas parallel components would lead to attractive interaction. The dipole-dipole interaction seems to be weaker than the elastic one. For instance, steps on vicinal Ag(111) have weak dipoles as was shown in a theoretical study [161]. Entropic interaction is due to the condition that steps may not cross and leads to an effective repulsive potential, the weakest interaction type. This contribution is always present and results from the assumption that cavities under the surface are unstable. Experiments and theory investigating steps on surfaces were recently reviewed [162].

Surfaces vicinal to fcc(111) and the miscut about the $\langle \bar{1}\bar{1}2 \rangle$ direction, that is, with steps running along the dense $\langle \bar{1}10 \rangle$ direction, still can differ. Due to ABC stacking, the fcc $\{11\bar{2}\}$ planes are not mirror planes and the minifacets at the steps

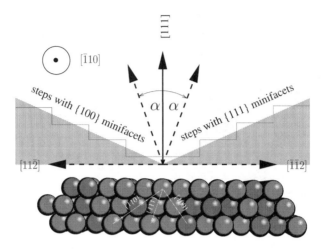

FIG. 46. Side view of a fcc(111) sphere model, illustrating the fabrication of pairs of vicinal surfaces with equal miscut but different minifacet orientation.

FIG. 47. ⟨110⟩ steps with closed minifacets on a Cu(111) surface with 11° miscut angle; 1000 Å, −1.4 V, 0.5 nA [164]. (Unpublished work by A. R. Bachmann et al.)

are either of {100} or {111} type, depending on whether the miscut is clockwise or counterclockwise (Fig. 46).

Figure 47 shows STM images of the 11° miscut samples with {111} minifacets. The parallel steps run from top to bottom. The upward direction of the steps is from right to left. Regular patterns of monoatomic steps as shown in Fig-

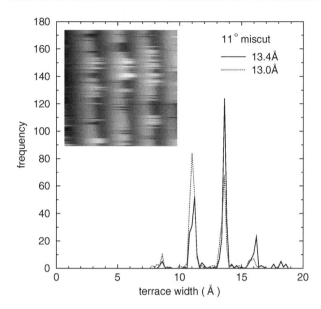

FIG. 48. Terrace width distribution histogram of $\langle 110 \rangle$ steps on $11°$ miscut surface. The inset shows a small frame of 50 Å, $-1.4$ V, 0.5 nA [164]. (Unpublished work by A. R. Bachmann et al.)

ure 47 have been observed on surfaces with {100} and {111} step types and with $5°, 7°, 9°$, and $11°$ miscuts [163]. In STM images the step edges exhibit so-called frizziness, which indicates that atoms move along step edges and that the STM under-samples the topography in time. At room temperature this is a typical feature of materials with relatively low binding energy (e.g., Pb, Ag, and Cu). The frizziness at the {111} steps was found to be stronger than with {100} steps on the $9°$ miscut surfaces [164].

Figure 48 shows a histogram of the terrace widths calculated from the STM frame shown in the inset. The maxima of the terrace width distributions are separated by $a_\perp = 2.2$ Å. Thus, one may conclude that step edge atoms spend the predominant part of time in fcc compatible hollow sites, indicating that their jump rate is moderate compared to the sampling rate (here 2.5 kHz). With lower miscuts, that is, wider terraces, quantization is less pronounced. Analyses of terrace width distributions and of step correlation functions, extracted from STM frames, allow the determination of the step-step interaction potential and the dimensionality of the diffusion processes at steps [165, and references therein].

5.2.3 Alloy Surfaces.    Alloy surfaces are important for catalysis. For instance, $Pt_3Sn$ surfaces catalyse many dehydrogenation reactions better than pure Pt. Further motivation can be material economization, for example, the dilution

of pure Au with Pd without loss of desired properties such as chemical inertness. With respect to nanostructuring, the wider spectrum of preparable surface
structures is of importance. Due to the stress in the surface region highly specific
self-ordering structures, like strain relief patterns or faceting, can be prepared.

With alloy surfaces the phenomena of preferential sputtering and segregation
govern equilibrium and metastable structures. The component with lighter atoms
is most often preferentially sputtered and hence depleted in the surface region. In
a first approximation, the component that has the lower surface energy as pure
material will segregate because this will enhance surface energy. As a rough estimate, one may compare the melting temperatures of the pure materials instead of
the surface energies. A segregation database for transition metals based on LMTO
calculations is given Reference [166]. Restoring the original composition in the
surface region after sputtering requires a higher temperature than annealing of the
geometrical sputter damages because the first requires bulk diffusion and the latter
only surface diffusion.

Two principal categories of alloys can be distinguished with binary alloys
$A_x B_{1-x}$. The component that is less efficiently sputtered segregates towards the
surface. For instance, with $Au_3Pd$ surfaces the Pd is more efficiently sputtered
and Au segregates ($m_{Pd} < m_{Au}$ and $T_{Pd} > T_{Au}$). The $Au_3Pd$ crystallizes in
the fcc structure without strict chemical order. Standard preparation leads to
a pure unreconstructed Au(100) layer on the alloy with the lattice constant of
$a_{Au_3Pd} = 3.99$ Å [46]. An unreconstructed Au layer is special because pure low
indexed Au surfaces on Au ($a_{Au} = 4.08$ Å) always reconstruct in equilibrium. In
the Au(100) quasihexagonal ($5 \times 1$) reconstruction [167, 168] the surface atoms
are compressed with respect to the bulk. One can argue that the $Au_3Pd(100)$ surface provides a substrate that allows a stress-free Au(100) layer (Fig. 49).

A flat $Au_3Pd$ surface with nonzero Pd concentration in the first layer can be
prepared when there is sputtering when the crystal is being annealed. The Pd
atoms show up in the low energy ion scattering (LEIS) (see Section 2.2) signal and by chemical contrast in STM [46]. They are measured as protrusions with
larger apparent STM height than the Au atoms. There was no indication for chemical order in the alloy surface observed. Only the vertical composition profile of
the outermost three layers shows slight oscillations as shown in a $I(V)$-LEED
study [169].

In alloys of the other category, the preferentially sputtered component is segregating towards the surface. Here $Pt_3Sn$ serves as an example, in which Sn is
segregating and preferentially sputtered ($m_{Sn} < m_{Pt}$ and $T_{Sn} < T_{Pt}$). The $Pt_3Sn$
exhibits a strict chemical order of the $L1_2$ type, that is, fcc structure with Pt at the
corner sites and Sn at the face sites of the unit cell. The depletion in Sn in the surface region leads to a smaller lattice constant ($a_{Pt} < a_{Pt_3Sn}$). All three low-indexed
surfaces of $Pt_3Sn$ respond to this depletion by formation of metastable phases
with characteristic stress compensation features (Table III). A mesoscopic dislo-

FIG. 49. Unreconstructed Au(100) on Au$_3$Pd single crystal after 800 K anneal; 90 Å, −0.3 V, 1.7 nA [46].

TABLE III.    Summary of the Structures Observed by STM on Pt$_3$Sn Surfaces After Annealing at Moderate and High Temperature

|  | 600–800 K | 1000–1100 K |
|---|---|---|
| (111) | ($\sqrt{3} \times \sqrt{3}$) R30° (Pt$_2$Sn), mesoscopic subsurface dislocation network | p(2 × 2), adatom islands |
| (100) | multiple row structure, pyramids bordered by {102} and {104} facets | c(2 × 2), double steps, single atomic ad rows |
| (110) | hill-and-valley-like structure with {102} facets | (2 × 1), double steps, holes at Sn positions |

cation network, pyramids, or ripples are formed on the different surface planes (111) [91], (100) [89], and (110) [92]. All these structures could not be identified in earlier LEED analyses [90, 170, 171]. The pyramidal phase of Pt$_3$Sn(100) is shown in Figure 50. The facets of the pyramids are of {102} and {104} orientation. Despite the depletion in Sn in the surface region, in the outermost surface layer the Sn concentration is enhanced. For instance, the outermost layer of Pt$_3$Sn(111) has the composition and structure of Pt$_2$Sn. This Sn excess can be

FIG. 50. Pyramidal phase of the $Pt_3Sn(100)$ surface after 620 K anneal; 700 Å, 0.6 V, 1 nA [89].

attributed to the higher surface energy of Sn-rich surfaces. The surface energies of pure Sn and Pt are 0.61–0.62 J/m$^2$ and 2.3–2.8 J/m$^2$, respectively [172].

The equilibrium phase of the three $Pt_3Sn$ surfaces develops after annealing to temperatures as high as 1000 K. The STM measurements, and also LEED and LEIS [90, 170], revealed that only the mixed PtSn planes of $Pt_3Sn(100)$ and (110) are observed. In the bulk these planes are stacked in layers with alternating composition of PtSn and pure Pt. Thus, bulk-truncated PtSn terraces can only be separated by doubles or multiples of double steps. This is, in fact, observed in STM images. No steps of heights corresponding to odd numbers of layers have been found [89, 92]. In contrast, fcc(111) planes have homogeneous composition and stacking. Thus, monoatomic step heights are compatible with the bulk-truncated $Pt_3Sn(111)$ surface structure in this case. This is in line with the observations that is, STM topographies that show exclusively monoatomic steps.

All surfaces have in common that the Sn between the Pt is not imaged as a bump in the STM topography but as a depression. The origin of this strong chemical contrast is the large difference in the electronic structure. Band structure calculations reveal that in the region near the Fermi level the density of states at the Sn atoms is clearly lower than that at the Pt atoms (Fig. 51). To illustrate this effect, the STM topography of the $Pt_3Sn(110)$ surface is shown in Figure 52. The bumps are the Pt atoms and the Sn atoms are the weak depressions in between. Note that due to the "invisibility" of the Sn the apparently more densely packed rows run along $\langle 100 \rangle$. Holes in the form of monolayer deep depressions are ob-

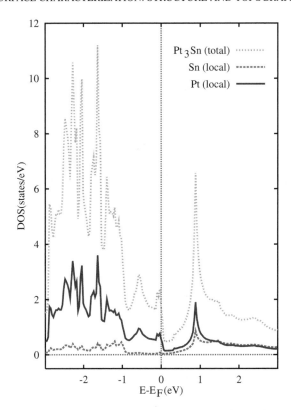

FIG. 51. Total (per unit cell) and local (per space-filling atomic spheres of equal size at both Pt and Sn sites) densities of states of Pt₃Sn, calculated by the tigh-binding linear muffin-tin orbitals method. At positive (sample-) bias voltages the unoccupied states above $E_F$ are imaged in the STM. Reprinted with permission from [89], © 1999, The American Physical Society.

served only at the Sn positions. As the sublimation of Sn is probably stronger than of Pt and the stability of single vacancies is usually low, an adatom gas depleted in Sn is the most plausible explanation for the holes in Figure 52 [92]. Discrimination between two atomic species by STM is not obtained with any system and tip. Initially, chemical contrast was obtained on PtNi alloys [173].

5.2.4 **Ultrathin Films.** The demand for artificial materials or so-called designer solids is increasing. Many applications such as solid state lasers and new generations of transistors require ever finer structuring of materials. It is very common for the properties of devices based on heterostructures to depend on the quality of the interfaces. The structures can be grown by chemical vapor deposition (CVD) or molecular beam epitaxy (MBE). Transmission electron microscopy

FIG. 52. An STM image of the Pt$_3$Sn(110) surface; 120 Å, 0.5 V, 0.8 nA. The Pt atoms are visible as protrusions (open circles), the Sn atoms are invisible (filled circles).

(TEM) and RHEED (see Section 3.2) are the standard methods used to check the crystallinity of the layers and interfaces. However, the origins for imperfect crystallinity of thicker films are not directly inferrable from such analyses. Here, understanding the evolution at the onset of growth is the goal. Besides preparation of nanostructures with novel physical properties in general, this is one of the major motivations for studying the topography and structure of incomplete layers by STM. An overview of the nucleation and aggregation of thin metal layers is given in Reference [174].

With regard to magnetoelectronics, structures of nonferromagnetic combined with (anti)ferromagnetic materials are of special interest (see also Section 4.6). As an example, we show in Figure 53 an STM image of 0.12 ML Cr on Cu(111) [175] grown by thermal evaporation. The structure is similar to Co/Cu(111) [176, 177] and Fe/Cu(111) [178]. Although Cr and Cu are not miscible in bulk, etching of the substrate occurs. Vacancy islands are observed in the vicinity of steps after initial growth. Such rearrangement contributes to interfacial roughness. It is driven by a decrease in total energy because material of lower surface energy is going to cover material of higher surface energy (Cr, Co, Fe). This process can be extenuated when the initial growth is performed at low sample temperature [179].

In Figure 53 the steps are decorated by approximately 50-Å wide Cr bands of bilayer height, measured from the upper terrace. Similar step decoration in the form of interrupted island bands was found with Co/Cu(111) [176], Fe/Cu(111) [178], and Ni/Ag(111) [180]. The islands that have agglomerated in

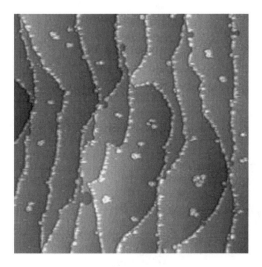

FIG. 53. 0.12 monolayers Cr on Cu(111). The height of the Cr bands at the steps is 4 Å, measured from the upper terrace. The height of the islands is 4 Å. The small indentations at the steps are caused by missing substrate material due to intermixing between Cr and Cu; 3000 Å, 2.7 V, 0.07 nA.

the middle of the terraces are also of bilayer height. Studying thicker Co films by STM and LEED revealed that one fraction of the islands occupies fcc positions and another fraction hcp positions on the Cu(111) substrate [177]. This twinning in nucleation causes problems in thick films, when the different types of islands cannot connect and canyon-like defects remain.

Another challenge is the production of 1D structures that might exhibit totally different magnetic properties [181]. Such lateral systems can be obtained by step decoration on vicinal surfaces.

Here we give an example using substrates with terraces of nanoscopic width. This will suppress the formation of the typical form of decoration bands of incompletely connected islands as observed on the low indexed (111) surfaces. There are striking differences in the morphology of vicinal surfaces with {100} and {111} step minifacets. The most remarkable effect of the Co is that on the surfaces with the {100} steps the step array is rearranged into a configuration where double steps prevail (Fig. 54). These rearrangements afford considerable mass transport in the substrate surface. The double steps do not exhibit frizziness. Prior to deposition only single steps in a regular step array similar to those shown in Figure 47 were observed. In principle, double steps can also result after exposure to oxygen [182]. On the basis of AES (see Section 4.3) and CO adsorption experiments we attribute the double steps shown in Figure 54 to incorporated Co. The Co agglomerates appear immobile, probably as a consequence of the stronger binding compared to Cu. Thus, kink diffusion becomes slow compared to the usual STM sampling rates of some kHz when Co is involved. Note also that at the double steps only immobile, that is, "frozen" kinks (labeled f in Fig. 54) are observed to display a width of multiples of $a_\perp$. The topography at the merging points (labeled

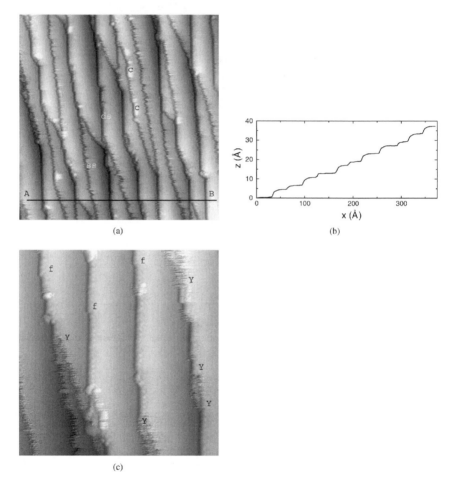

(a)

(b)

(c)

FIG. 54. Cu(111) surface with $5°$ miscut, {100} steps and low Co coverage. (a) 385 Å, $-0.3$ V, 0.7 nA, "ds" marks a double step, "ss" marks a single step and "c" marks very rare positions where single steps are unfrizzy, possibly due to contaminants; (b) line scan $z(x)$ of image (a); (c) 185 Å, $-0.3$ V, 0.7 nA. The "beady" step edges contain Co and have double height. The labels Y and f mark transition points from double to single steps and frozen kinks, respectively.

Y in Fig. 54) reminds one of a zipper and suggests that the incorporation of Co in {100} step minifacets also works like a zipper [183]. Most likely these double steps at the {100} minifacets represent a 1D alloy. This interpretation is according to the Co structures observed on the large counterparts, the low-indexed Cu(100) surface: According to an STM and a density-functional theory study Co occupies substitutional sites in the flat Cu(100) substrate [184].

**5.2.5 Adsorbate Layers.**    The background of applications behind the study of adsorbate layers is heterogeneous catalysis where surfaces are used to accelerate chemical reactions. Examples for reactions to be accelerated are the detoxication processes in exhaust gases, for example, $O_2 + 2CO \rightarrow 2CO_2$, and the industrial production of technical gases. The aim is to understand catalytic surface reactions on an atomic level, that is, the ability to improve catalysts systematically. One element in such analyses is the study of adsorbate layers by STM [185–187].

Because sticking coefficients are material dependent, adsorption allows marking and titration of elements. As an example, Figure 55 shows the topography of $Pt_3Sn(110)$ after CO adsorption. The CO is found on top of the Pt bumps but never between them, at the Sn positions [188]. Thus, in addition to catalytic applications, adsorption experiments provides a method to discriminate between elemental species.

In many catalysts sulfur acts as a poison, which necessarily motivates study of S layers on catalytically interesting surfaces. To produce S layers on surfaces,

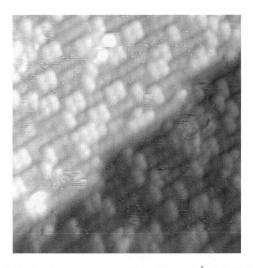

FIG. 55. The $Pt_3Sn(110)$ after exposure to 60 L CO; 120 Å, 0.4 V, 0.8 nA. The CO is found on top of the Pt bumps, never in between at the Sn positions. The carbon side binds to the Pt substrate atoms and the molecules are arranged in dimers and tetramers.

$H_2S$ adsorption can be used because upon dissociation the S stays at the surface whereas the H is believed to desorb or be incorporated in the subsurface. Adsorbates can appear as protrusions or depressions, or they can be invisible in STM topographies and thus theoretical predictions [189] must be employed for interpretation.

A somewhat uncommon approach to production of adsorbate layers is to exploit the impurity atoms in the crystal that segregate to the surface during annealing to higher temperatures ($T_{anneal} \approx 0.5 \cdot T_M$). For instance, most Pd crystals contain S impurities. Annealing to $>800$ K results in 0.15-0.34 monolayers S at the surface. The adlayer is partly disordered and the steps are decorated with $(2 \times 2)$ bands [190] (Fig. 56a). In the $(2 \times 2)$ structures the sulfur atoms prefer acute angles ($30°$). This is reflected in the shape of the step bands and the triangles. The origin of acute angles in the structures is the even valency of the sulfur.

Interestingly, neither the decoration bands nor the triangles are reproduced when the sulfur films were produced by 2-3 Langmuir exposure to $H_2S$ gas (Fig. 56b). Instead the $(2 \times 2)$ structures form patches with bands. A structure developing only with adsorption is the $(\sqrt{3} \times \sqrt{3})$ R30°. The other structures as the $(\sqrt{7} \times \sqrt{7})$ R19.1° and the disordered structures are also obtained by adsorption. When annealing to higher temperatures ($>400$ K) the $(\sqrt{7} \times \sqrt{7})$ R19.1° displaces the other structures to a large extent [191]. Thus, only the $(\sqrt{7} \times \sqrt{7})$ R19.1° structure is an equilibrated state.

**5.2.6 Organic Films.** The fields of applications behind layers of organic molecules are biolelectronical interfaces, electronic devices like transistors, light emitting diodes (LED), lasers, and electronic wiring. The advantage of organic light emitting diodes (OLEDs) [192, 193] compared to conventional LEDs on a semiconductor basis are the flexibility of the material and the relatively easy tuneability of the gap, that is, the color. The length of suitable molecules, like multinary phenyls, exceeds 20 Å such that the low conductivity through such a layer often makes STM difficult and such films are usually studied by AFM (Part IV) [194, 195].

**5.2.7 Nanostructures.** In 1959 Feynman brought up the idea of miniaturing machines down to the atomic level [196]. Surfaces are an especially nice playground for the realization of controlled fabrication of ordered structures with nanometer dimensions. The prequisite step involves overcoming the limits of conventional optical lithography ($\approx$200 nm). Today it has become possible to fabricate many of such nanostructures but they are not yet part of everyday products. Since STM and AFM are extensively used to image nanostructures on surfaces we illustrate in the following a few aspects of this field.

Without any precautions, islands typically agglomerate at rather random positions on the surfaces, governed by nucleation and agglomeration processes. Ordered nanostructures can be obtained when the film is grown on mesoscopic network structures arising from reconstructions or stress compensation. An early

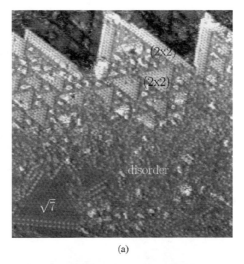

(a)

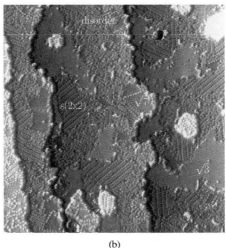

(b)

FIG. 56. Sulfur layers after segregation (a) and $H_2S$ adsorption (b) on Pd(111). The parameters are 400 Å, $-0.07$ V, 1.0 nA and 640 Å, $-0.05$ V, 2.0 nA. Visible are (2×2)triangles, (2×2)step decoration bands, disordered areas, and ($\sqrt{7} \times \sqrt{7}$) R19.1°. Areas with decorated domain boundaries (a) and disordered, $\sqrt{3} \times \sqrt{3}$R 30°, ($\sqrt{7} \times \sqrt{7}$) R19.1°; and striped (2 × 2) areas (b).

example is the growth of Ni islands at the "elbows" of the Au(111) ($\sqrt{3} \times 22$) reconstruction [197] shown in Figure 57. An example for ordered islands using a 2D defect pattern is given in Reference [198]. A strain relief network is formed after 800 K annealing for a monolayer Ag on Pt(111). A second Ag layer nucle-

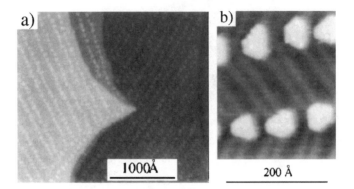

FIG. 57. The 0.1 monolayers Ni on Au(111). The islands grow at the elbow sites of the $(\sqrt{3} \times 22)$ reconstruction of Au(111) [197]. Reprinted with permission from D. D. Chambliss et al., *Phys. Rev. Lett.* 66, 1721 (1991), © 1991, The American Physical Society.

ates predominantly inside the fcc meshes and a regular pattern of Ag islands, that is, of zero-dimensional structures, results (Fig. 58).

The 1D counterparts are epitaxial bands that grow at steps on surfaces. These "wires" can be more or less smooth. For instance Cu grows on stepped W(110) and Mo(110) in bands at the lower sides of the steps. On W(110) the new Cu step edge of these bands appears rough compared to the original step edge (1D Stranski-Krastanov) whereas on stepped Mo(110) the Cu bands display smooth edges (1D layer-by-layer) [199]. Contrast between these different metals was managed in STM images by resonant tunneling via surface states and image states. This is a nice example of how identification of the elements by STM can be achieved via knowledge about the electronic structure of the materials. Equally fascinating are purely electronic low dimensional structures. Clean self-ordering steps as described in Section 5.2.2 represent a set of 1D nanostructures and can 1D lead to confinement and hybridization effects [200].

An elegant way to produce ordered nanostructures is atom lithography through a light field of standing light waves in front of the substrate. The resulting structures can be arranged according to any possible pattern of the light field, that is, rows, dots, zig-zag, honeycombs, pearl necklets [201, 202].

One of the special features of nanostructures is that the dimensions of such structures are of the order of the wavelengths of electrons in metals such that quantum effects dominate. The first and most impressive realization is the quantum corrals on Cu(111) arranged by Xe atoms at low temperature [203]. The Xe atoms could be attached and deattached to the STM tip in a controlled manner by means of voltage pulses. The electron waves of the Cu(111) surface state are partly reflected at the inside of the Xe boundary and a standing wave pattern forms

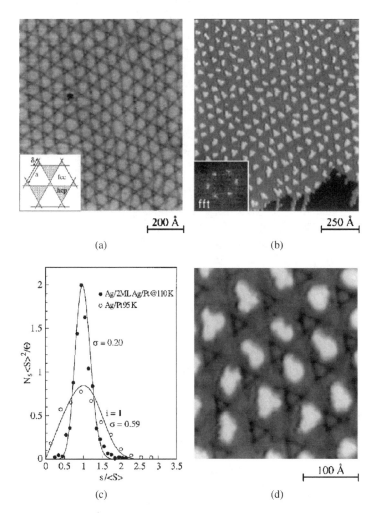

FIG. 58. Confinement nucleation of adatom islands on a dislocation network. (a) Ordered $(25 \times 25)$ dislocation network formed by the second Ag monolayer on Pt(111) on deposition at 400 K and subsequent annealing to 800 K. The inset shows a model of this trigonal strain relief pattern. (b) A superlattice of islands is formed on Ag deposition onto this network at 110 K (coverage = 0.10 monolayers). The inset shows the Fourier transform of the STM image. (c) Island size distributions for random and ordered nucleation. The curve for ordered nucleation is a binominal fit. The curve labeled $i = 1$ shows the size distribution from scaling theory for random nucleation on an isotropic substrate. Size distributions were normalized according to scaling theory ($s$ is the island size in atoms, $\langle S \rangle$ its mean value, and $N_S$ the density of islands with size $s$ per substrate atom). (d) Zoom into image (b) [198]. Reprinted with permission from H. Brune et al., *Nature* 344, 451 (1998), © 1998, Macmillan Magazines Ltd.

that can be imaged by STM. These were the first real space images of electronic states. The latest attainment by such structures is the mirage Co atom imaged by STM inside a corral of such type [204]. If one real Co atom is placed at a focus point of an elliptical corral, due to the Kondo resonance, some of its properties appear at the other focus, where no atom actually exist.

Standing electron wave patterns of surface states also can be observed in defects. This was realized with adatom islands on Ag(111) at low temperature [205]. Vacancy islands did not show electron confinement probably because of absorption losses via bulk transitions.

## 5.3 Scanning Tunneling Spectroscopy

Scanning tunneling spectroscopy (STS) allows local electronic properties to be determined. The tunnel current $I$ (Eq. (47)) depends on two variables, the voltage $U$ and the tip-sample-separation ($\propto z$). Hence, there are three modi of local spectroscopy $I(U)$, $I(z)$, and $U(z)$, with the third parameter being kept constant. The $U(z)$ spectroscopy is rarely used because $I$ is a dependent variable and it is difficult to keep it constant when the two other parameters vary.

5.3.1 **Local Work Function.** $I(z)$ and $U(z)$ spectroscopy allows one to determine the mean barrier height $\bar{\Phi}$ that is related to the local work function. The acquisition of such spectra is made at each scan point. Once, before taking the spectroscopic curve with the regulation loop deactivated, $z$ is adjusted according to the setpoint current. With Eq. (47) the average barrier height is

$$\bar{\Phi} = \frac{\hbar^2}{8m} \left( \frac{\Delta \ln(I/I_0)}{\Delta z} \right)^2 = \frac{\hbar^2}{8m} \left( \frac{\Delta \ln(U/U_0)}{\Delta z} \right)^2 \qquad (50)$$

$$\text{with} \quad \bar{\Phi} = \frac{\Phi_{\text{probe}} + \Phi_{\text{sample}}}{2}$$

Due to the image charge the barrier height is lowered by a few eV, compared with work functions of the free surfaces.

For free surfaces the local work function is usually lowered at steps on metals due to Smoluchowski smoothing [160]. The spatial width of such line dipoles can be determined only from STS work function maps. With steps on Au and Cu surfaces a reduction of the work function in an approximately 8 Å wide zone was observed [206]. Figure 59 shows an STM image and a work function map of 0.8 monolayers Au on Cu(111). Nonlocal methods, for example, traditional photoelectron spectroscopy, simply yield a lowered average work function for stepped surfaces.

5.3.2 **Local Density of States.** Acquisition of a characteristic $I(U)$ curve at constant $z$ piezolength, that is, with the feedback loop switched off, allows us to obtain qualitative information on the local density of states (LDOS) of a surface. The quantitative context of LDOS and characteristic $I(U)$ curves is not

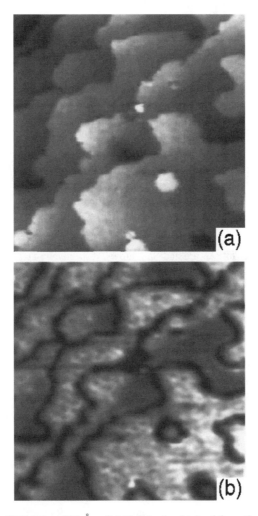

FIG. 59. (a) The STM image 580 Å, −2.0 V, 0.1 nA, obtained from the Au/Cu(111) surface with 0.8 ML of Au. (b) Local work-function image obtained simultaneously with (a). A higher brightness represents a higher local work function [206]. Reprinted with permission from J. F. Jia et al., *Phys. Rev.* B58, 1193 (1998), © 1998, The American Physical Society.

completely clear. There are plenty of suggestions as to how to calculate a curve from $I(U)$ in order to obtain maximal similarity with the LDOS. With metals and low gap voltages ($<1$ V) the local differential conductances are calculated to this purpose [207, and references therein], that is, $\rho_{\text{sample}}(E) \propto dI(U)/dU$.

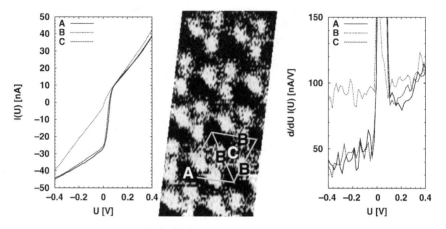

FIG. 60. Atomically resolved STS measurements of the $(\sqrt{7} \times \sqrt{7})$ R19.1 ° structure.
A,B,C positions refer to $S_{fcc}$, Pd and $S_{hcp}$ atoms, respectively. $I(U)$ curves (left). Drift
corrected $I$ image at $U_0 = -0.1$ V (middle). $(d/dU)I(U)$ curves (right) [209]. Reprinted
with permission from S. Speller et al., *Phys. Rev. B* 61, 7297 (2000), © 2000, The American
Physical Society.

This is the simplest, qualitative extension of Eq. (49) when ignoring the variation of the transmission $T$ with energy. With semiconductors and higher gap
voltages the normalized differential conductance $\rho_{sample}(E) \propto (dI/dU)/(I/U)$
is used [208]. Such $\rho_{sample}(E)$ curves reproduce pronounced features of the real
density of states.

We illustrate the spatially varying LDOS by means of the $(\sqrt{7} \times \sqrt{7})$ R19.1°
layer of sulfur on Pd(111) [209]. Figure 60 shows characteristic curves that are
taken at different positions in the S layer. Within the empty states ($<0$ V) a lower
current and $dI(U)/dU$ is found when measured on S atoms. This reduced electron density can be understood by means of the LDOS calculated by the full potential linear augmented plane wave (FLAPW) method [209].

### 5.3.3 Vibration Spectroscopy.

Vibrations of molecules on surfaces can be
excited by the tunneling electrons when the corresponding eigenenergies ($h\nu$) are
surpassed by the energy corresponding to the gap voltage ($eU$). The conductance
is then raised by this small fraction of inelastic tunnel processes. Such information is hence available by inelastic electron tunneling spectroscopy (IETS) from
$I(U)$ curves. Calculating $d^2I(U)/dU^2$ yields small peaks at the energies of the
vibration frequencies of the molecules [210]. As with other STS methods, the
traditional (nonlocal) IETS spectroscopy had already been developed much earlier [211].

## PART IV. ATOMIC FORCE MICROSCOPY

The AFM was developed a few years after STM [212]. A special feature is that no tunnel current is needed and thus insulators, for example, salts, glasses, layers of organic molecules, or crystals of biological molecules and complete biological objects can be studied. For AFM, the probe is mounted on a lever that bends under the force of the sample surface (Fig. 61). For technical reasons, in most AFMs the sample is attached to the fine piezoelements, and hence the sample is scanned against the tip, which is the opposite of what is done with traditional STM. The bending of the lever can be measured by means of a piggyback STM, an interferometer, or a light pointer. Nowadays, the light pointer principle is commonly used: The beam from a light emitting diode is reflected at the end of the cantilever onto a 2D position sensitive detector (PSD). The deflection is approximately proportional to the applied force. This technique allows discrimination between the normal force $F_z$ and lateral forces $F_x$ and $F_y$. Meanwhile, combined AFM/STM heads and AFMs operating in gases, vacuum, or liquid cells are commercially

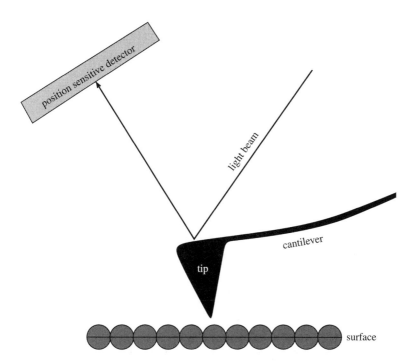

FIG. 61. Illustration of the AFM method (not to scale).

available. A millipedelike ultrahigh density data storage concept has been developed on the basis of AFM [213].

A variety of forces act between tip and sample atoms (see Table IV). The potential is Lennard-Jones-like

$$\Phi(d) = -\frac{A}{d^6} + \frac{B}{d^{12}}$$

a superposition of a short-range repulsive and a long-range van-der-Waals attractive part (Fig. 62a). This potential does not comprise magnetic, electrostatic, and hydrofluid contributions. The negative gradient of such a potential gives the force

TABLE IV.    Overview of Important Interactions Occuring Between Tip and Sample Atoms in the Atomic Force Microscope

| Interaction type | Nature | Range |
|---|---|---|
| Pauli exclusion | Short range, repulsive | $\approx 0.1$ nm |
| Coulomb repulsion (cores) | Short range, repulsive | $\approx 0.1$ nm |
| Chemical bond | Short range, attractive | $\approx 0.1$ nm |
| Van der Waals | Long range, attractive | Up to 100 nm |
| Electrostatic | Long range, attr. or rep. | $\approx$ Several 100 nm |
| Magnetic | Long range, attr. or rep. | $\approx$ Several 100 nm |
| Capillary forces | Attractive | Up to 10 nm |
| Hydrodynamic | Very long range, damping | $\approx 10\ \mu$m |

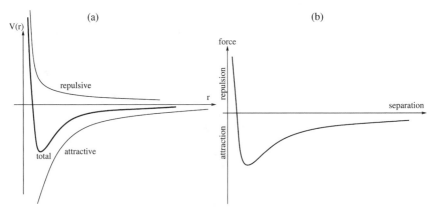

FIG. 62. Variation of the potential between tip and sample with distance (a) and the variation of the force between tip and sample with distance (b).

that governs the deflection of the lever (Fig. 62b). The derivative of the force-distance curve gives the force constant curve.

### 5.3.4 Force-Distance Curves.

Measuring force-distance curves (Fig. 63) of samples gives a completely different image because since the distance (abscissae) refers to the relative $z$ piezoposition and not directly to the position of the cantilever tip. There are two discontinuities in the force-distance loop that reflect points of instability, points (1) and (3) in Figure 63. During approach, no force is initially active. At point (1) the surface's (attractive) force gradient surpasses the lever's spring constant and the lever jumps towards the surface, into contact, to the repulsive force regime. As this jump takes place only at the tip and not at the $z$ piezo a sudden change in force without any $z$ change is noticed in the force-distance curve. Therefore, the shapes of the force curve inferred from the potential and the measured one differ (Figs. 62b and 63). In contact, the lever is pushed by the sample (points (2) to (3)). Upon further approach of the sample the normal force increases linearly according to the spring constant of the lever. When the sample is being retracted, the slope of the force curve is slightly different. The main reason is piezo creep. Close to point (3) the force is maximally attractive and the spring constant is surpassed a second time. The lever jumps back and is then again in an interaction free point (4), far from the surface.

The area between approach and retract curves (hysteresis) reflects the energy loss of the cantilever, for example due to deformation of the surface. The shape of force-distance curves changes when repulsive forces dominate (no jump-to and jump-off) or when measuring in liquids.

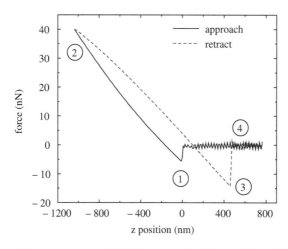

FIG. 63. Force–distance curve on Si(111) covered with a self-assembled octadecyltrichlorosilane layer.

The force constants of levers can be calculated from the geometrical dimensions and Young's module. Therefore, force-distance curves allow calibration of the AFM forces with respect to the voltages at the PSD.

The interaction force versus the tip-sample distance can be calculated partly from force-distance curve. The distance-values are transformed such that the contact line $F(z) = -k \cdot z + F_0$ becomes the ordinate of the interaction force curve.

**5.3.5 Contact Mode.** In contact mode a constant normal force, typically in the nN-range, is maintained via a regulation loop. Similar to STM this is done by adjusting the $z$-piezo accordingly. The force must be calibrated for every new cantilever via a force distance curve (see Section 5.3.4). With hard materials the measured surfaces of constant force represent the topography. However, contact measurements are destructive to a certain extent. Problems arise especially with soft materials, because material is moved or destroyed by the applied forces during the scan. This is already true for metals such as Au.

Alternatively, the contact mode can be operated in the error signal mode by slowing down the regulation speed and collecting the normal force signal directly. This mode is the analog to the CHM in STM. The images show higher contrast, but are not quantitatively related to the topography.

**5.3.6 Lateral Forces and Friction Maps.** Frictional forces show up as a systematic asymmetry between the structures of two images taken at backward and forward scan. This can be observed in the AFM (and STM) topography, for example, on HOPG.

The variations in friction between the tip and sample causes a stick and slip movement of the lever's tip. If the fast scan direction ($x$) is perpendicular to the lever axis this results in lever torsion. Deflection of the light beam by a twisted lever on the position sensitive detector is perpendicular to the usual deviation stemming from normal ($z$) forces. Thereby, discrimination of $F_x$ and $F_z$ is possible. Lateral force microscopy (LFM) measures the forces parallel to the surface plane. The feedback loop must be slowed down, as always when a force channel is measured.

Changes of $F_y$ and $F_z$ deflect the light beam in the same direction. The cantilever is then buckled, which must distort the topography signal. In certain cases, one may assume that the frictional force dominates the normal force and a $F_y$-map can be aquired. A clean topography signal is not possible in this scan geometry. On the other hand, if the surface is not atomically flat, the cantilever will additionally be twisted when the slope of the surface changes. To separate such topographical from frictional origins LFM and AFM images should be taken simultaneously.

However, terraces that exhibit the same appearance in topography can show different lateral force signals. This is attributed to a sort of material contrast or different nanoscopic roughness [214].

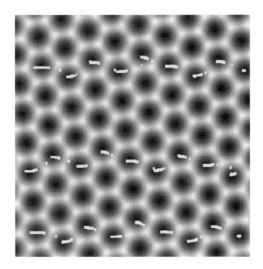

FIG. 64. Gray scale plot of the interaction potential on HOPG (image size 20 Å) and three typical calculated paths of the tip in this potential plotted by dots separated by equal time intervals ($\Delta t = 0.1$ ms, $v = 40$ nm/s) [215]. Reprinted with permission from H. Hölscher et al., *Phys. Rev.* B57, 2477 (1998), © 1998, The American Physical Society.

A nice illustration for the stick-and-slip process is the trajectory of the cantilever on HOPG. For many years it remained puzzling as to why only one type of carbon atom was visible by scanning methods. In the case of AFM this seems to be solved by comparing LFM maps with simulations [215]. Figure 64 shows the interaction potential and three typical calculated paths of the tip in this potential plotted by dots separated by equal time intervals. The bright positions show the atomic honeycomb surface lattice. In LFM, AFM and STM images a simple hexagonal structure similar to the black position is observed. Although the scan speed is constant, the tip moves discontinuously over the surface and stays most of the time in the minima (dark areas), that is, between the atoms. This stick-and-slip movement causes the force maps to represent "hollow-site resolution" instead of "atomic resolution."

5.3.7 Noncontact Mode. In the noncontact mode the influence of the force gradient between probe and surface on a vibrating cantilever is utilized to measure the topography. The tip-sample separation is large, typically between 1 and 100 nm. In contrast to the contact mode, the noncontact mode is less destructive. The cantilever oscillates, driven by a quartz. In air the lever vibration is damped more strongly and it is very difficult to avoid a jump-to-contact through a meniscus of the thin water film that is always present in air. Therefore, noncontact measurements generally require vacuum conditions. Even at the reversal points of the vibration the tip should not touch the surface, that is, the amplitude of the oscillation must be kept small. An intermediate mode when that is not fullfilled, that is, with intermittent contact, is called the tapping mode. This is often used in air or gases when imaging poorly immobilized or soft samples. Both methods are dynamic modi.

The force gradient $F' = dF/dr$, from the interaction between surface and tip (see Fig. 62a,b), is measured via the shift in resonance frequency. This shift serves as the feedback signal. The harmonic approximation is only valid with fairly small amplitudes because the potential in front of the surface is highly asymmetric. The force gradient is varying with distance from the sample, and alters the effective spring constant, $k_{eff} = k + \Delta k$ with $k$ being the spring constant of the free cantilever. At a certain tip-sample separation $r_1$ the total force gradient becomes $F' = k_{eff} \approx k - dF/dr(r_1)$. Accordingly, the resonance frequency shifts to $\omega_0^* = \sqrt{k_{eff}/m_{eff}}$ with $m_{eff}$ being the effective mass of the cantilever.

The quality factor of an oscillation is defined by $Q = \sqrt{\gamma m}/k$ with $\gamma$ being the damping coefficient. If the damping is small the $Q$ factor equals the quotient of the eigenfrequency of the free cantilever and the width of the resonance maximum at the so-called half power point (HPP) (where the amplitude has decayed to $A_{HPP} = A_{max}/\sqrt{2}$): $Q = \omega_0^{free}/\Delta\omega_{HPP}$. In air the $Q$ factors are typically between 100 and 1000. In UHV they are approximately 100 times larger.

### 5.3.8 Principles of Force Gradient Detection.

#### 5.3.8.1 Slope Detection.
Slope detection is often used for AFMs operated in air. The cantilever is then driven at its free eigenfrequency. After the approach the amplitude decreases by $\Delta A$ due to its being out of tune. The reduced amplitude is detected and used as the feedback signal (see Fig. 65). The $z$ piezoadjustment, necessary to keep the amplitude at the setpoint value (e.g., 80% of $A_0$),

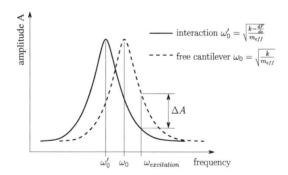

FIG. 65. Illustration of the slope detection principle by means of the resonance curve of the lever. In the vicinity of a sample the resonance frequency shifts and leads to a reduced amplitude at the excitation frequency.

gives the topography. As the lever is not operated in resonance, the phase lag of the oscillation against the excitation shifts away from $\pi/2$, the magnitude of this shift dependent upon local damping. This shift $\Delta\Phi$ is often coimaged as an additional information channel. Although is not completely clear what physical information this extra channel carries, it is very rich in contrast and is not dominated by the relatively large topographic variations. It might be interpretable as material or chemical contrast.

*5.3.8.2 Frequency Modulation Detection.* With a sharp resonance curve, as in UHV, the amplitude drops too quickly to zero under the presence of a force field. Regulation of $\Delta A$ is therefore not possible. Instead, in the frequency modulation (FM) mode the cantilever is vibrating freely with its respective resonance frequency $\omega^*$ including the local force gradient. Such an oscillation must be self-excited, that is, part of the oscillation signal itself is fed back, correctly phase-shifted, to a modulation piezo behind the cantilever. In this way maximal resonance is maintained and the amplitude is regulated to a constant value by an extra feedback circuit. However, the actual feedback signal is the frequency shift. Hence surfaces of constant force gradient are measured during the scan. The advantage is that the frequency shift is more directly related to the interaction force than the amplitude in slope detection is.

**5.3.9 Probes.** Due to its hardness and stiffness $Si_3N_4$ probes have been widely used for contact force microscopy. Microfabricated force sensors with an integrated tip are commercially available. They are etched out of Si wafers and the outermost layer can either be covered by a $Si_3N_4$ film or oxidized. For combined STM/AFM UHV applications this film can be sputtered off to make the cantilever conductive again. Depending on the dimensions, the eigenfrequencies vary between 10 (contact levers) and several 100 kHz (noncontact levers). The torsion eigenfrequencies are usually one order of magnitude larger. For noncontact applications bare Si tips are commonly used.

## 5.4 Topographies

Much progress has been made in understanding atomistic properties of surfaces by noncontact AFM [216]. In noncontact mode true atomic resolution was first obtained on $Si(111)7 \times 7$ [217], on InP(110) [218], and on NaCl [219]. Meanwhile, even subatomic features are observable by noncontact AFM [220]. In contact mode, atomic resolution is achievable but unlike with STM and noncontact AFM it is inconclusive if this resolution is real. True atomic resolution can be recognized by the correct imaging of lattice defects, for example, vacancies as depressions. Otherwise, apparent atomic resolution can arise from the corrugations of the tip's surface and the sample's surface being in phase. The image is then a superposition of many patches of the surface and vacancies cannot be seen.

The AFM is mostly used to image physical properties of biological samples and organic layers [194, 195]. There are a few examples with high-resolution topographs of organic layers [221, and references therein]. With softer materials of macromolecular size or biological cells resolution drops drastically, especially in liquids. Noncontact measurements are, in principle, nondestructive but the topographies can be influenced by long range forces and are not completely reliable.

The motivation for the study of topography and atomic structure is the relation of function to structure. Topography often is acquired as a control channel, simultaneously with other physical properties. In the following section we give a few examples.

## 5.5 Beyond Topography

When using frequency modulation detection of the force gradient, the amplitude is kept constant. The signal to compensate for the damping, which is necessary to maintain a constant amplitude, reflects the energy loss of the oscillation. A contrast in the damping image arises therefore at places where the oscillating system dissipates more or less energy. That is related to a local change in the quality factor of the oscillation. The damping image might be regarded as similar to the lateral force image in the contact mode (Section 5.3.6).

It was reported that on Au(111) the damping at steps especially close to dislocations is weaker compared to the terraces [214]. Furthermore, the damping lines were reported to deviate from the actual step lines. Rigid dislocations in the subsurface region, which might not necessarily be attached to steps, might explain these observations. On Si(111)7 × 7 the damping was found to be stronger at the steps than on terraces [214], resulting in bright step lines in the damping images.

Despite the general interpretation of damping as energy dissipation, the exact physical mechanisms are not yet fully understood. Damping might be attributed to local excitations of phonons or to tip-sample contacts at the reversal point near the surface. During contacts the chemical reactivity of the surface material will increase the damping. This effect can be modulated with topographical features because the size of the interaction area governs the extent of the chemical reactivity, for example, at steps the damping will in general be lowered.

Organic islands on a silicon substrate might serve as a good example of the difficulty of interpreting damping images [222]. The soft organic layer is decisively different from the relatively hard silicon substrate and one might expect a material specific contrast in the damping image. The noncontact topography in Figure 66b shows bright fractal-shaped islands of self assembled octadecyltrichlorosilane (OTS) molecules (Fig. 66a) on activated Si(111). The OTS was deposited by a Langmuir-Blodgett technique. As the assembling of larger OTS aggregates has already taken place in the solution the age of the solution is an important parameter and allows for control of size and density of the OTS islands.

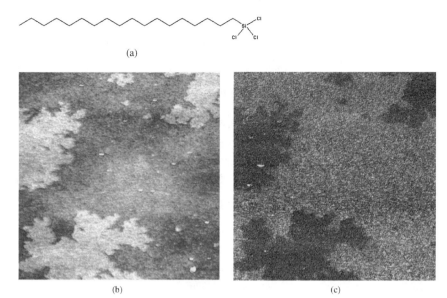

(a)

(b)                                                    (c)

FIG. 66. Octadecyltrichlorosilane molecule (OTS, $C_{18}H_{37}Cl_3Si$) (length $= 26$ Å) (a) AFM noncontact topography image of self-assembled OTS film on $Si_x$/Si(111); 1.0 $\mu$m, $\Delta f = -54$ Hz (at 740 Hz), amplitude $= 0.2$ V. The apparent height of the islands is 16 Å, (b) Simultaneously acquired damping image (c). The Si surface was activated by UV/ozone treatment prior to exposure. The time of exposure was 20 s and the age of the OTS solution 10 min.

The simultaneously acquired damping map is shown in Figure 66c. Here, the fractal-shaped OTS islands appear dark, indicating less energy dissipation on the islands as compared to that on the silicon substrate. A possible explanation is the different chemical reactivity of the substrate and the OTS. The saturated hydro-carbon chains of the OTS do not react with the silicon tip as strongly as with the activated silicon substrate. Therefore, adhesion forces between tip and sample will be stronger on substrate regions than on OTS regions.

In magnetic force microscopy (MFM) [223, 224] the tip of the flexible can-tilever is covered by a ferromagnetic material and magnetized prior to the mea-surement. The magnetic stray field of the sample leads to a force onto the mag-netized tip. This is a long-range interaction and it can be detected up to several 10 nm away from the surface. The measurement is performed in noncontact mode, that is, the frequency shift is used as a feedback signal. With soft-magnetic tips magnetic domains and/or domain walls can be imaged, that is, of a track on a hard disk [225] and of Co dots [226]. The lateral resolution typically reached in MFM is approximately 50 nm. In case of a hard-magnetic tip and a soft magnetic sample it is possible to "write" magnetically with the AFM tip [227].

Analogously, in electrostatic force microscopy (EFM) electrostatic forces can be used in the regulation feedback [228, 229]. Ferroelectric media can be polarized by a voltage applied to a conductive AFM probe in contact (writing). The surface polarization leads to a long range force that can be detected via interaction with the probe at low bias. This interaction force is used as feedback in noncontact mode (reading).

Scanning Kelvin force microscopy (KFM) uses the tip-sample-contact as a local Kelvin probe and allows acquisition of contact potential (CP) distributions or work function maps with nanometer resolution. Initially, such experiments have been performed in atmosphere [230, 231]. Because the work function of most surfaces is altered by oxidation and adsorption, this method was augmented for measurements under UHV conditions [232]. The tip-sample contact represents a capacity that varies during the cantilever oscillation. Additionally, both ac and dc voltage are applied. The result is a time-varying electrostatic force that is minimal, when the dc voltage $U_{dc}$ and the work function difference $\Delta\Phi/e$ compensate. During scanning the dc voltage is adjusted to the force minimum by a feedback loop and a CP map ($U_{dc}(x, y)$) is acquired. This is similar to conventional (nonlocal) Kelvin probes where the current instead of the force is minimized.

## 5.6 Force Spectroscopy

In the spectroscopy mode of AFM force-distance curves $F(z)$ are recorded at one or more scan points after the $z$ piezo has been adjusted to the force setpoint (contact mode). The classical shape, as illustrated in Section 5.3.4 and Figure 63, is obtained mostly with hard materials or simpler molecules. In air a meniscus of water is formed at the jump-to-contact. Due to the meniscus force the jump-to- and jump-off-contact separations differ largely and the area of the hysteresis loop becomes quite large. Force-distance curves can have various appearances. An overview is given in Reference [233]. Spectroscopy is rarely employed in the dynamic mode because with an oscillating probe the tip-sample separation is never well-defined. On the other hand the snap-on is avoided and the complete interaction potential can be inferred from the measured frequency versus distance curve employing simulations [234, 235].

When the tip is functionalized with a chemical species, chemical discrimination can be achieved (chemical force microscopy, CFM) [236, 237]. Covalently functionalized nanotubes can be prepared, allowing chemical contrast between areas with different SAM layers [238]. For biomolecular applications tips can be chemically modified by a layer of molecules that bind especially strongly to complementary molecules. Insight into mechanical properties of biomolecules, such as binding/recognition interactions, unfolding, and elasticity of complex biomolecules has been gained on the basis of force-distance curves [239–243].

## 6 COMPARATIVE CONSIDERATIONS

The different surface analytical methods are compared in Table V with respect to the method used, chemical information, structural information, sensitivity, probing depth and lateral resolution. The comparison is given in qualitative terms defining ranges rather than absolute limits. There are special instruments which surpass some of the specifications given. Some of the listings are physical limits, e.g., diffraction methods give long range structural information in principal, all local information is obtained by calculations based on models. All diffraction methods provide, however, the most accurate lattice constants. In contrast,

TABLE V.    General Characteristics of Surface Analytical Methods

| Method | Excitation detection | Information chemical | Information structural | Monolayer resolution | Effective probing depth | Lateral resolution |
|---|---|---|---|---|---|---|
| LEIS | Ion-ion | Elements Be–U | Short-range order | $10^{-2}$ | <1 nm | 1 $\mu$m$^2$ |
| RBS | Ion-ion | Elements Be–U | Order via channeling | $10-10^{-4}$ | 10 nm | 1 mm$^2$ |
| SIMS | Ion-ion | Elements H–U | None | $10^{-5}$ | 1 nm | 100 nm$^2$ |
| LEED RHEED | Electron-electron | Poor | Long-range order | poor | 1 nm | 100 nm$^2$ |
| AES | Electron-electron | Elements Li–U | None | $10^{-2}$ | 2 nm | 100 nm$^2$ |
| XPS | Photon-electron | Elements Li–U Chemical bonding | Short-range order via XPD | $10^{-1}$ | 3 nm | 10 $\mu$m$^2$ |
| STM | Tip-current | Poor | Short-range order | poor | <1 nm | 0.1 nm$^2$ |
| AFM | Tip-force | Poor | Short-range order | poor | <1 nm | 0.1 nm$^2$ |

TABLE VI.  Some Special Aspects of Surface Analytical Methods with Chemical Sensitivity

| Features | LEIS | RBS | SIMS | SNMS | AES | XPS |
|---|---|---|---|---|---|---|
| Elements not directly detected | H, He | H, He | None | None | H, He | H, He |
| Detection of H and He | Via recoils | Via recoils ERDA | Direct | Direct | Via line shapes ESCA | High resolution |
| Isotope detection | Low Z | Low Z | Yes All masses | Yes All masses | No | No |
| Variation of detection with Z | $\sim 10^2$ | $\sim 10^3$ | $10^4$–$10^5$ | $10$–$10^2$ | $\sim 20$ | $\sim 20$ |
| Organic sample Polymers | Elements at the surface, poor | Elements in depth range | Total mass, fragments, elements | Total mass, fragments, elements | Elements, line shapes | Elements, chem. shifts, valence band |
| Damage | small | Very small | Yes | Yes | small | Very small |
| Charging | Yes | Small | Yes | Yes | Yes | Small |
| Quantification | Using elem. Standards | Complete | Very poor | Good | Using elem. standards | Yes |
| Deviation | 5% | 2% | 5–100% | 5–10% | 1–5% | 1% |
| Depth profiling | Via sputtering | Via energy loss | Sputtering | Sputtering | Via sputtering | Line shape analysis, sputtering |

STM and AFM are local probes, long range information is limited by the scanning range and/or the management of adding up scanning areas, the final limit being the total scanning time with respect to the time stability of the sample. Table VI compares some aspects of the surface analytical methods providing chemical sensitivity. Again, the listings are rather qualitative. Completeness is not the goal of this representation, e.g., PIXE is not mentioned using proton excited X-rays for chemical analysis. All variations of photoelectron spectroscopy including ESCA (Electron Spectroscopy for Chemical Analysis) are summarized under the heading XPS. The features listed are in many discussions the key issues when decisions are to be made which instrument to chose for the problem at hand.

## References

*1.* M. Schleberger, S. Speller, and W. Heiland, *Exp. Meth. in the Physical Sciences* 30, 291 (1998).

*2.* H. Niehus, W. Heiland, and E. Taglauer, *Surf. Sci. Rep.* 17, 213 (1993).

*3.* J. Lindhard, *Kong. Danske Vidensk. Selsk. Mat.-Fys. Medd.* 34, 1 (1965).

*4.* M. Hou and M. T. Robinson, *Appl. Phys.* 17, 371 (1978).

*5.* E. Rutherford, *Phil. Mag.* 21, 669 (1911).

*6.* G. Moliere, *Z. Naturforsch.* 2a, 133 (1947).

*7.* J. F. Ziegler, J. P. Biersack, and U. Littmark, "The Stopping and Range of Ions in Solids." Vol. 1. Pergamon Press, New York, 1985.

*8.* E. Taglauer, in "Surface Analysis—The Principal Techniques." J. C. Vickerman, Ed., John Wiley & Sons, Ltd., p. 215, 1997.

*9.* For example: H. Goldstein, "Classical Mechanics." Addison-Wesley, Reading, MA, 1965.

*10.* M. T. Robinson and I. M. Torrens, *Phys. Rev.* B9, 5008 (1974).

*11.* M. Hou and M. T. Robinson, *Appl. Phys.* 17, 317 (1978).

*12.* J. P. Biersack and L. G. Haggmark, *Nucl. Inst. Meth.* 174, 507 (1980).

*13.* J. P. Biersack and W. Eckstein, *Appl. Phys.* 34, 73 (1984).

*14.* W. Eckstein, "Computer Simulation of Ion-Solid Interactions." Springer-Verlag, Berlin, 1991.

*15.* W.-K. Chu, J. W. Mayer, and M.-A. Nicolet, "Backscattering Spectrometry." Academic Press, New York, 1978.

*16.* L. C. Feldman and J. W. Mayer, "Fundamentals of Surface and Thin Film Analysis." North Holland, Amsterdam, 1986.

*17.* H. H. Andersen, "The Stopping Power and Ranges of Ions in Matter." Pergamon Press, New York, 1977.

*18.* L. R. Doolittle, *Nucl. Instr. Meth. in Phys. Res.* B9, 344 (1985).

*19.* Z. L. Liau, J. W. Mayer, W. L. Brown, and J. M. Poate, *J. Appl. Phys.* 49, 5295 (1978).

*20.* J. F. van der Veen, B. Pluis, and A. W. Denier van der Gon, in "Kinetics of Ordering and Growth on Surfaces" (M. G. Lagally, Ed.), p. 34, Plenum Press, New York, 1990.

*21.* J. W. M. Frenken and J. F. van der Veen, *Phys. Rev. Lett.* 54, 134 (1985).

22. S. A. E. Johansson and J. L. Campbell, "Pixe—A Novel Technique For Elemental Analysis." John Wiley, New York, 1988.

23. M. Aschoff, Ph.D. Dissertation (Thesis), Osnabrück, 1999 (http://elib.Uni-Osnabrueck.DE/dissertations/physics/M.Aschoff/).

24. H. H. Brongersma and G. C. van Leerdam, in "Fundamental Aspects of Heterogeneous Catalysis Studied by Particle Beams" (H. H. Brongersma and R. A. van Santen, Eds.), NATO ASI Series B265, 283 (1991).

25. H. D. Hagstrum, in "Inelastic Ion—Surface Collisions" (N. H. Tolk, J. C. Tully, W. Heiland, and C. W. White, Eds.), p. 1, Academic Press, New York, 1977.

26. I. S. Tsong, in "Inelastic Particle—Surface Collisions" (E. Taglauer and W. Heiland, Eds.), p. 258, Springer, Heidelberg, 1981.

27. R. Hoekstra, J. Manske, M. Dirska, G. Lubinski, M. Schleberger, and A. Naermann, in "Inelastic Ion Surface Collisons." Proc. of the IISC-11 (W. Heiland and E. Taglauer, Eds.), Nucl. Instr. Meth. B125, 53 (1997).

28. H. Brenten, H. Müller, and V. Kempter, Phys. Rev. Lett. 70, 25 (1993).

29. J. Los and J. J. C. Geerlings, Rep. Progr. Phys. 190, 133 (1990).

30. W. Heiland, in "Low Energy Ion-Surface Interactions" (J. W. Rabalais, Ed.), p. 313, John Wiley & Sons, New York, 1994.

31. E. C. Goldberg, R. Monreal, F. Flores, H. H. Brongersma, and P. Bauer, Surf. Sci. Lett. (1999).

32. R. L. Ericksen and D. P. Smith, Phys. Rev. Lett. 34, 297 (1975).

33. A. Zartner, E. Taglauer, and W. Heiland, Phys. Rev. Lett. 40, 1259 (1978).

34. J. C. Tully, Phys. Rev. B16, 4324 (1977).

35. E. Taglauer, W. Englert, W. Heiland, and D. P. Jackson, Phys. Rev. Lett. 45, 740 (1980).

36. D. P. Jackson, W. Heiland, and E. Taglauer, Phys. Rev. B24, 4198 (1981).

37. E. Taglauer, in "Ion Spectroscopies for Surface Analysis" (A. Czanderna and D. M. Hercules, Eds.), Plenum Publ. Co., New York, 1991.

38. H. H. Brongersma, M. Carriere-Fontaine, R. Cortenraad, A. W. Denier van der Gon, J. P. Scanlon, I. Spolveri, B. Cortigiani, U. Bardi, E. Taglauer, S. Reiter, S. Labich, P. Bertrand, L. Houssiau, S. Speller, S. Parascandola, H. Unlü-LAchnitt, and W. Heiland, Nucl. Instr. Meth. B142, 377 (1998).

39. J. Kuntze, S. Speller, W. Heiland, P. Deurinck, C. Creemers, A. Atrei, and U. Bardi, Phys. Rev. B60, 9010 (1999).

40. A. Niehof and W. Heiland, Nucl. Instr. Meth. B48, 306 (1990).

41. S. Speller, M. Schleberger, A. Niehof, and W. Heiland, Phys. Rev. Lett. 68, 3452 (1992).

42. Y. Fujii, K. Namuri, K. Kimura, M. Mannami, T. Hashimoto, K. Ogawa, F. Ohtani, T. Yoshida, and M. Asari, Appl. Phys. Lett. 63, 2070 (1993).

43. R. Pfandzelter, Surf. Sci. 421, 263 (1999).

44. M. Aono, C. Oshima, S. Zaima, S. Otani, and Y. Ishizawa, Jap. J. Appl. Phys. 20, L829 (1981).

45. H. Niehus, Nucl. Instr. Meth. 218, 230 (1983).

46. M. Aschoff, S. Speller, J. Kuntze, W. Heiland, E. Platzgummer, M. Schmid, P. Varga, and B. Baretzky, Surf. Sci. 415, L1051 (1998).

47. C. Kim, C. Hoefner, and J. W. Rabalais, *Surf. Sci. Lett.* 388, 11085 (1997).
48. C. Kim, C. Hoefner, V. Bykov, and J. W. Rabalais, *Nucl. Instr. Meth.* B125, 315 (1997).
49. C. Kim, J. Ahn, V. Bykov, and J. W. Rabalais, *Int. J. Mass Spectr. Ion Phys.* 174, 305 (1998).
50. C. Hoefner and J. W. Rabalais, *Phys. Rev.* B58, 9990 (1998).
51. Ch. Linsmeier, H. Knetzinger, and E. Taglauer, *Surf. Sci.* 275, 101 (1992).
52. A. W. van der Gon, R. Cortenraad, W. P. A. Jansen, M. A. Reijme, and H. H. Brongersma, *Nucl. Instr. Meth.* B161–163, 56 (2000).
53. L. Houssiau and J. W. Rabalais, *Nucl. Instr. Meth.* B157, 274 (1999).
54. P. Sigmund, *Phys. Rev.* 184, 383 (1969).
55. R. Behrisch (Ed.), "Sputtering by Particle Bombardment I and II." *Topics Appl. Phys.*, Springer, Berlin, 47 (1981); 52 (1983).
56. R. Behrisch and K. Wittmaack (Eds.), "Characteristics of Sputtered Particles Applications." *Topics Appl. Phys.*, Springer, Berlin, 64 (1991).
57. H. Oechsner, *J. Mass Spectr. Ion Proc.* 143, 271 (1995).
58. C. E. Young, M. J. Pellin, W. F. Callaway, B. Jørgensen, E. L. Schreiber, and D. M. Gruen, *Nucl. Instr. Meth.* B27, 119 (1987).
59. H. Gnaser, *J. Vac. Sci. Tech.* A12, 452 (1994).
60. H. Gnaser, *J. Vac. Sci. Tech.* A15, 445 (1997).
61. H. Oechsner, *J. Mass Spectrometry and Ion Proc.* 143, 271 (1995).
62. A. Benninghoven, F. G. Rüdenauer, and H. W. Werner, "Secondary Ion Mass Spectrometry." Wiley, New York, 1987.
63. www-hasylab.desy.de/science/groups/materlik-group/research.html B. N. Dev and G. Materlik, in "Resonant Anomalous X-ray Scattering" (G. Materlik, C. J. Sparks, and K. Fischer, Eds.), pp. 119–145, North Holland, Amsterdam, 1994.
64. H. Dosch, *Appl. Phys.* A61, 475 (1995).
65. J. Als-Nielsen, *Physica* A40, 376 (1986).
66. H. Dosch, "Critical Phenomena at Surfaces and Interfaces." *Springer Tracts in Modern Physics*, Springer, Heidelberg, 1995.
67. S. Krimmel, W. Donner, B. Nickel, and H. Dosch, *Phys. Rev. Lett.* 78, 3880 (1998).
68. W. Donner, H. Dosch, S. Ulrich, H. Ehrhardt, and D. Abernathy, *Appl. Phys. Lett.* 73, 777 (1998).
69. C. J. Davisson and L. H. Germer, *Phys. Rev.* 30, 705 (1927).
70. R. L. Park and H. E. Farnsworth, *Rev. Sci. Instr.* 35, 1592 (1964).
71. G. Scheithauer, G. Meyer, and M. Henzler, *Surf. Sci.* 178, 441 (1986).
72. M. Lagally, *Meth. in Exp. Phys.* 22, 237 (1985).
73. K. Heinz, *Rep. Progr. Phys.* 58, 637 (1995).
74. K. Heinz, M. Kottke, U. Löffler, and R. Döll, *Surf. Sci.* 357-358, 1 (1996).
75. www.fkp.uni-hannover.de/fkp/research/spaleed/spaleed.html
76. J. B. Pendry, "Low Energy Electron Diffraction." Academic Press, London, 1974.
77. M. A. Van Hove and S. Y. Tong, "Surface Crystallography by LEED." Springer, Berlin, 1979.
78. C. B. Duke, *Surf. Sci.* 299/300, 24 (1994).
79. J. B. Pendry, *Surf. Sci.* 299/300, 375 (1994).

80. J. M. MacLaren, J. B. Pendry, P. J. Rous, D. K. Saldin, G. A. Somorjai, M. A. Van Hove, and D. D. Vvedensky, "Surface Crystallographic Information Service." D. Reidel Publ. Co, Dordrecht, 1987; For new information see: http://www.nist.gov/srd/nist42.htm

81. A. Barbieri and M. A. Van Hove, Symmetrized Automated Tensor LEED Package, Version 4.1, Private communication (1999).

82. A. Barbieri and A. M. Van Hove, Private communication (1999).

83. S. Crampin and P. J. Rous, Surf. Sci. Lett. 244, L137 (1991).

84. J. Pendry, J. Phys. C13, 937 (1980).

85. A. M. Van Hove, W. H. Weinberg, and C.-M. Chan, "Low Energy Electron Diffraction" (G. Ertl and R. Gomer, Eds.), Springer Series in Surface Science, 6, Springer Verlag, Berlin, 1986.

86. A. Zangwill, "Physics at Surfaces." Cambridge University Press, Cambridge, 1988.

87. www.fkp.uni-erlangen.de/fkp/literatur/pubholo.html

88. P. K. Larson and P. J. Dobson, "Reflection High Energy Electron Diffraction." NATO ASI Series, 188, Plenum Press, New York, 1988.

89. M. Hoheisel, J. Kuntze, S. Speller, A. Postnikov, and W. Heiland, Phys. Rev. B60, 2033 (1999).

90. A. N. Haner, P. N. Ross, and U. Bardi, Surf. Sci. 249, 15 (1991).

91. J. Kunze, S. Speller, W. Heiland, A. Atrei, I. Spolveri, and U. Bardi, Phys. Rev. B58, R16005 (1998).

92. M. Hoheisel, S. Speller, J. Kuntze, U. Bardi, and W. Heiland, Phys. Rev. B (2001).

93. G. A. Prinz, Physics Today April, 58 (1995).

94. G. Bayreuther and S. Mengel, "Magnetoelektronik, Grundlagenforschung—Zukunftstechnologie?" VDI-Technologiezentrum, 1998.

95. P. Walser, M. Schleberger, P. Fuchs, and M. Landolt, Phys. Rev. Lett. B80, 2217 (1998).

96. M. Schleberger, P. Walser, M. Hunziker, and M. Landolt, Phys. Rev. B60, 14360 (1999).

97. E. Bauer, Z. Kristallogr. 110, 372 (1958).

98. D. A. Shirley, Phys. Rev. B5, 4709 (1972).

99. S. Tougaard and C. Jansson, Surf. Interface Anal. 20, 1013 (1993).

100. S. Tougaard and P. Sigmund, Phys. Rev. B25, 4452 (1981).

101. S. Tougaard, Surf. Interface Anal. 11, 453 (1988).

102. S. Tougaard, J. Vac. Sci. Technol. 8, 2198 (1990).

103. M. P. Seah and W. A. Dench, Surf. Interface Anal. 1, 2 (1979).

104. S. Tanuma, C. J. Powell, and D. R. Penn, Surf. Interf. Anal. 11, 577 (1988).

105. J. Lindhard, K. Dan. Vidensk. Selsk. Mat.-Fys. Medd. 28, 1 (1954).

106. R. Ritchie and A. Howie, Philos. Mag. 36, 463 (1977).

107. S. Tougaard and I. Chorkendorff, Phys. Rev. B 35, 6570 (1987).

108. S. Tougaard and J. Kraaer, Phys. Rev. B 43, 1651 (1991).

109. S. Tougaard, Solid State Commun. 61, 547 (1987).

110. S. Tougaard, Surf. Interface Anal. 25, 137 (1997).

111. M. Schleberger, Surf. Sci. 445, 71 (2000).

112. M. Schleberger, D. Fujita, C. Scharfschwerdt, and S. Tougaard, Surf. Sci. 331-333, 942 (1995).

113.  M. Schleberger, D. Fujita, and S. Tougaard, *J. Electr. Spectr.* 82, 173 (1996).

114.  E. N. Sickafus, *Phys. Rev.* B 16, 1436 (1977).

115.  M. Gryzinski, *Phys. Rev. A* 138, 336 (1965).

116.  M. Schleberger, D. Fujita, C. Scharfschwerdt, and S. Tougaard, *J. Vac. Sci. Technol.* B13, 949 (1995).

117.  E. Daugy, P. Mathiez, F. Salvan, and J. Layet, *Surf. Sci.* 154, 267 (1985).

118.  T. Yasue, T. Koshikawa, H. Tanaka, and I. Sumita, *Surf. Sci.* 287/288, 1025 (1993).

119.  P. Ho, G. W. Rudloff, J. E. Lewis, V. L. Moruzzi, and A. R. Williams, *Phys. Rev.* B22, 4784 (1980).

120.  R. Matz, R. J. Purtell, Y. Yakota, G. W. Rudloff, and P. S. Ho, *J. Vac. Sci. Technol.* A2, 253 (1984).

121.  R. Kern, G. L. Lay, and M. Manville, *Surf. Sci.* 72, 405 (1978).

122.  G. L. Lay, *Surf. Sci.* 132, 169 (1983).

123.  S. Ino and A. Endo, *Surf. Sci.* 293, 165 (1993).

124.  K. Sumitomo, T. Kobayashi, F. Shoji, K. Oura, and I. Katayama, *Phys. Rev. Lett.* 66, 1193 (1991).

125.  J.-J. Yeh, J. Hwang, K. Bertness, D. J. Friedman, R. Cao, and I. Lindau, *Phys. Rev. Lett.* 70, 3768 (1993).

126.  A. Cros, J. Derrien, F. Salvan, and J. Gaspard, *Proc. ECOSS-3*, unpublished (1980).

127.  M. Schleberger, A. Cohen Simonsen, S. Tougaard, J. L. Hansen, and A. N. Nyland-sted, *J. Vac. Sci. Techn.* A15, 3032 (1997).

128.  A. Cohen Simonsen, M. Schleberger, S. Tougaard, J. L. Hansen, and A. N. Nyland-sted Larsen, *Thin Solid Films* 338, 165 (1999).

129.  Y. W. Mo, D. Savage, B. Schwarzentruber, and M. Lagally, *Phys. Rev. Lett.* 65, 1020 (1990).

130.  M. Asai, H. Ueba, and C. Tatsuyama, *J. Appl. Phys.* 58, 2577 (1985).

131.  E. Richmond, *Thin Solid Films* 252, 98 (1994).

132.  P. Grütter, W. Zimmermann-Edling, and D. Brodbeck, *Appl. Phys. Lett.* 60, 2741 (1992).

133.  S. Sheiko, M. Möller, E. Reuvekamp, and H. Zandbergen, *Phys. Rev.* B48, 5675 (1993).

134.  S. S. P. Parkin, N. More, and K. P. Roche, *Phys. Rev. Lett.* 64, 2304 (1990).

135.  P. Grünberg, R. Schreiber, Y. Pang, M. B. Brodsky, and H. Sowers, *Phys. Rev. Lett.* 57, 2442 (1986).

136.  P. Zahn, I. Mertig, M. Richter, and H. Eschrig, *Phys. Rev. Lett.* 75, 2996 (1995).

137.  S. Toscano, B. Briner, H. Hopster, and M. Landolt, *Phys. Rev. Lett.* 73, 340 (1994).

138.  R. Kilper, St. Teichert, Th. Franke, P. Häussler, H. G. Boyen, A. Cossy-Favre, and P. Oelhafen, *Appl. Surf. Sci.* 91, 93 (1995).

139.  P. Häussler, Private communication.

140.  J. J. de Vries, J. Kohlhepp, F. J. A. den Broeder, R. Coehoorn, R. Jungblut, A. Rein-ders, and W. J. M. de Jonge, *Phys. Rev. Lett.* 78, 3023 (1997).

141.  C. Dufour, A. Bruson, G. Marchal, B. George, and Ph. Magnin, *J. Magn. Magn. Mat.* 93, 545 (1991).

142.  G. Binnig, H. Rohrer, C. Gerber, and H. Weibel, *Phys. Rev. Lett.* 50, 120 (1993).

*143.* Absolute distance measure is not directly possible, as the nanoscopic tip shape is unknown in general. However, the distance can be measured by the $z$ piezovoltage necessary to drive the tip into a soft crash with the sample.

*144.* W. Moritz and D. Wolf, *Surf. Sci.* 88, L29 (1979); 163, L655 (1985).

*145.* G. Binnig, H. Rohrer, C. Gerber, and H. Weibel, *Surf. Sci.* 131, L379 (1983).

*146.* H. P. Bonzel and R. Ku, *Sci. Technol.* 9, 633 (1972).

*147.* T. Gritsch, D. Coulman, R. J. Behm, and G. Ertl, *Phys. Rev. Lett.* 63, 1086 (1998); *Surf. Sci.* 257, 297 (1991).

*148.* J. K. Gimzewski, R. Berndt, and R. R. Schlittler, *Surf. Sci.* 247, 327 (1997); *Phys. Rev. B* 45, 6844 (1992).

*149.* A. Zangwill, "Physics at Surfaces." Cambridge University Press, Cambridge, 1988.

*150.* S. Speller, J. Kuntze, T. Rauch, J. Bömermann, M. Huck, M. Aschoff, and W. Heiland, *Surf. Sci.* 366, 251 (1996).

*151.* R. Koch, M. Borbonus, O. Haase, and K. H. Rieder, *Phys. Rev. Lett.* 67, 3416 (1991).

*152.* J. Kuntze, The Atomic Structure of the Clean and Adsorbate Covered Ir(110) Surface, Ph.D. Dissertation (Thesis) (1999). (http://elib.Uni-Osnabrueck.DE/dissertations/physics/J.Kuntze/thesis.ps.gz).

*153.* J. J. Schulz, M. Sturmat, and R. Koch, *Phys. Rev. B* 62, 15402 (2000).

*154.* P. Hanesch and E. Bertel, *Phys. Rev. Lett.* 79, 1523 (1997).

*155.* E. D. Williams, *Surf. Sci.* 299/300, 502 (1994).

*156.* C. Alfonso, J. M. Bermond, J. C. Heyraud, and J. J. Metois, *Surf. Sci.* 262, 371 (1992).

*157.* M. Giesen, *Surf. Sci.* 370, 55 (1997).

*158.* L. Barbier, L. Masson, J. Cousty, and B. Salanon, *Surf. Sci.* 345, 197 (1996).

*159.* J. W. M. Frenken and P. Stoltze, *Phys. Rev. Lett.* 82, 3500 (1999).

*160.* R. Smoluchowski, *Phys. Rev.* 60, 661 (1941).

*161.* R. Stumpf and M. Scheffler, *Phys. Rev. B* 53, 4958 (1996).

*162.* H.-C. Jeong and E. D. Williams, *Surf. Sci. Rep.* 34, 171 (1999).

*163.* A. Bachmann *et al.*, to be published.

*164.* A. R. Bachmann, A. Mugarza, J. E. Ortega, A. Närmann, and S. Speller, to be published.

*165.* M. Giesen, *Prog. Surf. Sci.*, in press (2001).

*166.* A. V. Ruban, H. I. Skriver, and J. K. Norskov, *Phys. Rev. B* 59, 15990 (1999).

*167.* B. M. Ocko, D. Gibbs, K. G. Huang, D. M. Zehner, and S. G. J. Mochrie, *Phys. Rev. B* 44, 6429 (1991).

*168.* L. Bönig, S. Liu, and H. Metiu, *Surf. Sci.* 365, 87 (1991).

*169.* J. Kuntze, S. Speller, W. Heiland, A. Atrei, G. Rovida, and U. Bardi, *Phys. Rev. B* 60, 1535 (1999).

*170.* U. Bardi, L. Pedocchi, G. Rovida, A. N. Haner, and P. N. Ross, in "Fundamental Aspects of Heterogenous Catalysis Studied by Particle Beams" (H. H. Brongersma and R. A. van Sannen, Eds.), p. 393, Plenum Press, New York, 1991.

*171.* A. Atrei, U. Bardi, G. Rovida, M. Torrini, E. Zanazzi, and P. N. Ross, *Phys. Rev. B* 46, 1649 (1992).

*172.* L. Vitos, A. V. Ruban, H. L. Skriver, and J. Kollár, *Surf. Sci.* 411, 186 (1998).

*173.* M. Schmid, H. Stadler, and P. Varga, *Phys. Rev. Lett.* 70, 1441 (1993).

*174.* H. Brune, *Surf. Sci. Rep.* 31, 121 (1998).

*175.* S. Speller, S. Degroote, J. Dekoster, G. Langouche, J. E. Ortega, and A. Närmann, unpublished.

*176.* J. de la Figuera, J. E. Prieto, C. Ocal, and R. Miranda, *Surf. Sci.* 307–309, 538 (1994).

*177.* J. de la Figuera, J. E. Prieto, G. Koska, S. Müller, C. Ocal, R. Miranda, and K. Heinz, *Surf. Sci.* 349, L139 (1996).

*178.* M. Klaua, H. Höche, H. Jenniches, J. Barthel, and J. Kirschner, *Surf. Sci.* 381, 106 (1997).

*179.* S. Speller, S. Degroote, J. Dekoster, G. Langouche, J. E. Ortega, and A. Närmann, *Surf. Sci. Lett.* 405, 542 (1998).

*180.* S. Morin, A. Lachenwitzer, O. M. Magnussen, and R. J. Behm, *Phys. Rev. Lett.* 83, 5066 (1999).

*181.* F. J. Himpsel, J. E. Ortega, G. J. Mankey, and R. F. Willis, *Adv. in Phys.* 47, 511 (1998).

*182.* G. Witte, J. Braun, D. Nowack, L. Bartels, B. Neu, and G. Meyer, *Phys. Rev. B* 58, 13224 (1998).

*183.* A. R. Bachmann, A. Mugarza, J. E. Ortega, A. Närmann, and S. Speller, in preparation.

*184.* F. Nouvertné, U. May, M. Bamming, A. Rampe, U. Korte, G. Güntherodt, R. Pentcheva, and M. Scheffler, *Phys. Rev. B* 60, 14382 (1999).

*185.* G. Ertl and H. J. Freund, *Physics Today*, Jan. 32 (1999).

*186.* J. Wintterlin, S. Völkening, T. V. W. Janssens, T. Zambelli, and G. Ertl, *Science* 278, 1931 (1997).

*187.* F. Besenbacher, I. Chorkendorff, B. S. Clausen, B. Hammer, A. M. Molenbroek, J. K. Nørskov, and I. Stensgaard, *Science* 279, 1913 (1998).

*188.* M. Hoheisel *et al.*, to be published.

*189.* I. S. Tilinin, M. K. Rose, J. C. Dunphy, M. Salmeron, and M. A. van Hove, *Surf. Sci.* 418, 511 (1998).

*190.* J. Bömermann, M. Huck, J. Kuntze, T. Rauch, S. Speller, and W. Heiland, *Surf. Sci.* 357/358, 849 (1996).

*191.* S. Speller, T. Rauch, J. Bömermann, P. Borrmann, and W. Heiland, *Surf. Sci.* 441, 107 (1999).

*192.* J. R. Sheats, H. Antoniadis, M. Hueschen, W. Leonard, J. Miller, R. Moon, D. Roitman, and A. Stocking, *Science* 273, 884 (1996).

*193.* Z. Shen, P. E. Burrows, V. Bulovi, S. R. Forrest, and M. E. Thompson, *Nature* 276, 2009 (1997).

*194.* A. Ikai, *Surf. Sci. Rep.* 26, 261 (1997).

*195.* J. A. DeRose and R. M. Leblanc, *Surf. Sci. Rep.* 22, 73 (1995).

*196.* http://www.zyvex.com/nanotech/feynman.html

*197.* D. D. Chambliss, R. J. Wilson, and S. Chiang, *Phys. Rev. Lett.* 66, 1721 (1991).

*198.* H. Brune, M. Giovannini, K. Bromann, and K. Kern, *Nature* 394, 451 (1998).

*199.* T. A. Jung, R. Schittler, J. K. Gimzewski, and F. J. Himpsel, *Appl. Phys. A* 61, 467 (1995).

*200.* J. E. Ortega, S. Speller, A. R. Bachmann, A. Mascaraque, E. Cr. Michel, A. Närmann, A. Mugarza, A. Rubic, and F. J. Himpsel, *Phys. Rev. Lett.* 84, 6110 (2000).

*201.* Th. Schulze, B. Berger, R. Mertens, M. Pivk, T. Pfau, and J. Mlynek, *Appl. Phys.* B 70, 671 (2000).

*202.* U. Drodofsky, J. Stuhler, Th. Schulze, M. Drewsen, B. Brezger, T. Pfau, and J. Mlynek, *Appl. Phys.* B 65, 755 (1997).

*203.* M. F. Crommie, C. P. Lutz, and D. M. Eigler, *Science* 262, 218 (1993).

*204.* H. C. Manoharan, C. P. Lutz, and D. M. Eigler, *Nature* 403, 512 (2000).

*205.* J. Li, W.-D. Schneider, R. Berndt, and S. Crampin, *Phys. Rev. Lett.* 80, 3332 (1998).

*206.* J. F. Jia, K. Inoue, Y. Hasegawa, W. S. Yang, and T. Sakurai, *Phys. Rev.* B 58, 1193 (1998).

*207.* G. Hörmandinger, *Phys. Rev.* B 49, 13897 (1994).

*208.* J. A. Stroscio and R. M. Feenstra, in "Scanning Tunneling Microscopy" (J. A. Stroscio and W. J. Kaiser, Eds.), Academic Press, New York, 1993.

*209.* S. Speller, T. Rauch, A. Postnikov, and W. Heiland, *Phys. Rev.* B 61, 7297 (2000).

*210.* B. C. Stipe, M. A. Rezaei, and W. Ho, *Science* 280, 1732 (1998); *Phys. Rev. Lett.* 82, 1724 (1999).

*211.* R. C. Jaklevic and J. Lambe, *Phys. Rev. Lett.* 17, 1139 (1966).

*212.* G. Binnig, C. F. Quate, and C. Gerber, *Phys. Rev. Lett.* 56, 930 (1986).

*213.* P. Vettiger, M. Despont, U. Drechsler, U. Dürig, W. Häberle, M. I. Lutwyche, H. E. Rothuizen, R. Stutz, R. Widmer, and G. K. Binnig, *IBM J. Res. Develop.* 44, 323 (2000).

*214.* S. Molitor, Ph.D. Dissertation (Thesis), Osnabrück; http://elib.Uni-Osnabrueck.DE/dissertations/physics/

*215.* H. Hölscher, U. D. Schwarz, O. Zwörner, and R. Wiesendanger, *Phys. Rev.* B 57, 2477 (1998).

*216.* "Proceedings of the Second International Workshop on Noncontact Atomic Force Microscopy" (R. Benewitz, Ed.), *Appl. Surf. Sci.* 157 (2000).

*217.* F. J. Giessibl, *Science* 267, 68 (1995).

*218.* Y. Sugawara, M. Ohta, H. Ueyama, and S. Morit, *Science* 270, 1646 (1995).

*219.* M. Bammerlin, R. Lüthi, E. Meyer, A. Baratoff, J. Lü, M. Guggisberg, Ch. Gerber, L. Howald, and H.-J. Güntherodt, *Probe Microscopy* 1, 3 (1997); or http://monet.unibas.ch/gue/uhvafm/pub_mb/nacl_pm.html

*220.* F. J. Giessibl, S. Hembacher, H. Bielefeldt, and J. Mannhart, *Science* 289, 422 (2000).

*221.* T. Uchihashi, T. Ishida, M. Komiyama, M. Ashino, Y. Sugawara, W. Mizutani, K. Yokoyama, S. Morita, H. Tokumoto, and M. Ishikawa, *Appl. Surf. Sci.* 157, 244 (2000).

*222.* M. Reiniger, B. Basnar, G. Friedbacher, and M. Schleberger, *Surface Interface Analysis*, accepted.

*223.* P. Grütter, H. J. Mamin, and D. Rugar, in "Scanning Tunneling Microscopy II" (R. Wiesendanger and H.-J. Güntherodt, Eds.), Springer Verlag, Berlin Heidelberg, 1995.

*224.* Y. Martin and H. K. Wickramasinghe, *Appl. Phys. Lett.* 50, 1455 (1987).

*225.* T. D. Howell, D. P. McCown, T. A. Diola, and Y. S. Tang, *IEEE Trans. Magn.* 26, 2298 (1990).

*226.* M. Hehn, K. Ounadjela, J.-P. Bucher, F. Rousseaux, D. Decanini, B. Bartenlian, and C. Cahppert, *Nature* 272, 1782 (1996).

*227.* A. Born and R. Wiesendanger, *Appl. Phys.* A 68, 131 (1999).

228. R. Lüthi, H. Haefke, K.-P. Meyer, E. Meyer, L. Howald, and H.-J. Güntherodt, *J. Appl. Phys.* 74, 7471 (1993).
229. L. Eng, M. Bammerlin, C. Loppacher, M. Guggisberg, R. Bennewitz, R. Lüthi, E. Meyer, Th. Huser, H. Heinzelmann, and H.-J. Güntherodt, *Ferroelectrics* 222, 153 (1999).
230. M. Nonnenmacher, M. P. O'Boyle, and H. K. Wickramasinge, *Appl. Phys. Lett.* 58, 2921 (1991).
231. J. M. R. Weaver and D. W. Abraham, *J. Vac. Sci. Technol.* B 9, 1559 (1991).
232. S. Kitamura and M. Iwatsuki, *Appl. Phys. Lett.* 72, 3154 (1998).
233. B. Capella and G. Dietler, *Surf. Sci. Rep.* 34, 1 (1999).
234. H. Hölscher, W. Allens, U. D. Schwarz, A. Schwarz, and R. Wiesendanger, *Phys. Rev. Lett.* 83, 4780 (1999).
235. B. Gotsmann, B. Anczykowski, C. Seidel, and H. Fuchs, *Appl. Surf. Sci.* 140, 314 (1999).
236. C. D. Frisbie, L. F. Rozsnyai, A. Noy, M. S. Wrighton, and C. M. Lieber, *Science* 265, 2071 (1994).
237. A. Noy, D. V. Vezenov, and C. M. Lieber, *Annual Reviews of Materials Science* 27, 381 (1997).
238. S. S. Wong, E. Joselevich, A. T. Woolley, C. L. Cheung, and C. M. Lieber, *Nature* 394, 52 (1998).
239. M. Radmacher, *Physics World* 12, 33 (1999).
240. M. Rief, F. Oesterhelt, B. Heymann, and H. E. Gaub, *Science* 275, 1295 (1997).
241. M. Rief, M. Gautel, F. Oesterhelt, J. M. Fernandez, and H. E. Gaub, *Science* 276, 1109 (1997).
242. F. Oesterhelt, D. Oesterhelt, M. Pfeiffer, A. Engel, H. E. Gaub, and D. J. Müller, *Science* 288, 143 (2000).
243. H. Li, M. Rief, E. Oesterhelt, and H. E. Gaub, *Appl. Phys.* A 68, 407 (1999).
244. D. Fujita, M. Schleberger, and S. Tougaard, *Surf. Sci.* 357, 180 (1996).

# 2. APPLICATION OF PHOTOELECTRON SPECTROSCOPY IN INORGANIC AND ORGANIC MATERIAL SYSTEMS

Sudipta Seal*

Advanced Materials Processing and Analysis Center (AMPAC) & Mechanical, Materials and Aerospace Engineering (MMAE) University of Central Florida, Orlando, Florida, USA

Tery L. Barr

Materials Engineering and Laboratory for Surface Studies (LSS) University of Wisconsin, Milwaukee, Wisconsin, USA

## 1 INTRODUCTION

Surfaces and interfaces in material systems play an important role in the design of their unique physical and chemical properties. The most effective way to define a surface or interface in a system is to refer to its many degrees of abruptness. A change in any property in a system may be used to define an interface. How to form a surface? One can achieve it by the process of comminution, that is, by subdividing a huge block of material into small parts, which may consist of individual atoms or molecules. There are two basic approaches to the formation of an interface. In the comminution approach, a crystal is separated along a predetermined plane of atoms and then allowed to relax to its equilibrium point. In the processing approach, this crystal is treated as a molecule of increasing size, where the equilibrium configuration is calculated for each incremental increase in size. A detailed study and thorough review on surfaces and interfaces is described elsewhere [1].

The explosive commercial development of ultrahigh vacuum techniques in the 1960s prompted both academic and industrial interest in understanding the fundamental concepts of surfaces and interfaces in both inorganic and organic material systems. The primary techniques used to study surfaces and interfaces are commonly known as surface analytical techniques. Of the many recent analytical techniques, electron spectroscopy of chemical analysis (ESCA) (also known as X-ray photoelectron spectroscopy (XPS)) is being devoted to the field of surface science. This technique has been available for the last 30 years and is still being improved at a fairly rapid pace. More than 1000 ESCA units are in operation throughout the world and are obvious testimony to the viable utility of this spectroscopy as one of the premier methods for surface chemical analysis. This chapter will be devoted

---

*To my wife Shanta, my parents Prof. Bijoy Kumar Seal and Mary Seal, my brother Pradipta, my in-laws Mr. Subroto Ghosh and Mrs. Alokanada Ghosh and all my other friends, colleagues and students.

111

EXPERIMENTAL METHODS IN THE PHYSICAL SCIENCES
Vol. 38
ISBN 0-12-475985-8
ISSN 1079-4042/01    $35.00

to an introduction of the fundamental principles and use of the ESCA technique in various materials systems. Other general reviews and readings on this subject regarding theory and applications of ESCA are highly encouraged [2–10].

## 2 FUNDAMENTALS AND A BRIEF HISTORY OF ESCA

The photoelectron effect was first discovered by Henrich Hertz [11] in early 1887 in order to verify the implications of Maxwell's theory and relations. Hertz noticed a spark of light on metal contacts in electrical units when exposed to light. The dawn of a new era actually came in 1905. Albert Einstein brilliantly utilized Planck's new quantum energy concept to explain how low radiation intensity and high frequency can actually eject electrons from a metal piece. The converse failed to produce any electrons. Max Planck received the Nobel Prize on quantization of energy [12] in 1918 and Einstein received the Nobel Prize on photoelectric effect in 1921. The single relationship proposed so long ago by Einstein is still today the fundamental basis of photoelectron spectroscopy,

$$hv = \frac{1}{2}mv_e^2 (\text{or } E_K) + E_B + q\phi \tag{1}$$

$h$ = Planck's constant, $v$ = frequency of the incoming photons, $m$ = photoelectron mass, $v_e$ = velocity of the electrons, $E_K$ = kinetic energy, $q$ = reference charge, $\phi$ = the work function of the emitting materials and $E_B$ = the binding energy, expressed here against the Fermi level of the chosen materials. One of Planck's comments in nominating Einstein for the Prussian Academy in 1913 is quite humorous. Planck spoke of Einstein, "That he may sometimes have missed the target in his speculations, as for example, in his hypothesis of light quanta, cannot really be held against him." Although subsequent research continued [13] to generate suitable photoelectron spectra, lack of both high vacuum instruments and sophisticated electronics caused most scientists of the 1930s and 1940s to ignore the prospect of a "useful" photoelectron spectroscopy instrument.

Most of these problems were eventually overcome by a Swedish scientist, Kai Siegbahn (son of Manne Siegbahn, 1924 Nobel Prize winner) and his research group [14]. His group recorded the first three-photoelectron spectrum in 1955 using a giant Helmholtz coil. In so doing, they found the *famous chemical shifts* in two chemically different sulfurs in a thiosulfate. Realizing the potential impact of his observations, Siegbahn rephrased this new technique, electron spectroscopy for chemical analysis, often known as ESCA. In 1981, he was awarded the Nobel Prize for his outstanding contributions in the field of physics of photoelectron spectroscopy. Although the term ESCA is widely used today, it is still also called XPS, X-ray photoelectron spectroscopy. Although in practice the terms are interchangeable, ESCA encompasses both XPS and Auger emission spectroscopy (X-ray induced: X-AES).

## 3 A BASIC ESCA SYSTEM

Any conventional and commercial ESCA generally consists of either Al $k\alpha$ (1486 eV) or Mg $k\alpha$ (1254 eV) X-ray source or both. A variety of electron analyzers are employed to collect and energy sector the ejected photoelectrons, with particular use of either hemispherical or cylindrical mirror arrangements. A variety of detectors are used to collect the final electron emissions. The detectors include channeltrons, microchannel plates, and resistive anode plates, which are then often coupled with various arrangements of phosphor screens and highly sensitive cameras.

Figure 1 illustrates the basic principle of an ESCA process. In this illustration, an X-ray photon radiating an electron in the K shell causes an emission of a 1s photoelectron. The resulting K shell vacancy is filled by an electron from the next higher level, which leads to X-ray fluorescence or the radiationless deexcitation process of Auger emission. The kinetic energy of the outgoing photoelectron is related to the electron binding energy (a key cornerstone of the experimental XPS) and is given by Eq. (1). Unfortunately, directly focused X-rays may eventually damage some samples.

One cannot perform very high-resolution spectroscopy in a conventional ESCA and, therefore, the development of monochromatic ESCA has been given special

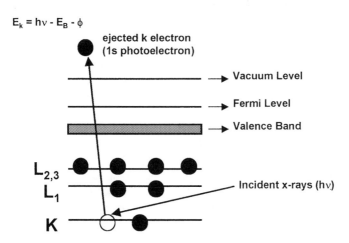

FIG. 1. A schematic representation of a photoemission process.

emphasis. The Siegbahn group discussed [15], proposed [16], and brought to general practice the use of monochromatically scattered X-rays in the spectrometer. This concept led to the development of the use of geometrically arranged silica (quartz) crystals [17], bent to produce the optimum Bragg angle, such that a Rowland circle arrangement will permit the focus on the sample of a diffracted segment of the principal Al $k\alpha$ X-ray peak. The aforementioned beam damage is substantially curtailed [17] by this monochromatic alignment. In addition, line widths may be reduced by several times, and the general background may be sufficiently reduced to realize an improvement in signal-to-background of several orders of magnitude. Monochromators in the Hewlett-Packard ESCA system were first marketed from 1971 to 1976 [17] and other manufacturers have followed the same trend, developing the high-resolution acute sensitivity mode of ESCA or XPS spectrometers.

# 4 INFORMATION AVAILABLE FROM ESCA

An ESCA analysis of a surface will provide both qualitative and quantitative information on all elements and compounds (except H and He) present at $>0.1$ atomic%. However, ESCA will probe only the top 50–60 Å of the surface. From elemental chemical shifts (see in what follows), one can determine molecular environment, including oxidation state and bonding. For organic systems, such features as aromatic vs unsaturated structures can be determined from a combination of principal peaks and shake-up ($\pi$ to $\pi^*$) satellites. By using angle dependent ESCA (photoelectrons with varying escape depths), the chemistry of the surface to subsurface layer can be studied in both inorganic and organic systems. If one wants to study the surface of bulk chemistry, an elemental depth profile for several layers of the sample can be achieved using ion etching (often known as sputtering (a destructive process)). The overall advantage in using ESCA (using both core level and valence band spectra) is in the fingerprinting of materials and identifying of bonding orbitals that permits the physics and chemistry of the samples to be studied.

# 5 BINDING ENERGY AND CHEMICAL SHIFT

It became common in the early development of X-ray photoelectron spectroscopy (XPS or ESCA) for users to classify some of the shifts detected for elemental covalent bonding energy peaks following changes in the chemistry of that element as a chemical shift. This section will describe how $E_B$ can be affected and how measurement of $E_B$ can be used to analyze the materials chemistry. Using current nomenclature and/or arguments, the problem of accurately measuring chemical shifts may be described as follows. Measured core level binding energies are a reflection of both their pre-measurement initial state value and those relaxation features induced by the measurement itself [2, 18–20].

The binding energy of a hole state is usually defined as the difference between the initial and final state of the species containing the core hole. The simplest approach to binding energy calculations assumes that Koopmans' theorem [21] is valid, that is, that the electron orbitals remain frozen during the photoemission process. This assumption neglects the contribution that occurs from the change in electron screening arising from other electrons in the solid. However, in the early days there was much controversy over whether it was initial or final state effects that were dominant in affecting the magnitude of the chemical shift, with the acceptance of Koopmans' theorem implying the former.

As demonstrated in the selected examples that follow, it is very difficult to achieve detailed closure on the arguments concerning the degree to which one may directly relate measured binding energy shifts to changes in the chemistry.

## 5.1 The Initial vs Final State Binding Energy Problem

The essential features of the dilemma rest on several points. First, as is well known, the measured core level chemical shifts for many oxides $(M_xO_y)$ symbolized as $(M^X)$ relative to their respective metal elemental $(M^o)$ states are often quite small, that is,

$$\Delta E_m(M^o \rightarrow M^{+X}) = E_m(M^{+X}) - E_m(M^o) \qquad (2)$$

The resulting measured chemical shifts vary substantially between various metal oxides, but are known to be generally <4 or 5 eV, and often are <1 eV. In a number of select cases the shifts may even be negative. Perhaps the alkaline earth oxides represent a good series of values that express these negative progressions. Although some values are in dispute (due to such hard to control features as charging), a representative set of chemical shifts may be expressed as:

| MgO | CaO | SrO | BaO |
|------|------|------|------|
| $-1.0$ eV | $-0.6$ eV | $-0.4$ eV | $-0.2$ eV |

Nothing is seriously wrong with the size or the progression of these values unless one considers the following.

The binding energies realized by any measured core level peak may be considered to be composed of (several) terms that are part of the (initial state) premeasured status of the system and those due to certain postmeasurement (final state) effects. In fact, it is not unreasonable to express a measured binding energy as

$$E_m(M) \approx E_1(M) - E_R(M) \qquad (3)$$

where $E_1$ is the (collective) initial state contribution and $E_R$ stands for the final state (relaxation) effects. The choice of sign here is somewhat ambiguous but we can assume that so-called relaxation effects result from various postdisturbance adjustments of the $N - 1$ remaining electrons that tend to collapse around any core hole. This collapse will always act effectively to increase the kinetic energy of the photoelectron that is, to decrease or subtract from the resulting initial state binding energy.

With this in mind we may use Eq. (3) to rewrite Eq. (2),

$$\Delta E_m(M^o \rightarrow M^{+x}) = \Delta E_1 - \Delta E_R \tag{4}$$

where in Eq. (4),

$$\Delta E_1 = E_1(M^{+x}) - E_1(M^o) \tag{5}$$

and

$$\Delta E_R = E_R(M^{+x}) - E_R(M^o) \tag{6}$$

However, we can also modify Eq. (6) by noting that the relaxation experienced by species M may be divided into two parts:

(i)  that part due to the individual units themselves (intra-atomic) $E_{IRA}$; and
(ii) that part experienced by the M units due to the presence of different surrounding species (extra-atomic), $E_{ERA}$.

Now it is also reasonable to assume that if our shift of interest involves two chemical states of the same M atom then:

$$\Delta E_R(M^o \rightarrow M^{+x}) = \Delta E_{EAR}(M^o \rightarrow M^{+x}) \tag{7}$$

Thus, the measured chemical shift energy of M for the process $M^o \rightarrow M^{+x}$ will differ from its initial state value because of any resulting extra-atomic relaxation shift.

It is well known [20] that although individual $E_R$ may be large, sometimes >20 eV, $\Delta E_{EAR}$ values are always much smaller. Although this is true, it is also known that in view of the sizes common for measured chemical shifts, $\Delta E_{EAR}$ values are often significant and sometimes may be several eV in size.

Methods have been developed to calculate $E_R$ and $\Delta E_{EAR}$ values directly (i.e., the $\Delta E$ SCF developed by Broughton and Ferry [22]) or to try to estimate them accurately (see, e.g., Hedin and Johansson [23]), but none of these approaches have been effectively extended to large atoms or molecules.

In order to fill the resulting gap, several empirical methods have been proposed. One of the most intriguing appears to be the Wagner Auger parameter ($\alpha$) method [18, 24]. In this approach it has been noted that comparative measurements of both selected photoelectron and Auger peaks lead to a term $\alpha$ that depends *only* on relaxation effects. In fact:

$$\Delta \alpha = \alpha(M^{+x}) - \alpha(M^o) \tag{8}$$

and

$$\Delta \alpha = \Delta E_m + \Delta E_{km \text{ (Auger peak)}} \tag{9}$$

By following a somewhat tortured path and introducing several approximations we find that

$$\Delta \alpha = 2\Delta E_{EAR} \tag{10}$$

Based upon the results achieved as well as physical arguments it is generally apparent that the relaxation experienced by a metal ($M^o$) following core electron ejection exceeds that of the same metallic element in the form of one of its oxides $M^{+x}$ Eq. (2), that is,

$$\Delta \alpha(M^o \rightarrow M^{+x}) \text{ is always } < 0!$$

Thus it is equally true that:

$$\Delta E_{EAR} \text{ is always } < 0!$$

And numerically $\Delta E_{EAR}$ must be added to $\Delta E_I$ to achieve $\Delta E_m$!
Now the alkaline earth oxides based on $\Delta \alpha$ values:

|  | MgO | CaO | SrO | BaO |
|---|---|---|---|---|
| $\Delta E_{EAR}$ | $-2.2$ eV | $-2.15$ eV | ? | $-1.85$ eV |

Thus all of the $\Delta E_{EAR}$ values are larger than all of the forementioned measured chemical shifts, $\Delta E_m(M^o \rightarrow M^{+x})$. Therefore, it must be true that the initial state chemical shifts $\Delta E_1(M^o \rightarrow M^{+x})$ for these oxides are negative. Based on physical intuition, however, this seems impossible. It should be realized that the initial state shift $\Delta E_1$ involves more things than just the change in oxidation state (primarily, structural changes). However, it is doubtful if the latter are always sufficiently negative to account for the persistent negative sign of $\Delta E_1$. This suggests that certain problems still exist in the fundamental arguments for these properties and, therefore, everyone should be careful not to ascribe absolute status to any interpretation.

Based upon the forementioned definitions and observations, one may worry that many measured chemical shifts will have little relationship to the realized (initial state) variation in chemistry and, perhaps even more disturbing, that most of the measured values will exhibit significant disparities with the $\Delta E_1$ (A–B). In point of fact, however, the measured $\Delta E_M$ (A–B) and, in many cases, the progressions exhibited by $\Delta E_M$ (A–B) for a particular class of compounds are often found to be reasonably close to $\Delta E_1$.

The reasons for the small $\Delta E_R$ (A–B) and, particularly, for the small $\Delta(\Delta E_R)$ are, as we will explain in what follows, fairly diverse. In fact, we will find that it is very useful also to consider in detail the reasons (beyond the scope of this chapter) why all $\Delta E_R$ may be uniformly large for certain categories of chemical shifts.

## 5.2 Fundamentals of Charging Shift

Charging occurs in XPS or ESCA because a nonconductive sample does not have sufficient delocalized conduction band electrons available to neutralize the macroscopic charged centers that build from clustering of the positive holes created by the ejection of the photoelectrons and/or Auger electrons [25]. As a result, a positive outer potential builds at or near the material surface, producing a retardation, or drag, on the outgoing electrons [26]. This retardation appears in the spectrum as a positive (nonchemical) shift that appears to subtract from the uncharged kinetic energy or, correspondingly, adds onto normal binding energies of the outgoing electrons. Compared to most of the features in various spectroscopic experiments, charging shifts are often relatively slow in their establishment and thus in the time frame of a normal experiment, charging has generally both dynamic and static components [27]. From a practical point of view then, charging may result from either the inherent insulating nature of the components of the sample, or the orientation of the sample. Removal of this charging shift may be achieved by simply neutralizing the forementioned retardation potential using a low energy electron flood gun. Figure 2 shows a typical charging shift spectrum (with variations in flood gun voltage) for the C (1s) of a fiberglass material.

    5.2.1 X-ray Satellites [2].   When conventional ESCA is utilized with direct focus X-ray units (usually Al K$\alpha$ or Mg K$\alpha$) the resulting XPS or ESCA spectra will produce and register photoelectrons from all aspects of the X-ray line source. In the case of the two major sources employed, this means that not only are photoelectrons ejected by the principal lines (centered at 1486.6 eV for Al K$\alpha$ and 1253.6 eV for Mg K$\alpha$), but they are also produced by any satellite lines associated with these X-ray sources. A typical example is shown in Figure 3, where the Al K$\alpha$ satellites at 9.6 and 11.5 eV (with respect to the position of the principal line) have been produced. These two satellite peaks have ~5 and 3.5% of the intensity of the principal photoelectron line. Often these satellite lines do not create any problems, and they may even be viewed as an additional useful feature for calibration processes; however, they may introduce some measurement problems, particularly if these satellites overlap weak photoelectron peaks. One must, therefore, always be cognizant of their presence. It should be noted that one of the advantages in employing an X-ray monochromator is the removal of these X-ray satellites.

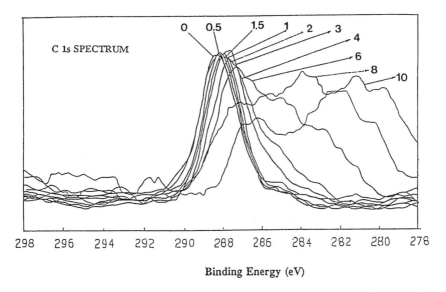

FIG. 2. A typical ESCA (performed in VG ESCA) C (1s) spectrum of fiberglass, showing charging shifts with different flood gun voltage settings.

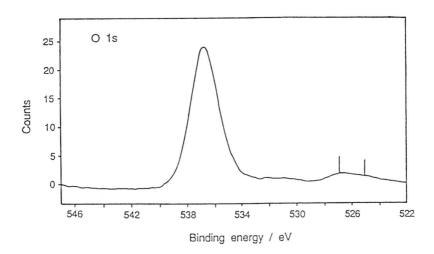

FIG. 3. ESCA O (1s) spectrum achieved with conventional Al K$\alpha$ for $AlPO_4^{-8}$ molecular sieve. Note X-ray satellite peaks located at 9.6 and 11.5 eV down field with respect to the principal O (1s) line.

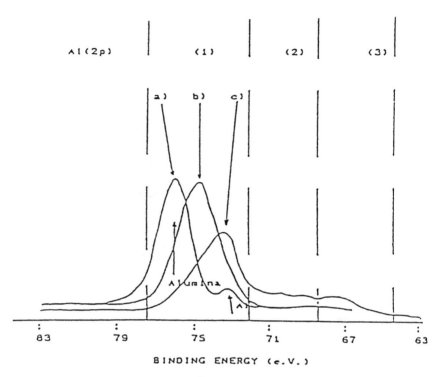

FIG. 4. Flood gun (FG) study of ~20 Å alumina on Al°: (a) FG off, (b) FG = 0.4 Ma, (c) FG = 0.4 Ma and 9.0 eV [2].

### 5.2.2 Differential Charging.

Differential charging may arise due to a number of factors, of which four of the most common, interrelated forms, are considered herein [27, 28]:

(1) the photoelectron spectroscopy sampling depth;
(2) the depth of field of any neutralization device (such as electron flood gun) that may be utilized to adjust any charging; and most importantly,
(3) the sample morphology, including layer systems, and the presence of clusters with distinctly different conductivity.

Figure 4 is an example of a charging shift and its progressive removal through application of a flood gun current and voltage. Application of various flood gun voltages definitely establishes the presence of differential charging in the alumina layer being considered. Figure 4 suggests that some of the charging species shift to different degrees with the applied voltage [29].

## 5.3  Charge Referencing in ESCA

Exposure of solid surfaces to ambient air invariably results in a modest over-layer of carbonaceous byproducts collectively labeled adventitious carbon (AC). The employment of the adventitious carbon to establish binding energy scales has been the standard in more ESCA analysis than all other referencing methods combined. In order to define the use of adventitious carbon (AC) [30–34], we must first try to describe what it is and how it behaves. Although each of these studies exhibits slight differences, the striking feature of the many thousands of XPS observations of adventitious carbon is their apparent chemical and physical similarity. Any attempt at analyzing the key details of AC requires a short review of the properties of the possible candidates.

### 5.3.1  Graphitic Systems.

The ESCA spectra of graphite is characterized by a number of distinct features whose variations (if any) from the spectra of hydrocarbons (and other carbonaceous systems) are more a matter of degree than of distinct differences. The first feature to note for graphite is its very narrow C (1s) line, which due to good conductivity exhibits no charging shift (see Fig. 5). The narrowness of this line is both a result of this conductivity and the macroscopic extent of the structural integrity of the graphite. Thus, related materials with somewhat compromised structures, such as conventional carbon blacks, usually do not exhibit charging but do generally show a slight broadening of their C (1s) line due to lack of structural singularity. (All of this is expressed in the differences seen in Fig. 5a,b.) The binding energy of the C (1s) for graphite has been variously expressed at several values, generally <285 eV. We have generally reproduced the value for graphite at 284.4 ± 0.2 eV [35, 36]. In addition to its singular core level line, graphite is found to produce a broad but distinct valence band signature that can be explained in terms of its charge density and bonding.

### 5.3.2  Hydrocarbon Polymers.

As in the case of graphitic materials, the ESCA spectra of most hydrocarbon polymers are so similar that it is difficult to distinguish one polymer from another. In fact, for many years, practitioners of ESCA were not even inclined to acknowledge any differences between the C (1s) binding energies of graphitic systems and any hydrocarbon polymers [37]. They thus clustered C (1s) values for all systems at ∼284.8 ± 0.3 eV (Fig. 5b). The features that seemed to play the largest roles in this indistinguishability were the lack of conductivity and resulting lack of a fixed Fermi edge (charging) exhibited by many of the hydrocarbon polymers. In recent years, very careful studies have led to the conclusion that these C (1s) binding energies are, in fact, distinct [38]. The range from ∼284.8 to ∼285.3 eV has been assigned to polymeric hydrocarbons (with some evidence for a progressive increase in value as the bonding in the polymeric chain becomes more complex) [38], whereas that for graphitic materials is as already stated, at 284.4 ± 0.2 eV.

SEAL AND BARR

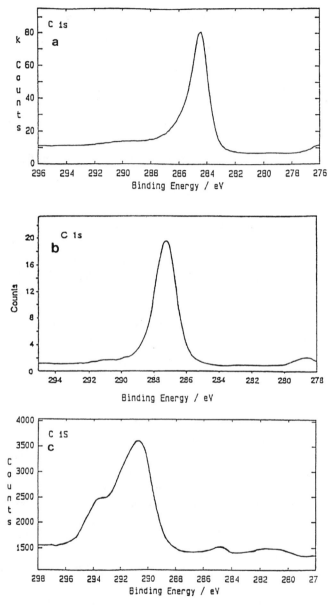

FIG. 5. Representative C (1s) spectra for select systems: (a) graphite; (b) pure hydrocarbons; and (c) adventitious carbon (AC) found on amphibole silicate. Note that only (a) is coupled to the spectrometer Fermi edge; (b) and (c) demonstrate the need for AC referencing. Reprinted with permission from T. L. Barr and S. Seal, *JVST* 13(3), 1239 (1995). Copyright (1995), American Vacuum Society.

These distinctions are very small and unfortunately may be undetectable for mixed (polymer/graphite) systems. This has led to the development of auxiliary detection methods based on the ESCA generation of valence bands and loss spectra. These procedures for the distinction of hydrocarbons are similar to those described in the preceding for graphite [36, 38].

### 5.3.3 Adventitious Carbon in General.

Having briefly described the ESCA characteristics of graphitic materials and hydrocarbon polymers, we must now see if we can use these features to provide a careful (and hopefully useful) chemical description of AC.

The complexity of the C (1s) spectra invariably registers the modest presence of multifaceted C—O containing type units. On the other hand, the principal C (1s) peak for AC generally exhibits fairly singular, chemical features that are predominantly expressive of either C—H or carbon-only type bonds. Despite this singularity, it is common for the principal peak to be broader than that realized by a pure graphitic system, or even that for a reasonably pure hydrocarbon polymer. This suggests that a substantial part of this broadening is due to the nature of the structural and chemical diversity of the AC rather than its particular chemistry. Noting, as we already have, that for proper utility AC should be composed of outer-surface-oriented, well-dispersed carbonaceous units, which are probably not formed into a contiguous layer. It should not be surprising that the ESCA line structure exhibits some degree of broadening [34]. Chemically, one cannot make a ready identification from the common C (1s) binding energy realized, ~284.8 eV [30, 31] (which is often made uncertain due to charging), Figure 2c, and the nature of the formation of the AC precludes any significant carbonaceous valence band. *All of this confirms that AC is predominantly formed from adsorbed hydrocarbon polymers and not graphitic carbons* [36, 37]. Support for this statement is also contained in studies by the authors of the ESCA properties exhibited by petroleum catalysts in which following use these systems exhibit various types of coke deposits formed on or near their surface [36, 39, 40]. If this coke growth continues to grow, eventually it may form a contiguous sheet that due to its fairly graphitic nature will exhibit reduced (perhaps zero) charging and its own valence band [36, 39]. All of the latter features are inherently different from the characteristic patterns of AC.

### 5.3.4 Adventitious Carbon: The Sputtering Effect.

A major problem of employing AC to establish a binding energy scale arises when there is not enough AC for ESCA detection. This commonly occurs following ion sputter etching. (It should be noted that conventional $Ar^+$ ion etching can be employed to verify our previous contentions of the general order of layering that occurs for air exposed surfaces. Thus, ion etching of an air-exposed metal foil will first remove most C=O species, then C—O—C, followed by the balance of the AC. If one performs this process carefully, one will note that the integrity of most air-induced oxide layers often remains intact during these steps, although any hydroxides or hydrates generally are removed with the AC species [36].)

Unfortunately, when the AC layer is gone, it is often difficult to continue to establish the binding energy scale, although sometimes some of the emerging components may be substituted by referencing back to their values when AC is present [31, 36].

Sometimes researchers will resort to reinsertion of AC to try to circumvent this problem. This is, however, a process that can be very deceiving because, depending on the materials and sputtering energies involved, some freshly sputtered surfaces are extremely reactive centers. Thus, we have found cases where carbon deposition has led to carbide production. An example for silicon is shown in Figure 6. Other examples of materials for which this has occurred include Fe, Ti, Mo,

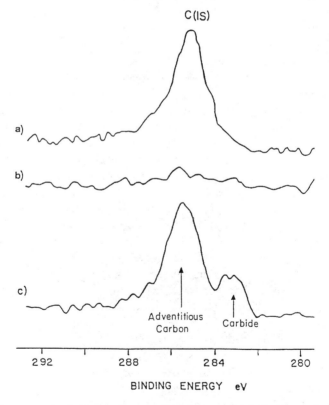

FIG. 6. Representative behavior of pre- and post-sputtered AC. In this case a sample of Si$^{\circ}$: (a) pre-sputtered outer surface; (b) following sputtering; and (c) after re-exposure to air. Reprinted with permission from T. L. Barr and S. Seal, *JVST* 13(3), 1239 (1995). Copyright (1995), American Vacuum Society.

and W [36, 37, 39]. Obviously, the generation of the species described herein is subject to the constraints of the original distribution of the AC. The fact that these features occur as often as they do is thus a testimony to the surprising singularity of AC.

### 5.3.5 Other Methods for Binding Energy Referencing and Determination.

Few researchers are very content with the process of employing AC to produce a binding energy scale. Many researchers are particularly unhappy with the fact that it represents the use of an uncontrolled "contaminant" for this purpose. For this, and related reasons, it has been common practice over the last 20 years to try to develop some other method for binding energy referencing. This is based upon a known ingredient that is introduced by the researcher in a controlled environment. A variety of methods have been tried, ranging from the introduction of small quantities of graphite to the implanation of modest metal dots [31, 36, 41]. All of these methods have the appearance on paper of being excellent candidates for energy referencing, but in practice all have suffered in ways that generally not only compromise the specific analysis, but also often make the methods inferior to AC. In general, the problem with these alternatives may be summarized in two phrases—"configurations" and "Fermi edges."

### 5.3.6 A Brief Summary of Binding Energy Determinations.

Over and above the problems of charging, it should be apparent that the major difficulties with ESCA binding energy measurements involve fixing and determining a Fermi edge. The Siegbahn research group determined very early in the development of ESCA that this was an easy feature to circumvent for any solid material that is a good conductor or narrowband gapped semiconductors [43]. In order to avoid the necessity for simultaneously determining the work function $\Phi$ of the system being examined, the Siegbahn method promotes the coupling of the Fermi edge of the sample with that of the spectrometer. Measurements are subsequently made against the latter by subtracting the spectrometer work function (an easily determined and generally constant parameter). This method, which works so well for the forementioned materials, is in fact a major reason for the difficulties in establishing the binding energies for nonconductors. Despite the fact that this Fermi edge coupling is generally not achieved for insulators or wide band gapped semiconductors, the spectrometers continue to generate measured results against the Fermi edge of the spectrometer [37, 40, 43]. This means that practitioners are constantly faced with the dilemma of shifting and discarding a variable series of measured results.

Methods have been proposed, with a few attempted, to provide an adjustable field to the measurement process that, when correctly employed, may promote the forementioned Fermi edge coupling. Perhaps the most well-documented example of this was the use by Stephenson and Binkowski [44] of the Hewlett-Packard (HP) flood gun to couple the Fermi edge of $SiO_2$ to that of the HP ESCA. The study was a major success but its extension into a readily acceptable, general procedure has eluded us so far. Therefore, we are left today with the somewhat uncomfortable situation that no method for determining the ESCA binding energies of nonconductive materials has been found to supersede energy scale referencing, and the primary application of this method is still through the use of AC-adventitious carbon [36, 41].

When properly configured, AC is assumed to be uniformly dispersed throughout the surface region of our sample of interest. As the AC species are neither contiguous nor in finite islands, it may be assumed that the carbonaceous valence region has a zero energy level but no independent band structure or Fermi edge. Because of dispersion, this AC valence zero is coupled to the Fermi level of the matrix material (sample) just as if AC is an integral component of the matrix. Thus, in effect, the C (1s) binding energy of AC scaled to, for example, 284.6 eV is an integral electronic state of the AC-laced matrix system. If the matrix system is a nonconductor and charges with photoelectron ejection, all of its energy states, including that for AC, will effectively charge shift the same amount. Therefore, adjusting all binding energies in the matrix subject to referencing the C (1s) binding energy to 284.6 eV effectively establishes valid binding energies for all peaks in the matrix. Most of these aspects are depicted in Figure 7.

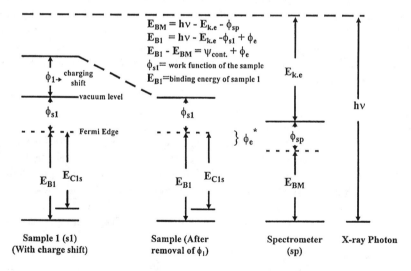

FIG. 7. Generalized binding energy scale diagram where the C (1s) is amalgamated into the level scheme for the sample and due to its nonconducting character, neither AC nor the sample Fermi edge is coupled to that for the spectrometer. Reprinted with permission from T. L. Barr and S. Seal, *JVST* 13(3), 1239 (1995). Copyright (1995), American Vacuum Society.

Two addendum features must be noticed: (a) in this model AC is coupled to the Fermi edge of the sample and not necessarily to that of the spectrometer; and (b) if AC is modified to achieve its own Fermi edge, this coupling may disappear unless the sample is a good conductor [36, 42].

# 6 QUANTIFICATION

Quantification is one of the most important and, in many cases, *the* most important feature of ESCA. Surprisingly, although this capability was recognized in the pioneering work in Uppsala [43], the development and explanation of the degree and methodology of realizing quantitative information in ESCA is still a very active and necessary area of research [45]. Thus, the fact that ESCA seems to detect material to a depth $\eta$, which in one sense is considered a surface analysis and in another a subsurface examination, makes the exact definition of its quantification very difficult. In particular, we must keep in mind that the predisposition of a factor labeled as an *escape depth* $\lambda$ can be a very beguiling fact. Thus, it is often suggested that one employ an average depth factor and the *average* ESCA observation is rendered to a depth of (for example) 30 Å. A major problem is, of course, that the actual depth of detection $\eta$ of an element A may be expressed as:

$$\eta = \lambda_M (E_{AX}) \cos \theta \tag{11}$$

where $M$ defines the particular matrix containing A, and the appropriate photoelectron is emitted from the X core level with energy $E$ at angle $\theta$ with respect to the surface normal. Thus, for an ideal sample the actual escape depth will vary with the energy of the source, the photoelectron binding energy, and the angle of emission.

As already suggested, quantification in electron spectroscopy has been the active research area of a number of the best practitioners in the field. Efforts in ESCA were originally championed by Wagner [46, 47], who has researched empirically derived, atomic, sensitivity factor scales [47]. Further development has been provided by Wagner et al. [48], Nefedov et al. [49], Seah (primarily for Auger spectroscopy) [50] and Powell [51] (primarily for ESCA).

The feature measured in an ESCA experiment and directly related to quantification is expressable, as expected, in terms of an electron current, or intensity factor, generally labeled herein as $I_{AX}$, where both A and X were defined in the preceding, and in [2, 52–54], that is,

$$I_{AX} = \sigma_{AX}(hJ)D(E_{AX}) \int_X^{Y\pi} \gamma = 0 \int_X^{y2\pi} \phi$$
$$= 0x L_{AX}(\gamma) \int_X^{Y\infty} y = -\infty \int_X^{y\infty} w$$
$$= -\infty J_0(wy) T(wy\gamma\phi E_{AX})$$

$$X \int_X^{y\infty} z = 0 N_{AX}(xyz)e^{-Z/\lambda_M(E_{AX})\cos\theta} dz dw dy d\phi d\gamma \qquad (12)$$

where $\sigma_{AX}(h\nu)$ is the quantum mechanical photoelectron cross section for the emission of an electron from the X core level of atom A due to the consumption of a photon of energy, $h\nu$, $D$ is the photoelectron's detection efficiency, $L_{AX}(\gamma)$ is the atomic angular asymmetry of the photoelectron intensity, $J$ illustrates the properties of the X-ray line in the detection plane, $T$ is the analyzer transmission function of the spectrometer, $N_A$ is the atomic density, and $z$ is the detected distance into the sample. As pointed out by Seal [54], this equation is still specifically constructed for the so-called homogeneous, semi-infinite slab model and therefore must be modified if the sample is nonuniform.

Calculations of absolute intensities are generally not attempted. In almost all cases, particularly in practical analysis, the analyst is interested in *relative* knowledge. Thus, questions such as, "What is the silicon to aluminum ratio in that zeolite?" or "How much more potassium is there on the surface of thin film A than B?" are indicative of typical requests. Fortunately, in this form of analysis, one may generally simplify the complex equation relating intensity and concentration. The relative precision may also be substantially improved, and it may be possible to employ the following:

$$\frac{I_{AX}}{I_{BT}} \cong \frac{\sigma_{AX} N_{AX}}{\sigma_{BT} N_{BT}} \qquad (13)$$

For this relationship to apply, it is necessary that the different photoelectrons AX and BT be emitted under nearly identical circumstances, that is, from the same place in the same (or, at least, very similar) materials and with kinetic energies such that

$$E_{AX} \sim E_{BT} \qquad (14)$$

When these criteria are not observed, it may be necessary to include in Equation (12) all terms that are energy dependent, particularly $\lambda_M$. A reasonable use of Eq. (12) was accomplished when we generated the Si/Al ratio for a series of zeolites employing the Si (2p) and Al (2p) peaks, as is displayed in Table I. Even in this case, however, a slight compensation was included for the fact that the more energetic Al (2p) is sampled at a slightly greater depth than the Si (2p).

In many cases, particularly in applied studies, the relative analysis is actually what we refer to as relative/relative. Thus, for example, the A/B ratio for sample 1 may be compared to the A/B for sample 2. Now, if all features are

TABLE I.    ESCA Detected Relative Quantification for Select Zeolites:
[Si/Al] [2]

| | Representative materials examined [Si/Al] | | |
|---|---|---|---|
| Material | Bulk (reputed) | Surface (determined) | %Na$^+$ |
| NaA | 1.0 | 1.1 | 100 |
| NaY | 2.4 | 2.6 | 100 |
| CaX | 1.25 | 1.3 | <5 |
| CaY | 2.4 | 2.7 | 10 |
| NH$_4$Y | 2.4 | 2.8 | 10 |
| Kaolinite | 1.0 | 1.1 | 0 |
| Bentonite | 2.0 | 2.2 | 0 |
| $\alpha$-SiO$_2$ | | 0 | |
| Silica gel | | 0 | |
| $\alpha$-Al$_2$O$_3$ | | 0 | |
| $\alpha$-Al$_2$O$_3$ | | 0 | |

Note: Each example listed is one of several commercial systems examined. All are reputed to be of excellent purity. Our ESCA survey scans suggest that claim to be correct.

ideal,

$$\frac{(I_A/I_B)_1}{(I_A/I_B)_2} \cong \frac{(N_A/N_B)_1}{(N_A/N_B)_2} \tag{15}$$

The use of Eq. (14) may actually relax the requirements of homogeneity and the concept of an infinite slab. If the samples are subject, for example, to inhomogeneities, then it is only important that they be equally so. Consider, for example, the results for several zeolites reported in Table II. These results are for relatively smooth, similarly prepared wafers pressed from uniformly sized powders.

In general, however, lack of surface uniformity and various types of inhomogeneities have played a major role in the interpretation of surface analysis. Fadley et al. [55, 56] have provided fairly detailed descriptions of the modifications expected for Eq. (11) when overlayers, patches, noncontinuous overlayers, and variable concentration gradients result. One must be cautious when applying these models, as they depend upon a certain regularity of variable morphology that may be difficult to produce or even verify. For the correct cases, however, the methods should be very valuable.

So far we have discussed some of the important features of the ESCA process. To learn more about this technique one should refer to other key texts that helped

TABLE II.    ESCA Detected Relative/Relative Quantification for Zeolites:
[Si/Al]$_A$/[Si/Al]$_B$ [2]

|  | [NaX]/[NaY] | [NaX]/[NaY] |
|---|---|---|
|  | Compositional formulas | Surface (XPS)[a] |
| Na | 1.57 | 1.47 |
| Si | 0.77 | 0.77 |
| Al | 1.57 | 1.53 |
| 0 | 1.01 | 1.08 |

[a] Based upon several measurements of same and different systems, ±05.

establish the foundation of this method of surface analysis for the last two decades
[2, 6, 9, 57–61].

The ESCA system is widely utilized in the study of both inorganic and organic
material systems to understand the fundamentals of materials chemistry. Subsequent sections will describe some of the selected application of ESCA (results
from VG, HP, PHI 5400 model ESCA will be presented) in general materials
science problems.

# 7 APPLICATION OF ESCA IN OXIDES

One of the most critical uses of photoelectron spectroscopy is to try to achieve
accurate and proper registration of the key chemical changes in interfacial regions
between thin films (e.g., heterojunctions or quantum wells), or between a matrix
and a dopant (e.g., a composite or catalyst). The most prevalent compounds in
these areas are generally oxides. It is obvious that proper spectral interpretation
in these areas may be made difficult by such things as interfacial [62] and small
cluster shifts [2], as well as certain final state-effects [18], and perhaps any charging shifts due to lack of conductivity [42]. Binding energy tables such as those in
Briggs and Seah [18] should still provide very useful data, because these tabula-
tions feature many of the most extensively employed compounds in materials areas. As examples one need only consider alumina and several related compounds
and go into detailed discussions. As described in what follows, this consideration
leads to some very disquieting generalizations about the measurement of binding
energies that *may affect, to some degree, the photoelectron spectroscopic investigation of almost every nonconductive compound* [63].

## 7.1 Binding Energy Interpretation of Different Forms of Alumina

### 7.1.1 Bulk or Thick Forms of Alumina.
Alumina is a designation commonly employed to describe a series of $Al_2O_3$ structures, and also sometimes misused to indicate various hydrated and partially hydrated species, such as AlOOH and $Al(OH)_3$. According to the Briggs & Seah table [18], all of these systems should produce Al (2p) binding energies of between 73.85 and 74.3 eV, with some indication that the species containing OH units may yield the largest Al (2p) values. It turns out that because of the extensive use of aluminas in such areas as catalysis [64], these binding energy values are among the most reproduced analytical results in surface science. For example, one may note that in Reference [39], one of the authors of this chapter provides another table with Al (2p) binding energies for $Al_2O_3$ listed as 74.0 *based on the same reference scale as that employed by Wagner* in Briggs & Seah [18]. Other versions of these values have recently appeared in a new text by the author [2]. It is equally important to note that on this binding energy scale, Wagner reports an O (1s) for $Al_2O_3$ at 531.0 eV, whereas one of the authors of this chapter suggests 530.75 eV, that is, both results fall within the same range of precision.

It is crucial to note, at this point, that *all of these results are obtained from the surface of various morphological arrangements of what we label herein as the "bulk" form of alumina, that is, powders, pressed wafers, thick crystals, or alumina films, where the latter are at least 50 Å thick.* Further, one should note that although a variety of techniques are used in these measurements to deal with the fact that all bulk aluminas are insulators, the reported binding energies are referenced to a C (1s) energy of $\sim$284.8$\pm$0.3 eV for adventitious hydrocarbons. A representative example of the Al (2p) photoelectron peak in question is presented in Figure 8.

### 7.1.2 Thin Films of Alumina on Aluminum.
How about carefully grown alumina films on top of reasonably good conductors? The classic examples of this are, the anodic or thermal growths of alumina on $Al°$, or, more fundamentally, systems that result from the careful adsorption or deposition of small (Langmuir) quantities of $O_2$ onto various single crystal faces of $Al°$.

Substantial effort has also been employed by numerous groups using photoelectron spectroscopy to investigate the chemical properties of thin films of various aluminas on aluminum. Although perhaps not as numerous as the thousands of catalytic and composite driven studies involving bulk alumina, these thin-film/chemisorption studies are often characterized by the use of even greater care in the obtaining and maintenance of materials integrity and precision of measurement. Although it is important to note this distinction, it is equally important to be aware that the rather pristine nature of the process seems to be *unnecessary* in realizing the features of *principal* concern in this discussion. Thus, whereas the use of these pristine features may be necessary to generate some of the *detailed*

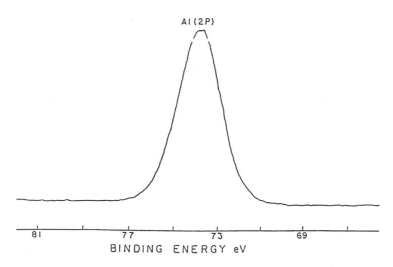

Al (2P)

BINDING ENERGY eV

81       77       73       69

FIG. 8. Al (2p) for $Al_2O_3$ with flood gun removal of charging. Reprinted with permission from T. L. Barr, S. Seal, L. M. Chen, and C. C. Kao, *Thin Solid Films* 253, 277 (1994). Copyright (1994) with permission from Elsevier Science.

chemical and structural features, the *general* binding energy patterns may be easily rendered during an XPS study of, for example, the anodic oxidation of reasonably clean, polycrystalline aluminum [65]. In any case, as already mentioned, no matter what the method of production, all of the resulting thin-film (alumina evolving) systems exhibit the same "problem," that is, their resulting aluminum and oxygen (core level binding energies for the Al−O system of more than three or four "molecular" layers) differ substantially from the previously described bulk results, (typified by Al (2p) of 75.8 ± 0.5 eV). Representative examples of these thin-film results are outlined in what follows.

### 7.1.3 Controlled Thin-Film Growth Studies.

The "pristine" adsorption of $O_2$ onto the surface of Al(111) was studied by McConville et al. [66]. This was a very involved photoelectron study that was reviewed in detail by Bradshaw [67]. McConville et al. report the Al°(2p) 3/2 to 1/2 split with the Al°($2p_{3/2}$) at ∼72.5 eV. Several intermediate (chemisorption (?)) oxidized Al peaks were also detected, producing, among others, Al (2p) peaks split from the metal by ∼1.5 eV. Then, finally, after coverage of several monolayers of purported $Al_2O_3$, an Al (2p) was generated split from that of Al° by ∼2.7 eV.

### 7.1.4 Thermal Oxidation Studies.

Another example of a detailed study of the oxidation of aluminum metal was provided in the Cornell Ph.D. Dissertation by Eric Johnson [68], and in a subsequent review of various $Al_2O_3$ studies the he coauthored with Cocke and Merrill [65]. They reported prominent Al (2p) peaks

## Al(2p) XPS Binding Energies

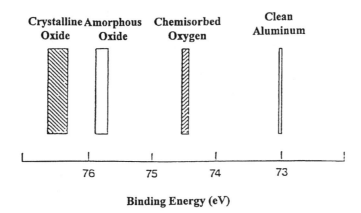

FIG. 9. Approximate Al binding energy shifts for different states of Al oxide. Reprinted with permission from T. L. Barr, S. Seal, L. M. Chen, and C. C. Kao, *Thin Solid Films* 253, 277 (1994). Copyright (1994) with permission from Elsevier Science.

shifted from that for the elemental metal by ∼1.4, ∼2.7, and ∼3.3 eV. The peak shifted by 2.7 eV was suggested to be that for amorphous alumina and that with 3.3 eV shift was due to crystalline alumina (see, e.g., Fig. 9).

**7.1.5 Binding Energies and Charging During Al Oxidation.** Part of the raison d'être for the present study was the realization that because few of the photoelectron spectroscopic systems employed in the previously mentioned investigations of oxide growth on metals were equipped with an electron flood gun, trying to register the *existence* of a charging shift, let alone attempting to *remove* it was naturally problematic. This is somewhat disquieting, in lieu of the well-known insulating nature of all aluminas. Perhaps part of the rationale for this neglect was the supposition that somehow the Fermi edge coupling of part of a thin oxide layer to another material (metal) (as indicated by the retention of valid (uncharged) binding energies by the metal) would compensate for this need. All of this, of course, ignores the ready prospect of a differential charging that gradually evolves with growth of the insulating layer [2, 39, 42]. The simple analogy of a conductive wire surrounded by a thin but partially effective insulating sheath seems somehow to have escaped notice. The operative word here is apparently "thin," a property whose definition varies for different metals and different "forms" of their oxides. It should be obvious, however, that for a wide band gapped, "true" insulator like alumina, charging may initiate for a very thin layer. Thus, in 1983 we presented an important related result demonstrating that a film of $<30$ Å of $SiO_2$ (with a bandgap similar to $Al_2O_3$) on $Si°$ exhibits obvious differential charging [2, 27, 39, 42].

With this potential problem in mind, the authors and his colleagues have expanded on a study of the features of the general oxidation of metals by including some of the particulars of that process on aluminum. One part of this study, accomplished by Kao [2, 69], did demonstrate that if the oxidative layer (on top of $Al°$) was "thick enough" obvious charging resulted. The layer studied was found to be generally somewhat thicker than that used by McConville et al. [66] or in the previous investigation [68] but *still thin enough to detect the subsurface $Al°$ with monochromatic alka X-rays*. In fact, in the Kao case [2, 69] the charging could be differentiated with relative ease, and by employing an additional energy bias of the utilized electron flood gun, the Al (2p) binding energy pattern of the alumina could be spread into a peculiar "depth profile" [2], wherein it was found that only the layers of oxide farthest from the metal were experiencing charging.

In order to test these results further, layered aluminum-oxygen systems with somewhat less oxide than in the case of Kao [2, 69] but a little thicker than used in the study by McConville et al. [66] were grown. The Al (2p) and O (1s) spectra produced by one of these systems somewhat resembled those finally achieved by McConville et al. [66] (with the added proviso that the differences in the photon energies employed may sample different depths). In any case, as suspected, the binding energy positions realized by part of the oxide formed on this system were still readily susceptible to movement in the electron flood gun (see, e.g., the O (1s) in Fig. 10). Once again, just as in the case of the previously mentioned $SiO_2$ on $Si°$ case, it appeared that it was only the oxide part of the Al (2p) manifold that could be moved by changing the current and energy settings of the HP-1850 electron flood gun.

These results seemed to concur with our assumption that the unusually large oxide binding energies achieved by the previously described studies of the formation of films of aluminas on $Al°$ may be, in part, "susceptible" to lack of diligence regarding charging. Consideration of the totality of the results, however, suggested that any assumption that the large difference between the chemisorbed and bulk-referenced values were *entirely or even primarily* due to charging might be an overstatement. In order to study the possibility of other potential effects we tried, therefore, through cleaning and regrowth, to create even thinner oxide films on $Al°$ (see Fig. 11). In this case, lack of XPS resolution (for the polycrystalline $Al°$ foil) seems to have precluded the appearance of the spin-orbit split of the metal $Al°$(2p), which was found to peak at 73.0 eV. Some suggestion of the detection of intermediate binding energy states were produced in our results, (see Fig. 11) near the appropriate peak positions suggested in the McConville study [66, 67], however, the resolution of these peak was very poor. The different depths of analysis of both 100 and 1486 eV photons may be partially the cause of the slight spectral mismatch (due to change in the sampling depth), as well as the difference in spec-

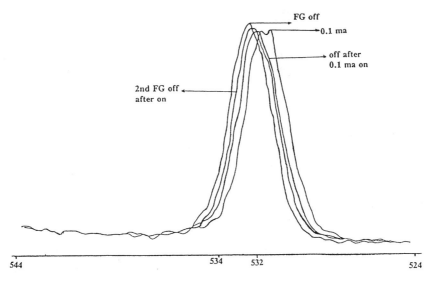

FIG. 10. O (1s) binding energy study of ESCA peaks for moderately thick oxide film on Al (HP ESCA and HP-2100A flood gun). Reprinted with permission from T. L. Barr, S. Seal, L. M. Chen, and C. C. Kao, *Thin Solid Films* 253, 277 (1994). Copyright (1994) with permission from Elsevier Science.

trometer resolution. Perhaps the most telling results, however, are the Al (2p) and related peaks due to the "outer" layer of our oxide (see Fig. 11a). One should note the similarity of these values reported by McConville et al. [66], Cocke et al. [65] and Johnson [68]. Particularly note the ΔAl (2p) ≅ 2.7 eV [70–72]. Once again an electron flood gun study was applied to this system, but in this case little involvement was affected, until application of a relatively large current, and some driving energy. Even in the case of the latter, however, there was only a slight shift in the Al (2p) of the oxide only, as shown in Figure 11b. (At the same time the Al°(2p) remained fixed at 73.0 eV.) In this regard it is important to note that not only the binding energies of the Al (2p), and corresponding (not shown) O (1s), remained largely impervious to the flood gun, but the small adventitious C (1s) (also not shown) also shifted only slightly. All these indicated a detachment of the Fermi edge (mode of charging) of perhaps the outermost alumina [2].

7.1.6 Conflicting Approaches to Oxide Binding Energies. Based on the forementioned results, there are apparently two conflicting approaches to oxide-binding energy measurement. In order to circumvent these dilemma, we must analyze the results achieved by the two methods and see if we can determine if one of the methods is producing correct and the other incorrect binding energies for the aluminas.

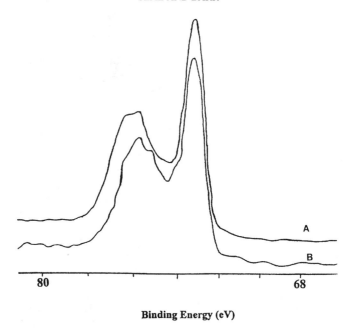

**Binding Energy (eV)**

FIG. 11. HP ESCA results for Al (2p) of a relatively thin oxide film. A – Flood Gun Off and B – Flood Gun On. Reprinted with permission from T. L. Barr, S. Seal, L. M. Chen, and C. C. Kao, *Thin Solid Films* 253, 277 (1994). Copyright (1994) with permission from Elsevier Science.

*7.1.6.1 Thin Film of Oxide on Metal (Al) Approach.* This method is based on the Siegbahn binding energy measurement approach of *coupled Fermi edges* [73]. First, one should note that if properly utilized the Fermi edges of *all* of the spectrometers employed in all of the measurements should be properly registered and coupled to the Fermi edge of any metallic sample, for example, Al°. Thus, for example, within the scope of the precision achievable by most measuring systems, Al (2p) values of 72.9 ± 0.2 eV should be easily obtained for thin films of $Al_2O_3$. Based upon these criteria, we also should note that up to a certain thickness, the thin films of oxide (e.g., $Al_2O_3$) grown on top of this metal appear to be Fermi edge-coupled to the metal, for example, Al°, and through it to the Fermi edge of the spectrometer [2, 42]. *Thus, we are prepared to state that, within the precision of the equipment employed, the resulting binding energies produced by this method are accurate measurements for the system, as energetically configured.*

*7.1.6.2 ESCA Analysis of Bulk Oxides.* This method is characterized by less utilization of the Fermi edge coupling, and instead substantial reliance on *binding energy referencing*, with particular use of adventitious carbon. First of all, a number of studies have been (correctly) conducted using conductive metal species as references (e.g., Au deposits [26]), and the results are essentially the same as those achieved with nonconductive carbon. Further, a number of conductive carbonaceous systems have been studied with ESCA (most notably graphite), and when Fermi edge coupled to the spectrometer all of these studies have produced C (1s) binding energies of 284.7 ± 0.5 eV. In addition, nonconductive carbonaceous material has been mixed with conductive and the ESCA of the latter also produces essentially the same C (1s) values. In fact, thousands of polymeric and composite ESCA studies have led to the conclusion echoed by Briggs [74] that *all C−C, C=C and C−H bonding units, no matter how complex, produce a C (1s) of essentially the same binding energy,* ∼285.0 eV. Thus, it is our conclusion that ESCA binding energies reported referenced to adventitious carbon with C (1s) ∼284.9 eV are moderately precise and often quite accurate. Further, it should be noted that the Al (2p) of ∼74.0 ± 0.3 eV that results from this assumption for the surface of a bulk aluminum oxide mixed with and referenced to adventitious carbon is consistent with values achieved with conductive references (e.g., those of the other Group IIIA and IVA oxides) [2, 75]; and also with the binding energies achieved when alumina is a simple progenitor of more complex oxides [2, 39, 75]. Further, we reported on a detailed series of studies of the ESCA-generated valence band spectra for alumina and related oxides [2, 39]. The patterns followed by these bands are all consistent with the forementioned Al (2p) value (∼74.0 eV) for alumina. In fact, it is our conclusion that *the XPS-generated binding energies for the surface of bulk aluminum oxides as presented in the preceding in Table III, for example, Al (2p) ∼74.0 eV, are reasonably precise and accurate values for the system, as energetically configured.*

We have thus reached an intriguing impasse where two vastly different sets of results are both djudged to be valid for what should be the same property of, apparently, the same system. This needs a closer look.

### 7.1.7 Interpretation of These Binding Energies.

In Table III, we present a tabulation of representative binding energies for select oxide compounds determined by adventituous carbon referencing. The results in column IV, thus, presupposes that the compounds in question are, if thick enough, nonconductors, and, therefore, the binding energies of these species are adjusted to a common value of the C (1s) peak for the (supposedly common) adventitious hydrocarbon species (in this case 284.9 eV was employed) adsorbed onto the surface of these (before measurement) air-exposed compounds.

The binding energies for the oxides in question may and in some cases also have been determined for a thin film of that compound on top of its elemental metal. In this case the compound film is thin enough that the key ESCA peaks of

TABLE III.    Binding Energies for Select Metals and Their Key Oxides

| Element | Core level | B.E. $\pm$ 0.2 eV for elemental metal ($M^\circ$) | B.E. $\pm$ 0.2 eV for key metal oxide[a] $M_xO_y$ (1) EC (1s) = 284.9 eV | B.E. $\pm$ 0.2 eV for O (1s) $M_xO_y$ using EC (1s) = 284.9 eV | Type of oxide[b] |
|---|---|---|---|---|---|
| Mg | 2p | 49.5 | 50.2 | 531.2 | SC |
| Al | 2p | 72.8 | 74.1 | 531.5 | SC |
| Si | 2p | 99.4 | 102.9 | 532.9 | SC |
| Ca | 2p$_3$ | – | 347.5 | 529.8 | I |
| Tl | 2p$_3$ | 453.3 | 458.5 | 529.7 | I |
| Fe | 2p$_3$ | 706.6 | 710.8 | 529.9 | I |
| Co | 2p$_3$ | 778.2 | 780.9 | 529.9 | I |
| Ni | 2p$_3$ | 852.7 | 854.6 | 530.0 | I |
| Cu | 2p$_3$ | 932.5 | 933.7 | 529.8 | I |
| Zn | 2p$_3$ | 1021.7 | 1022.0 | 530.7 | SC |
| Ga | 3d$_5$ | – | 20.2 | 530.9 | SC |
| Ge | 3d$_5$ | 30.0 | 32.7 | 531.2 | SC |
| Zr | 3d$_5$ | 178.9 | 182.2 | 529.9 | I |
| Mo | 3d$_5$ | 227.9 | 232.4 | 530.4 | I |
| Rh | 3d$_5$ | 307.2 | 308.4 | 530.0 | I |
| Pd | 3d$_5$ | 335.4 | 336.9 | 530.1 | I |

Reprinted from Barr, Seal, Chen, and Kao, *Thin Solid Films* 253, 277. Copyright (1994) with permission from Elsevier Science.

[a] In general, results for maximum valent (common) oxides are given. Where this is not practical the most common oxide is chosen.

[b] SC, semicovalent; I, ionic; B.E., binding energy.

the elemental species may still be detected through the film. Evidence suggests that in this case the two species (elemental and compound) are often Fermi edge coupled by short range (Schottky) flow of electrons and despite the fact that the compounds are often nonconductors, they are thin enough so that no charging shifts are *yet* produced. Thus binding energies determined by this method represent an example of a detailed use of the classic Siegbahn Fermi edge coupling method for binding energy determination [73].

It is perhaps not surprising that there are some differences in the resulting values for these two methods. What is surprising is that for many compounds these differences are fairly small (generally <0.4 eV), whereas for a select few the differences are much larger, for example, for $Al_2O_3$, Fermi edge-coupled thin-film Al (2p) $= 75.6$ eV, where as Column IV Al (2p) $= 74.1$ eV.

It should be apparent that the Fermi edge coupling procedure is per force the more *precise* method of measurement (based on the use of the word "precision" by analytical chemists). It is also true, however, that one must seriously doubt that the majority of the supposedly inert hydrocarbon or graphite adsorbed during the formation of the thin-film oxide on *certain select* elemental metals and metalloids is somehow subject to a substantial alteration that seemingly produces hydrocarbons with unprecedented increases in binding energies of as much as 1.5 to 2 eV, for example, in the case of $Al_2O_3$ thin film the adventitious carbon is found at $\sim$286.5 eV.

It should be noted that the act of onset of low-temperature oxide growth on these elements is known generally to extract electrons out of the elemental metal into the evolving oxide film [76]. *This fact (establishing the Cabrera-Mott field)* [77, 78] *may result in an effective shift of the Fermi edge of the evolving oxide.* This appears to be the case for the oxides referred to in Table III as semicovalent (SC). These oxides generally have wide, unobstructed bandgaps between the O (2p) and their conduction bands. Most of the common ionic (I) oxides, however, have bandgaps partially filled with relatively discrete (d) bands that apparently mitigate this effect. Thus, as demonstrated in Figure 12, the Fermi edge-coupled binding energy result for the SC oxide $Al_2O_3$ may, in fact, not be that for the intrinsic-only midgap oxide, and if one shifts the Fermi edge for these oxides back to that midgap point (as approximately dictated by choosing C (1s) for the adventitious hydrocarbon at 284.9 eV), binding energy values almost identical to those reported in column IV and V of Table III will be produced.

As implied, it appears that this effect has a great deal of generality beyond $Al_2O_3$. Thus, we have found in the case of other SC oxides, such as $SiO_2$ and $GeO_2$ [79], that this same property arises, that is, thin films of the oxides grown on their elemental metals (apparently in any structural state) produce metalloid oxides XPS results ($M^{+x}$, O (1s)) and also adventitious carbon C (1s) values that exceed those achieved for the corresponding $M^{+x}$, and O (1s) for thick samples of the oxides referenced to C (1s) at 284.9 eV. When the thin-film results are shifted to bring their C (1s) values to 284.9 eV, however, these values agree with the corresponding bulk metal oxide values to $\pm 0.3$ eV. We feel that this shifting generality can be extended to encompass most, if not all, of the oxides that we are herein labeling as (SC) or semicovalent, that is, those with O (1s) $> 530.5$ eV [2, 80]. In the case of those other oxides that we label herein as ionic, these adjustments are unnecessary as the two measurement methods produce binding energy values, even for C (1s), that are essentially the same.

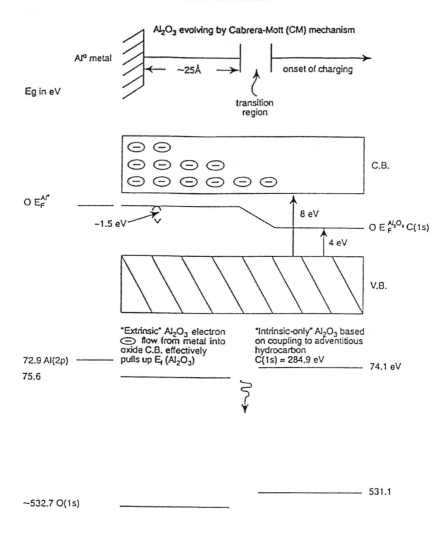

FIG. 12. Rendition of depth of oxidation-induced shift in bands and core levels for alumina due to space charge bias (vs fixed (intrinsic) Fermi energy). Reprinted with permission from T. L. Barr, S. Seal, L. M. Chen, and C. C. Kao, *Thin Solid Films* 253, 277 (1994). Copyright (1994) with permission from Elsevier Science.

# 8 APPLICATION OF ESCA IN COMPOSITE SYSTEMS

The principal theme of this section is concerned with the involvement of ESCA or XPS in various metal matrix-based composite systems. This section will be devoted to boron-, nitrogen-, aluminum-, and titanium-based composite systems and some of the ESCA results have been published elsewhere by the authors.

## 8.1 B—N—Ti System

As is known, nitrides of various elements play an important role in industry, as well as in science and technology, as a result of their interesting and useful resilient properties. Cubic boron nitride (c-BN) displays remarkable hardness (>50 GPa), wear resistance, and chemical inertness. The addition of certain third elements, for example, Ti, offers the promise of further enhancement of favorable wear properties of these boron nitrides as well as distinct improvements in their aesthetic features. The inclusion of a transition metal in the coating introduces metallic bonding, which may increase the probability of better adhesion between coating and many substrates. Further, TiN has found much acceptance in the industry in its own right as a surface coating because of its high hardness, good wear, and corrosion resistance [81–84]. One can synthesize Ti—B—N composite coatings by variety of deposition methods, such as CVD [85], plasma CVD [86, 87], sputtering [88–92], ion beam methods [92], and ion-assisted electron beam plasma vapor deposition (PVD) [87].

Some of the most important properties of these coatings occur due to the presence of the resulting binary and ternary phases, for example, $TiN_x$, $TiB_2$, BN, and $TiB_xN_y$ [93]. The chemical and microstructural characteristics of these thin film composite coatings, their surfaces, and the coating/substrate interfaces have been primarily analyzed using Auger electron spectroscopy (AES), XPS, electron energy loss spectroscopy (EELS) and cross-sectional transmission electron microscopy (XTEM) methods [94–99].

### 8.1.1 ESCA or (XPS) Results on B—N—Ti.
Detailed XPS peak binding energies of the Ti (2p), N (1s), and B (1s) of standard TiN, $TiB_2$ and BN samples are listed in Table IV and displayed in Figure 13a,b and Figure 15a. As expected, significant amounts of adventitious carbon (C) and oxygen (O) were found to be present, as expected, on these surfaces. The compositional ratios of the key compounds calculated from XPS peaks of TiN, $TiB_2$ and BN samples were Ti/N = 0.98, Ti/B = 0.4 and B/N = 1.2, respectively. The deviations from the ideal formulae values are due in part to the expected effects created by surface oxidation, that is,

$$2BN + 3O_2 = B_2O_3 + N_2$$

and also in part due to the presence of select defect sites.

TABLE IV.    Binding Energies (±0.2 eV) for Standard Systems—Binding
Energies Are Referenced to C (1s) = 284.6 eV

| | Ti (2p)$_{3/2}$ (eV) | N (1s) (eV) | B (1s) (eV) | O (1s) (eV) |
|---|---|---|---|---|
| | $^1$Ti$^{+4}$, $^2$TiN, | $^1$TiN, $^2$TiN$_{x(\text{unsat})}$, | $^1$TiB$_2$ | $^1$Ti$-$O |
| | $^3$TiB$_2$$^a$ | $^3$BN, $^4$BN$_y$O$_z$, | $-$ | |
| | | | $^2$BN | $^2$B$-$O |
| TiN (s) | $^1$458.6, $^2$455.57 | $^1$397.3, $^2$396.5 | $-$ | $^1$530.2 |
| TiB$_2$ (s) | $^1$458.6, $^3$454.45 | $-$ | $^1$187.42 | $^1$530.5, $^2$532.3 |
| BN (s) | $-$ | $^3$398.2, $^4$399.1 | $^2$190.38 | $^2$532.4 |

Reprinted with permission from S. Seal, T. L. Barr, N. Sobczak, and E. Benko,
*J. Vac. Sci. Technol.* A15, 505, Copyright (1997), American Vacuum Society.
$^a$s = standard samples, ($^1$Ti$^{+4}$, $^1$458.6, i.e., the binding energy of Ti$^{+4}$
is 458.6 eV).

### 8.1.2 Effect of Heat Treatment on the Ti$-$B$-$N Surface Chemistry.

The XPS method was employed to investigate the chemical states of Ti$-$B$-$N
at various stages of heat treatment (1000 °C for 1 hr and 1400 °C for 2 hr, respectively). The XPS spectra of Ti-BN (1000 °C for 1 hr) revealed the expected
presence of TiO$_2$ on the surface due to preferential air-induced oxidation of the Ti.
After heating at 1400 °C for 2 hr, the XPS (Ti (2p), N (1s), B (1s), O (1s)) spectra exhibited distinct changes. At this higher temperature condition, a number
of distinct new phases, such as TiB$_2$, TiN, BN, plus perhaps some oxynitrides
[101, 102], as well as the expected oxides of B and Ti, were suggested by XPS
to be formed on the surface (Table V and Figs. 13, 14, and 15). This suggests
that at higher temperatures B and N species have diffused towards the surface and
reacted with Ti, forming TiN and TiB$_2$ compounds, and also with O$_2$ to produce
B$_2$O$_3$ (Figs. 13d, 14, and 15b). The presence of a peak at 399.13 eV in the N (1s)
spectrum (Fig. 15b) suggests that oxynitrides are also formed in the air-induced
oxidized layer. Moderate amounts of TiO$_2$ (Ti (2p)$_{\text{B.E.}}$ = 458.8 eV) and B$-$O
containing species (B (1s)$_{\text{B.E.}}$ = 192.4–193.1 eV) were found on the top surface
layer. The boron oxide peak has a higher binding energy than that of BN due to
the more ionic nature of the B$-$O bonds (compared to B$-$N). A similar rational
explains the titanium binding energy for TiO$_2$ ($\sim$458.6 eV) relative to that for TiN
($\sim$455.5 eV) [103].

The XPS binding energies of 1000 °C and 1400 °C heat-treated BN + TiN
systems are also listed in Table V and presented in Figures 13e,f, and 15c,d. Two
types of chemical species for N atoms were found in TiN film (Table V):

(1)  One was designated as TiN, with apparently stoichiometric bonds with

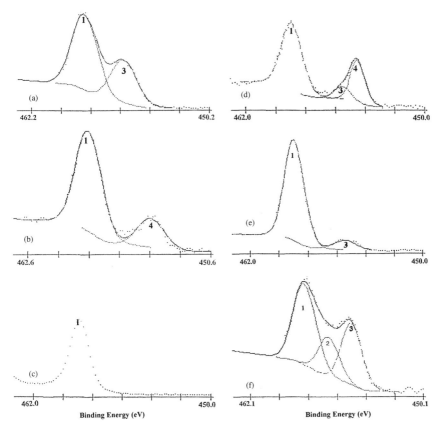

FIG. 13. The ESCA deconvoluted Ti (2p) spectra: (a) TiN; (b) $TiB_2$; (c) and (d) Ti + BN (treated at 1000 °C and 1400 °C); and (e) and (f) TiN + BN (treated at 1000 °C and 1400 °C). Possible peak identifications: (1) $TiO_2$, (2) $TiN_x$, (3) TiN and (4) $TiB_2$. Reprinted with permission from S. Seal, T. L. Barr, N. Sobczak, and E. Benko, *J. Vac. Sci. Technol.* A15, 505 (1997). Copyright (1997) with permission from American Vacuum Society.

titanium (Ti atoms surrounded by N atoms: N $(1s)_{B.E.}$ = 397.7 and 397.7 eV for 1000 °C and 1400 °C, respectively) and

(2) The other was designated as $TiN_x$, with nonstoichiometric N−Ti bonds (i.e., N/Ti > 1 and different bonding due to the excess nitrogen) (N $(1s)_{B.E.}$ = 396.8 and 396.2 eV for 1000 °C and 1400 °C, respectively) (Fig. 15c,d).

The fraction of N with nonstoichiometric bonds with Ti may be one of the principal factors for determining film quality. It was shown that coatings with the

SEAL AND BARR

TABLE V.   Binding Energies ($\pm 0.2$ eV) for Representative Ti$-$B$-$N Deposition Systems—Binding Energies Are Referenced to C (1s) = 284.6 eV

| | Ti (2p)$_{3/2}$ (eV) | N (1s) (eV) | B (1s) (eV) | O (1s) (eV) |
|---|---|---|---|---|
| | $^1$Ti$^{+4}$, $^2$TiN, $^3$TiB$_2$ $^4$Ti-Oxynitrides, $^5$TiN$_{x\,(\text{unsat})}$ | $^1$TiN, $^2$TiN$_{x\,(\text{unsat})}$, $^3$BN, $^4$BN$_y$O$_z$, $^5$Ti-Oxynitrides | $^1$TiB$_2$, $^2$BN, $^3$TiB$_x$N$_y$, $^4$BN$_y$O$_z$, $^5$BO, $^6$B$_2$O$_3$ | $^1$Ti$-$O, $^2$B$-$O, $^3$Si$-$O, $^4$Oxynitrides, $^5$OH/C$-$O$-$C |
| $^x$Ti+BN$^a$ | $^1$458.97 | Nil | Nil | $^1$530.3, $^3$532.66 |
| $^y$Ti+BN$^a$ | $^1$458.8, $^2$455.5, $^3$454.5 | $^1$397.25, $^3$398.1, $^4$399.13 | $^1$187.57, $^2$190.38, $^3$189.44, $^4$191.8, $^5$192.4, $^6$193.1 | $^1$530.22, $^2$532.8, $^4$530.5–531.6 |
| $^x$TiN+BN | $^1$458.8, $^2$455.5 | $^1$397.7, $^2$396.8, $^5$400.2, $^5$398.3 | Nil | $^1$530.07, $^3$532.8, $^4$530.8–531.5 |
| $^y$TiN+BN | $^1$458.2, $^2$455.6, $^5$456.5 | $^1$397, $^2$396.2, $^5$398.5 | Nil | $^1$529.92, $^4$530.7–531.8, $^5$533.8 |

Reprinted with permission from S. Seal, T. L. Barr, N. Sobczak, and E. Benko. Copyright American Vacuum Society (1997).

$^a$$x$ and $y$ denote 1000 and 1400 °C heat treatment, respectively.

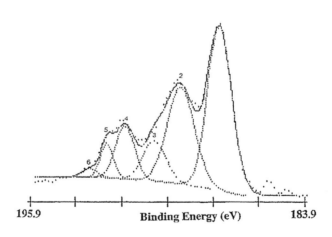

195.9                                    Binding Energy (eV)                            183.9

FIG. 14.   The B (1s) ESCA spectrum of Ti + BN treated at 1400 °C for 2 hr. Possible peak identifications: (1) TiB$_2$; (2) TiB$_x$N$_y$; (3) BN; (4) BN$_y$O$_z$; (5) B$-$O; and and (6) B$_2$O$_3$. The peaks chosen in this figure and the next one are based upon the expected products and their probable positions. Reprinted with permission from S. Seal, T. L. Barr, N. Sobczak, and E. Benko, *J. Vac. Sci. Technol.* A15, 505 (1997). Copyright (1997) with permission from American Vacuum Society.

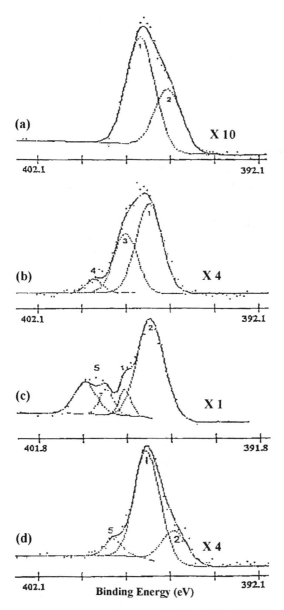

FIG. 15. The N (1s) ESCA spectra of (a) TiN, (b) Ti + BN (1400 °C), (c) and (d) TiN + BN treated at 1000 °C and 1400 °C, respectively. Possible peak identifications: (1) TiN; (2) TiN$_x$; (3) BN; (4) B-oxynitrides; and (5) Ti-oxynitrides. Reprinted with permission from S. Seal, T. L. Barr, N. Sobczak, and E. Benko, *J. Vac. Sci. Technol.* A15, 505 (1997). Copyright (1997) with permission from American Vacuum Society.

best tribological properties contained the smallest fraction of "excess" nitrogen bonds with titanium [104, 105]. No $TiN_x$-type compounds were found in the Ti + BN system. Only nitrogen seems to be sensitive to the nonstoichiometry and, therefore, Ti + BN may exhibit better tribological properties than TiN + BN due to the absence of $TiN_x$ in the former case. In addition, the intensity of N in the 1000 °C treated BN + TiN sample is 4 times lower than the one treated at 1400 °C. This suggests that the mobility of N atoms is larger at higher temperature during the formation of titanium nitrides.

  **8.1.3 Detection of Contamination.**   In all of the currently described coated samples (Table V), a slight surface presence of silicon in the form of Si−O, Si−C, and $Si_3N_4$ was suggested by a multifaceted Si (2p) spectra. The presence of this silicon apparently occurred during the heat treatment of these samples in quartz tubes, indicating that Si may migrate from the tube to the sample surface.

  **8.1.4 XPS Quantification from Surface to Bulk Through Sputter Etching.**   Using depth profiling, compositional analyses were performed from the surface to a depth of more than 8000 Å, where the material reached was deemed to be that of the substrate. From Figure 16 we observe a decrease in the silicon concentration following modest sputtering, suggesting it to be a surface impurity. The Ti content was found to increase to a depth of 60 Å and remained flat until 4700 Å, followed by a sharp decrease as we approach the substrate/coating interface. The N content increased with depth of sputtering and a significant B concentration was first found at 2300 Å; its concentration increased exponentially until we reached "pure substrate" (i.e., BN). From the XPS quantification and the binding energy of the detected Ti (2p) (458.9 eV), $TiO_2$ was found to have been produced on the surface of the implanted Ti film. After $Ar^+$ ion sputtering for 6 min the thin layer of $TiO_2$ was removed and the Ti (2p) binding energy suggested primarily TiN. After 1 min of sputtering the presence of oxynitrides is evident, but disappeared after 50 min of sputtering, apparently resulting in a phase separation into stoichiometric and non-stoichiometric TiN, plus some interfacial $TiO_2$ and Ti−O−N. Earlier studies proposed the presence of suboxides at the interface between the oxide and nitride [106]. In our case we find XPS evidence of TiN films oxidized to $TiO_2$ with a suggested sublayer of Ti-oxynitrides. A similar feature was also suggested by other researchers [107]. For an oxynitride, the N (1s) peak is generally found between 399.1–400.2 eV, whereas the Ti ($2p_{3/2}$) is in the range of 455.6–455.8 eV [105]. After sputtering off 2300 Å, boron (B) was detected (see Fig. 16). With further sputtering the B and N content increased while the Ti content in the film decreased. This suggests the near completion of any Ti and N reaction. However, in this region, a mixture of $TiB_2$ (B $(1s)_{B.E.}$ = 187.5 eV), TiBN (B $(1s)_{B.E.}$ = 189.8 eV), and BN (B $(1s)_{B.E.}$ = 190.5 eV) phase was found to be present, suggesting that with depth the Ti−N reaction is preceded by a reaction between Ti and B.

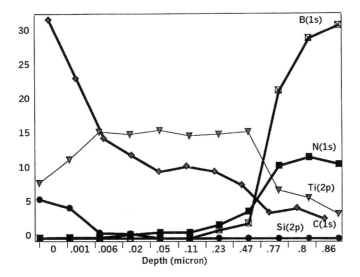

FIG. 16. A typical XPS sputter depth profile of Ti thin film deposited on BN substrate, heated at 1000 °C for 2 hr.

### 8.1.5 Effect of Sputtering on the Ti Chemistry.

As a result of ion bombardment, the line shape and full width half maximum (FWHM) are changed in all of the Ti (2p) spectra. These changes are caused by the appearance of additional chemical states, which, based on analogy with previous studies [108], were characterized as suboxided $Ti^{+3}$, $Ti^{+2}$. Based on related studies these features were found to be due to sputter-induced reduction reactions. Deeper in the film (>8000 Å), the $TiB_2$ phase disappears, while the relative XPS quantification of the BN phase increases (Fig. 16). Some oxidation of BN apparently into B-oxynitrides is found deeper into the film. In addition, the relative amount of $TiN_x$ (or TiN) decreases with increasing depth of sputtering. At a depth of ~8600 Å in the film, the appearance of a distinct Ti (2p) binding energy peak at 454.1 eV suggests the presence of small amounts of unreacted $Ti^0$.

From the XPS analysis we found that the inner region of the coating was primarily TiN and $TiN_x$, followed by a $TiB_2$ layer, while the region near the substrate/coating interface consists of BN, $B_2O_3$, and some unreacted $Ti^0$. By analyzing the phase composition of this film we also found a region at the 4700 Å level of titanium nitride, which appears to be extensively nitrogen deficient (i.e., based on our XPS results Ti/N ~ 2.1). This may be caused by the high-temperature (1000 °C) treatment of the substrate. At this temperature the mobility of the adatoms is higher and therefore it is possible that an initially formed TiN phase is, at this depth, partially transformed into a nitrogen-deficient phase [109, 110]. It should be noted that in studies such as in this case, one should not attach too much significance to the specific numbers or other details. Sputter profiling is known to cause structural damage and also often alters the newly exposed (sub)chemistry, particularly through the process of "sputter reduction." Thus, for example, it has been noted in Reference [2] that one of the structural forms of $TiO_2$ may (under certain sputtering conditions) exhibit selective sputter reduction. It may be expected, however, that the corresponding nitrides are more resistant to sputter reduction than the oxides, particularly in the case of BN. The structural integrity of the mixed nitride system $Ti-B-N$ is, however, another matter, as these units are expected to be structurally altered by the $Ar^+$ ion beam. Correspondingly, we do not anticipate that the sputtering process will gradually alter the integrity of *the TiN film in the Ti−B−N system.*

## 8.2 ESCA or XPS Studies of Al−Ti−N Systems and Its Oxidation Behavior

Why study $Ti-Al-N$? The $Ti-Al$ alloy nitrides possess relatively high hardness, good wear resistance, and oxidation and corrosion resistance at elevated temperatures [111–120]. They are often used as cutting and forming tools and are a better choice over TiN coatings because the latter oxidizes rapidly in air at temperatures >550 °C [121]. The information available on the ternary $Ti-Al-N$ phase diagram is limited. Several ternary compounds such as, $Ti_2AlN$ [123, 124], $Ti_3AlN$, and $Ti_3Al_2N_2$ [123] have been reviewed in great detail. The high oxidation resistance of $Ti-Al-N$ coatings is due to the formation of a stable protective $Al_2O_3$ layer in the outer scale as a result of selective oxidation of aluminum over titanium at the selected temperatures, thus protecting the underlying material. In a note, McIntyre et al. [125] showed that there is indeed multiple oxide layer formation on oxidation because the oxidation process in ternary alloy thin films is complex. Studies indicate that the oxide scales grown on a base of heated $Ti-Al-N$ consists of two sublayers, in which the outer one is alumina rich whereas the lower one is titania rich [125].

The complex nature of the bonding structure in transition metal nitrides incorporates a mixture of covalent, metallic and ionic components [126]. The nature on this bonding leads to high hardness, chemical inertness and, good electrical conductivity of these mixed nitride thin films. In what follows, we describe the surface chemical alteration of both a control and oxidized $Ti-Al-N$ films (i.e., two different types of deposition: room temperature (A) and liquid $N_2$ temperature (B)), by measuring the chemical shifts in the Ti (2p), Al (2p), O (1s) and N (1s) XPS spectra.

**8.2.1 Al (2p) Chemistry.** The as-deposited Al (2p) peaks for the $Ti-Al-N$ films (A and B) are at $73.7 \pm 0.1$ eV ($Ti-Al-N$ formation) along with a presence of a surface-exposed thin $Al_2O_3$ layer (indicated by a shoulder $\sim 74.3 \pm 0.2$ eV).

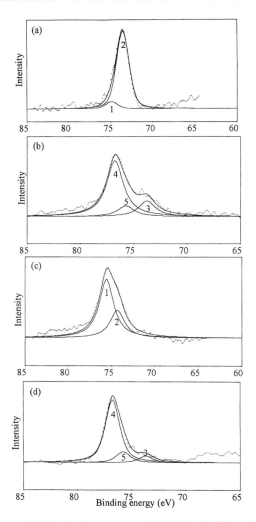

FIG. 17. Deconvoluted XPS Al (2p) spectra of (Ti,Al)N thin films; (a) ambient as deposited; (b) liquid $N_2$ deposited short-term oxidized at 850 °C; (c) ambient deposited short-term oxidized at 850 °C; and (d) liquid $N_2$ deposited long-term oxidized at 850 °C. Possible peaks: 1) $Al_2O_3$; 2) Ti—Al—N; 3) chemisorbed $Al_2O_3$; 4) crystalline $Al_2O_3$; and 5) amorphous $Al_2O_3$. Reprinted with permission from S. Seal et al., *J Vac. Sci. Technol.* A18(4), 1571, 2000, in press. Copyright (2000), American Vacuum Society.

The relative surface concentration of the Al content in both the films is higher than ambient. The Al (2p) B.E. in Ti—Al is 71.5 eV, which is attributed to the fact that following alloy formation the electron transfer is from Ti to Al [126], whereas the B.E. of Al (2p) metal in Ti—Al—N increases to $73.7 \pm 0.1$ eV (Fig. 17a), where all the electron transfer is from Al to N due to the electronegativity effect [127–129]. Both (A and B) films are oxidized at 850 °C for shorter and longer duration. In the short term oxidation, an increase in the binding energy of the Al (2p) occurs ~2 eV (A) to 3 eV (B) from that of the elemental metal, suggesting that Al might have migrated (in form $Al_2O_3$) to the surface towards the oxide/vapor interface. The main Al (2p) peak increased from 73.7 to 75.4 eV, indicating the oxidation of Ti−Al−N to an amorphous $Al_2O_3$ layer (relative to the position of coupling of the $Al°$ and $Al_2O_3$ Fermi edges, see preceding text) [129].

In the case of film B (Fig. 17b), following short-term oxidation, Al (2p) photolines are observed at higher binding energies, (B.E. ~75.5 ± 0.2 eV and B.E. ~76.5 ± 0.2 eV.) Various researchers have attributed these binding energies as those for amorphous (B.E. ~75.5 ± 0.2 eV) and crystalline (B.E. ~76.5 ± 0.2 eV) $Al_2O_3$ [129–131]. Various other studies document this similar chemical shift in Al (2p), indicating a transition of amorphous to crystalline behavior during aluminum oxidation [132]. Studies by Biaconi et al. [133] indicate that the amorphous to crystalline ($\gamma$-$Al_2O_3$) transition for very thin amorphous layers (~30 Å) occurs at 350 °C. In thicker amorphous layers (~300 nm), a modified $\gamma$-$Al_2O_3$ (called $\gamma$) is also reported to be formed [134, 135]. This species is regarded as an intermediate step between amorphous and crystalline $Al_2O_3$. The crystalline structure has a face-centered cubic lattice of oxygen ions containing interstitial $Al^{3+}$ ions statistically distributed in the available sites, where they are on average ≈70% octahedral and ≈30% tetrahedral. Only the oxygen ions are arranged regularly. It appears that the amorphous to crystalline transition involves a substantial movement of $Al^{3+}$ from the tetrahedral to the octahedral sites either directly or by reorganization of the oxide ions. This rearranged spinel structure possesses high bond strength and thus may be responsible for the higher binding energy for crystalline $Al_2O_3$. Similar observation has been made by several researchers [130–136]. In the case of film B, the transformation of amorphous $Al_2O_3$ into crystalline $Al_2O_3$ seems to be a function of oxidation time. Therefore, film A does not exhibit any crystalline $Al_2O_3$ formation, mostly all amorphous $Al_2O_3$ (Al (2p) B.E. ~1.3 eV shifted from that of Al metal Fig. 17c). As the time of oxidation of film B increases to 7.5 hr, almost all of the amorphous $Al_2O_3$ starts transforming to crystalline $Al_2O_3$, such as also indicated by the reduction in FWHM from 2 to 1.8 eV (see Fig. 17d). From the quantitative standpoint, the Al/O ratio in the film decreases as the film is oxidized, whereas for A in film B that ratio shows a comparatively high value (less oxygen). This can be due to the fact that improved nanocrystalline features of film B form an early protective $Al_2O_3$ film, preventing further oxygen uptake.

8.2.2 Ti (2p) Chemistry.   The main Ti (2p) photemission peak for as-deposited Ti−Al−N films (A and B) lies in the range of 456.7 to 456.3 eV, indicating the formation of alloy nitride along with a surface-exposed $TiO_2$ layer (B.E. 458.7 ± 0.3 eV). The Ti (2p) binding energy for TiN is at 455.6 eV [137].

Both films are deviated from the ideal stoichiometry, which is attributed to the difference in sputter yield of titanium and aluminum and its final recombination with nitrogen atoms in the chamber. The XPS Ti (2p) line of film A shows a stepwise chemical phase change when oxidized at both short and longer duration. As a result, the clean Ti (2p) in Ti−Al−N (B.E. ∼456.7 eV) converts to both Al−Ti−O−N (B.E. ∼457.8 eV) and $TiO_2$ (B.E. ∼458.6 eV).

Further oxidation for longer durations causes the total conversion of the oxynitride to pure titania (Ti 2p peak at 458.3 eV), accompanied by a decrease in the Ti/O ratio. Several studies have proposed an oxidation of suboxides at the interface between the oxide and the nitride [138]. Robinson et al. found that TiN films oxidized to $TiO_2$ with a sublayer of oxynitrides, whereas Ernsberger et al. labeled the oxynitrides as the principal component resulting from TiN oxidation [139, 140]. All of these observations are consistent with the result detected on our oxidized Ti−Al−N thin films.

In the case of the as-deposited ambient film, the amount of Ti−Al−N formation is more than that of the $TiO_2$ as calculated from the ratios of the relative peak intensities after peak deconvolution. Titanium attains enough oxygen activity along with the high temperature that it acts as a driving force for faster oxidation. The intensity of the Ti−Al−N (Ti 2p) peak starts to drop, suggesting that $TiO_2$ is now growing at a faster rate and forming on top of this nitride layer, thereby suppressing the intensity of the nitride layer. The FWHM of the Ti 2p manifold increases to 3.1 eV and indicates the formation of mixed oxide, nitride and oxynitiride species (Table VI). During longer oxidation periods, a reduction in the Ti (2p) FWHM to 1.5 eV indicates only the oxide $TiO_2$ (Table VI). When two oxides ($Al_2O_3$ and $TiO_2$ in this case) are forming, they will produce mixed oxides and create relative chemical shifts in their respective O 1s photoelectron peak binding energies due to their relative ionic and covalent behavior [2].

The oxidation behavior of film B (liquid $N_2$ deposition) at different durations is fairly simple as compared to that exhibited by film A. The short-term oxidation shows only a main titania peak (B.E. 460.7 ± 0.1 eV) with a small hump at the right-hand side of the peak (mixed oxynitride ∼B.E. 457.7 ± 0.1 eV). No stepwise chemical phase change is observed in this case. At longer durations, the binding energy of Ti (2p) indicates a surface with only $TiO_2$ formation. As discussed earlier, the formation of nanoparticles in the liquid $N_2$ deposited films seems to be responsible for the higher rate of oxidation, as a result of the higher surface area and hence greater surface reactivity of the growing oxide. When these films (B) are oxidized successively to shorter duration it is seen that the rapid oxidation of $TiO_2$ completely wipes out the Ti−Al−N Ti (2p) signal, indicating that the surface is now merely composed of oxides, and is contrary to the behavior seen in Ti (2p) spectra of film A under similar conditions (shorter oxidation). This is also reflected in the relatively low FWHM (1.44 eV at short-term oxidation and 1.56 eV at long-term oxidation) values for these peaks, indicating the presence

152                                SEAL AND BARR

TABLE VI.    XPS Binding Energies and FWHM Values in Ti (2p) Spectrum
for Ambient Deposition Ti—Al—N Thin Films (±0.2 eV)

| Sample Ti—Al—N | Phase | Ti (2p) Raw data B.E. (eV) | FWHM (eV) (total) |
|---|---|---|---|
| | Ambient A | | |
| | TiO₂ (s)ᵃ | 458.8 | |
| As deposited | Ti—Al—N (l)ᵃ | 456.7 | 2.75 |
| Oxidized | TiO₂ (mh)ᵃ | 458.6 | |
| (short term) | Ti—Al—N (l)ᵃ | 457.8 | 3.1 |
| | Mixed oxinitride (m)ᵃ | – | |
| Oxidized | TiO₂ | 458.3 | 1.5 |
| (long term) | | | |
| | Liquid nitrogen B | | |
| As deposited | TiO₂ (m)ᵃ | 458.2 | 2.76 |
| | Ti—Al—N (l)ᵃ | 456.6 | |
| Oxidized | TiO₂ (l)ᵃ | 460.9 | 1.44 |
| (short term) | Ti—Al—O—N (vs)ᵃ | 458 | |
| Oxidized | TiO₂ | 460.9 | 1.56 |
| (long term) | | | |

Reprinted with permission from S. Seal et al., *J. Vac. Sci. Technol.* A18(4), 1571 (2000). Copyright American Vacuum Society (2000).

ᵃIntensity: l = large; m = medium; mh = medium high; s = small; vs = very small.

of only $TiO_2$ (see Table VI). The Ti/O atomic concentration values for both the ambient and liquid $N_2$ temperature films decrease with increase in oxidation time, showing enhanced surface oxide film formation, but the metal/oxygen ratios are less than for Al/O.

8.2.3 Oxygen Chemistry.   The O (1s) spectra (Figs. 18a,b and 19a,b) of A and B films oxidized periodically for short and longer durations at 850 °C show that the relative $Al_2O_3/(Al_2O_3 + TiO_2)$ ratios for both films increase continuously, indicating that with treatment the aluminum is migrating out to the surface and forming a greater amount of aluminum oxide on the newly formed surface. This is due to the preferential formation of $Al_2O_3$ as a result of its larger (nega-

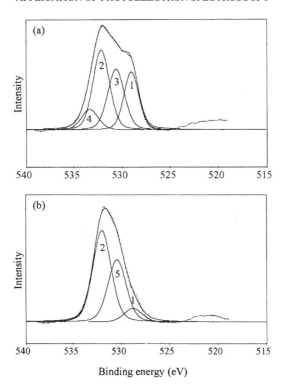

FIG. 18. Deconvoluted XPS O (1s) and spectra of ambient (Ti,Al)N thin films. (a) TiAlN film short-term oxidized at 850 °C; and (b) TiAlN film long-term oxidized at 850 °C. Possible peak identification: 1) $TiO_2$; 2) $Al_2O_3$; 3) mixed oxynitrides; 4) absorbed OH; 5) mixed oxides. Reprinted with permission from S. Seal et al., *J Vac. Sci. Technol.* A18(4), 1571, 2000, in press. Copyright (2000), American Vacuum Society.

tive) free energy of formation than for $TiO_2$. The $Al_2O_3/(Al_2O_3 + TiO_2)$ ratios in the case of film B (liquid $N_2$ film) are higher than for ambient films and can be attributed to the presence of nanoparticles, which possess more surface area and thereby result in surface reactivity that is enhanced and hence more prone to oxide formation. The FWHM increases from the control to the oxidized version in a short term and then decreases at longer-term oxidation, indicating formation or dissociation of mixed oxides. The FWHM of the O (1s) peaks for samples oxidized for shorter times at 850 °C in both the films (A and B) ranges from 5.1 eV (film A, see Fig. 18a) to 4.6 eV (film B, see Fig. 19a). The O (1s) peaks for film A oxidized for shorter times at 850 °C are 533.2 eV (hydroxide species), 532.1 eV ($Al_2O_3$), 530.52 eV (Al−Ti−N−O), and 529.07 eV ($TiO_2$) while for film B they are (short-term oxidation) 533.3 eV (hydroxide species), 532.4 eV

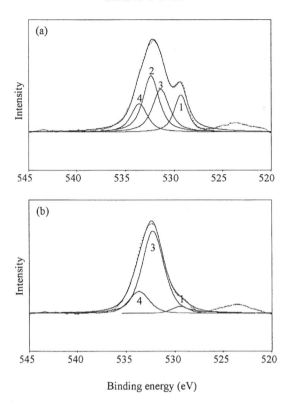

FIG. 19. Deconvoluted XPS O (1s) and spectra of liquid $N_2$ (Ti,Al)N thin films. (a) TiAlN film short-term oxidized at 850 °C; and (b) TiAlN film long-term oxidized at 850 °C. Possible peak identification: 1) $TiO_2$; 2) amorphous $Al_2O_3$; 3) crystalline $Al_2O_3$; and 4) absorbed OH. Reprinted with permission from S. Seal et al., *J Vac. Sci. Technol.* A18(4), 1571, 2000, in press. Copyright (2000), American Vacuum Society.

($Al_2O_3$-amorphous), 530.52 eV ($Al_2O_3$-crystalline), and 529.3 eV ($TiO_2$). A similar trend is observed in the Al (2p) spectra as described in the earlier section. Increase in the FWHM of the O (1s) peak may contribute to the presence of defects in the former film. In this case, there may be two crystallographically nonequilibrium oxygen ions and hence one would expect two O (1s) signals. On prolonged oxidation (long term), the $TiO_2$ signal (from O (1s)) starts to decrease due to enhanced $Al_2O_3$ formation. The mixed oxide formation (disappearance of oxynitride phases) is evident in film A (O (1s) ~ 530.526 eV, Ti—Al—O, not oxynitride, because N was not observed as discussed later). In this case, both Ti and Al form bonds with O in the mixed oxide system. The Ti—O bonds are more covalent than Al—O. As a result, a mixed oxide system valence electronic shell is no longer uniform but is polarized to reflect the difference in covalency and ionicity of Ti—O and Al—O bonds. The oxygen valence density oriented towards titanium is weakly held, which reflects the covalency in the Ti—O bond. The electrons are shared with Ti and the situation is quite different for Al, where electrons are held closer to the O ions (much more ionic). This creates inherent polarization of the oxygen electron valence shell reflected in the system. The core shell O (1s) electrons are, on the other hand, spherically distributed and photons of sufficient energy entering this shell will eject the electrons uniformly in all directions. The escaping O (1s) electrons then come in contact with the polarized valence shell, interact effectively, and shift in energy to recognize the polarity, resulting in selective chemical shifts in the mixed oxide system [2].

   **8.2.4 N (1s) Chemistry.**   The as-deposited ambient Ti—Al—N thin film (A) shows sufficient nitrogen concentration ($\approx 11\%$) as compared to that in the liquid $N_2$ deposited film (B) (3.5%). The surface nitrogen concentration decreases as the films (both ambient and liquid $N_2$) are subjected to oxidation. Figure 20b,d shows a relative comparison between oxidized films A and B at different times. This suggests that as the films are getting oxidized at subsequent intervals, the oxides are growing at a rate that almost covers the underlying nitride layers. The N (1s) binding energy of the as-deposited film (A) indicates the presence of a Ti—Al—N phase (B.E. $\sim$395.9 eV) with a possible presence of TiN ($\sim$396.7 eV, Fig. 20a). This negative shift for TiN (in N), is supported by the positive shift in the Ti (2p) spectrum (B.E. for Ti (2p) in Ti—Al—N > Ti (2p) in TiN). On the contrary, the as-deposited film B not only shows the presence of Ti—Al—N, (B.E. $\sim$395.6 eV), but also a peak at a 3.5 eV up-field shift (Fig. 20c). This peak at 399.1 eV can be attributed to the formation of a possible oxynitride phase due to high surface reactivity of film B. Similar binding energy assignments for oxynitride phases are listed in the literature and other studies (397.3 eV to $TiN_{0.5}O_{0.5}$, 398.4 to 399.9 eV to oxynitride) [138].

   For short-term oxidation, both films (A and B) show a small presence of the oxynitride phase (399.1 to 399.4 eV). The amount of N content is much smaller in the latter film due to quicker oxide growth. After a long-term oxidation at 850 °C, both film surfaces (A and B) are deprived of N due to the heavy oxide growth and hence the binding energy values could not be extracted even after deconvolution. After short-term oxidation both films indicate the formation of mixed oxides and oxynitrides, which dissociate on prolonged oxidation times. The nitrogen signal is weak in film B due to heavy oxidation as a result of nanoparticle formation. This is attributed to the relatively higher $Al_2O_3/(Al_2O_3 + TiO_2)$ ratios in film B than in film A.

   The $N_2$ films produced in liquid tend to serve better than ambient films with respect to their mechanical and oxidation resistance properties, and ESCA has been instrumental in revealing the details of the divergent surface chemical changes responsible for these different material properties. Similar applications of ESCA

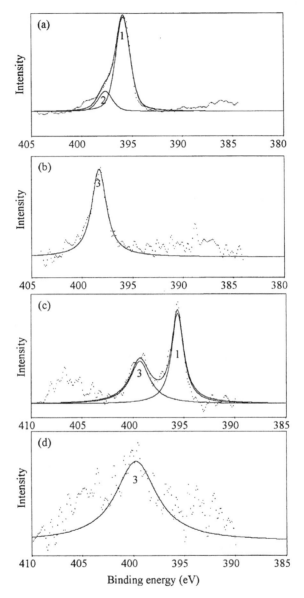

FIG. 20. Deconvoluted XPS N (1s) spectra of Ti—Al—N thin films. Ambient temperature: (a) as-deposited TiAlN film; (b) TiAlN film short-term oxidized at 850 °C; liquid $N_2$ temperature: (c) as-deposited TiAlN film; and (d) TiAlN film short-term oxidized at 850 °C. Possible peak identification: 1) Ti—Al—N; 2) TiN; and 3) oxynitride. Reprinted with permission from S. Seal et al., *J Vac. Sci. Technol.* A18(4), 1571, 2000, in press. Copyright (2000), American Vacuum Society.

in examining other composite material systems have also proved highly encouraging.

## 9 KEY SILICATE ANALYSES USING ESCA

Advances in catalysis, microelectronics, high-tech ceramics, and composite materials all require detailed chemical characterizations of complex solids. Many materials of interest contain Al, with the myriad aluminosilicates being of particular interest [141, 142]. The great success of NMR in these areas relies on the ease with which it can detect both octahedral and tetrahedral Al, even when both are present [143].

Early ESCA studies of aluminosilicates primarily originating in our group were heavily concerned with the surface chemical environment in zeolites [144, 145], the monitoring of such features as dealumination [141, 142], and the formation of byproducts. These studies have since been extended to a variety of other framework aluminosilicates [27]. All these materials are characterized by Al substituting for silicon in the tetrahedral framework. The ESCA method has been applied to sheet aluminosilicates [146] in which Al is often in octahedral coordination and to double-chain aluminosilicates [147] where both octahedral and tetrahedral Al atoms may be found. Again, two or more types of surface Al environment were simultaneously monitored. The research reports mention whether high-resolution/flood gun controlled ESCA is capable of differentiating between [4]Al and [6]Al on surfaces. Our success in this area will be described, including examples of a method to monitor simultaneously materials of simpler composition than chain silicates, but with well-defined Al coordination. Other materials will include aluminosilicates with mixed octahedral/tetrahedral structures, including yttrium aluminum garnets (YAGs).

### 9.1 General Comments

We shall first review oxide systems containing a single type of aluminum, in an attempt to determine the patterns in the ESCA binding energies that are particular to each structural type. Only materials containing tetrahedral (four-coordinate) and/or octahedral (six-coordinate) aluminum will be considered. Because of the well-recognized tendency of all ESCA peaks from a particular element to track the same chemical shift [37], one generally needs to report the binding energy of only one such peak. We will thus list only the Al 2p values.

As many oxides contain other elements (typically Si), the O 1s binding energies often reflect this complexity and require more elaborate interpretation, which is outside the scope of this chapter. Similar arguments apply to the valence-band spectra, which are useful for detailed oxide analysis [148, 149]. We include examples of the Al 2p linewidths, which often convey information about the integrity of the material. It is important not to over-interpret the numerical values, which may also reflect features related to morphological, rather than chemical integrity.

## 9.2 Structures Containing a Single Type of Aluminum

Nonsilicate aluminum oxide: The simplest aluminum oxide is alumina itself. Table VII lists its Al 2p ESCA binding energy along with values for other Al-containing systems. Of the several types of alumina [65], most have Al 2p binding energies very similar to those given in Table VII. The reported value ($\pm 0.2$ eV) has been reproduced by many workers [18], including this group [27, 144]. The Al$-$O bond in alumina is about as covalent as it can be in an octahedral environment, and consequently its Al 2p binding energy is relatively low (compared to more ionic Al$-$X bonds). Hydration of most metal oxides gives hydroxides in which the metal—oxygen (M—O) bond is made more ionic and the corresponding binding energies for M are significantly increased, usually by $>1.0$ eV [63]. Table VII shows that this increase occurs for aluminum, but that the largely amphoteric nature of gibbsite Al(OH)$_3$ mitigates this effect and the positive binding

TABLE VII.    The Al 2p ESCA Binding Energies ($\pm 0.2$ eV) for Key Oxides Containing Either Octahedral or Tetrahedral Aluminum [136]

| A/Octahedral Al | Si/Al | Al (2p) | O (1s) |
|---|---|---|---|
| Al$_2$O$_3$ | 0 | 73.8 | 530.6 |
| Al(OH)$_3$ | | 74.1 | – |
| Allophane | 1.0 | 74.3 | 531.75 |
| Kaolinite | 2.0 | 74.8 | 531.5 |
| Montmorillonite | | | 532.0 |
| (Pyrophyllite) | 4.0 | 74.15 | |
| Glaucophane | | | 531.35 |
| B/Tetrahedral Al | | | |
| Na$_2$Al$_2$O$_4$ | 0 | 73.2 | 530.2 |
| NaA | 1 | 73.5 | 530.5 |
| Sodalite | 1.05 | 73.6 | 530.9 |
| Analcime | 1.8 | 73.95 | 531.45 |
| Zeolite NaY | 2.5 | 74.2 | 531.75 |
| Zeolite NaX | 1.25 | 73.7 | 530.85 |
| Zeolite CaX | | 74.35 | 531.4 |

energy shift is only ~0.4 eV. One might expect this effect to be progressive—the Al 2p binding energy for boehmite (AlOOH) lies between those for $\gamma$-$Al_2O_3$, followed by $Al(OH)_3$ [2], but a more rigorous study is needed to confirm these relatively small shifts.

No simple tetrahedral aluminum oxide is known. The simplest compounds with [4]Al are the alkali-metal aluminates such as $Na_2Al_2O_4$ (Table VII), but they are almost impossible to prepare in a pure state. In addition, $Na_2Al_2O_4$ reacts easily in air with $CO_2$. Heating a mixture of $Na_2O$ and $Al_2O$ gives a complex mixture of compounds such as $Na_4Al_4O_{13}$, which contains four aluminate tetrahedral [144, 145]. In any case, the nature of the bonding is such that the much more ionic Na—O bond forces the Al—O bond to be more covalent than that exhibited by $Al_2O_3$. As a result, the Al 2p binding energy for $Na_2Al_2O_4$ (73.2 eV), derived by deconvoluting the peak from alumina produced by the reaction of $Na_2Al_2O_4$ with $CO_2$, is the lowest Al (2p) for any common oxide. The introduction of $K^+$ or $Cs^+$ should reduce the Al 2p binding energy even further.

## 9.3 Aluminosilicates

In framework silicates, the Al 2p binding energy of [4]Al changes progressively [27, 144, 145] with the replacement of Si by Al in the tetrahedral lattice. Thus, as the Si content of any tetrahedral framework is increased relative to the amount of Al, the combination of the reduced concentration of the interstitial cations and the increased population of Si—O units progressively forces the Al—O bond to be more ionic and as a result the Al 2p binding energy increases (Table VII). This effect has been explained in detail (see [27, 144, 145, 148, 150]).

The lower bound for the Al 2p binding energies for hydrothermally prepared [4]Al framework aluminosilicates is determined by the Loewenstein rule, which prohibits Al—O—Al bonding. This means that the Si/Al ratio must always be greater than 1.0. Thus, for zeolite Na—A and certain feldspathoids with this 1.0 Si/Al ratio, Al 2p is ca. 73.5 eV (Table VII) [37].

As expected, the type of countercation that compensates for the negative charge in the Al—Si tetrahedral lattices also plays a role in the binding energy shifts. We have seen that the presence of these cations tends to make the Al—O bond more covalent, thereby reducing the Al 2p binding energy. For reasons of charge compensation, multivalent cations such as $Ca^{2+}$ must straddle several aluminate centers, while often being smaller in size than their monovalent counterparts (such as $Na^+$). These factors reduce the compensatory effects induced by the cations in the Al—O bond. Thus, the Al—O bond in zeolite Na—X is thus much more covalent (and the Al 2p binding energy significantly lower) than in zeolite Ca—X (Table VII) [2].

Common framework silicates have only tetrahedral structural sites with O/Tet ~2. In all nonframework silicates O/Si > 2 and octahedral sites are always present [153]. The charge distribution in these complex oxides is such that the cations become more ionic (positive) as the coordination number increases. Thus, Al on octahedral sites is more positive than on tetrahedral sites, and on eight-fold sites more positive than on octahedral sites. As a consequence, in sheet silicates in which all Al is octahedral, the positive nature of the octahedral layer induced by the relatively negative silicate layer forces the Al—O bonds in kaolinite [148, 151] and allophane [152] to be more ionic than in alumina, and even more ionic (larger Al 2p) than in gibbsite. Charge compensation makes the corresponding Si—O bonds more covalent (lower Si 2p binding energy) than in silica [2]. This increase in the Al 2p binding energies found in 1–1 sheet minerals (such as kaolinite), should be enhanced in the case of the 2–1 minerals (such as pyrophyllite or the smectites). This is because each octahedral subsheet must compensate the charge of two relatively negative tetrahedral layers. The enhanced ionicity of the Al—O bond is reflected in the relatively large Al 2p value for montmorillonite (Table VII). The converse is not always continued for the lowering of the Si 2p binding energy where, in the case of montmorillonite, the influence of the aluminum on Si is smaller than in kaolinite [148].

## 9.4 Mixed Structures

### 9.4.1 Garnets.
We begin our discussion of mixed-coordination materials with an intriguing group of nonsilicates, yttrium aluminum garnets (YAGs). These are related to mineral garnets, which are orthosilicates with the general formula $\sim^{[8]}A_3{}^{[6]}B_2{}^{[4]}T_3$ oxide, as in $Ca_3Al_2(SiO_4)_3$. The garnet framework (Fig. 21) consists of alternating tetrahedral and octahedral sites in which each tetrahedral cation is linked to four octahedral cations and each octahedral cation is linked to six tetrahedral cations. Countercations lie within the interstitial framework cavities. In $^{[8]}Y_3{}^{[6]}Al_2{}^{[4]}Al_3O_{12}$ all framework sites are occupied by Al and the countercation is $Y^{3+}$. The general composition of a YAG is $Y_3{}^{[6]}Al_2{}^{[6]}Al_3O_{12}$, where $^{[4]}Al$ replaces the three silicons in the mineral garnet. Because of the presence of the mixed charged $AlO_4$ units, the charge of countercation A is higher by one than in the silicate garnets, so that effectively $Y^+$ is present in place of $Ca^{2+}$. A wide range of materials has been prepared by selective doping of all three (A, B and T) sites. These dopings have a profound influence on the properties of these materials, creating a variety of different aesthetic and, more recently, high-tech uses, particularly for lasers. Laser-oriented YAGs have been investigated by ESCA and other techniques. For our purposes here we restrict consideration to YAGs containing a mixture of $^{[4]}Al$ and $^{[6]}Al$.

The key results for a representative YAG are shown in Table VIII and Figure 22. Again, we give only the Al 2p binding energies, although those for O 1s and Y 3d are useful in confirming the integrity of the samples. A small amount of Mg, Si,

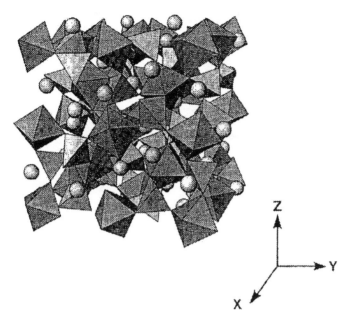

FIG. 21. Clinographic view of the garnet structure showing $AlO_6$ and $AlO_4$ polyhedra and interstitial $Y^{3+}$ couter cations (spheres) [136].

TABLE VIII.    The ESCA Binding Energies Obtained by Deconvolution of the Al 2p Manifold ($\pm 0.3$ eV) and Key $^{[6]}Al/^{[4]}Al$ Spectral Intensity Ratios ($\pm 0.15$ eV) for some Key Aluminosilicates [136]

| Material | Binding energy | | $^{[6]}Al/^{[4]}Al$ ratio | |
| | $^{[6]}Al$ | $^{[4]}Al$ | ideal | ESCA |
| --- | --- | --- | --- | --- |
| Hornblende | 74.5 | 73.5 | $U^a$ | 2.0 |
| Cummingtonite | 74.2 | 73.5 | $U^a$ | 1.3 |
| $Y_3Al_5O_{12}$ | 74.3 | 73.5 | 1.5 | 1.4 |
| $Y_3Al_5O_{12}{}^b$ | 74.2 | 73.4 | 1.5 | 1.5 |
| Sillimanite | 75.0 | 74.0 | 1.0 | 1.0 |

$^a$U = unknown and variable depending on geological source.

$^b$Doped with Pr and Mg.

Pr and Cr dopants, plus lesser impurities was insufficient to perturb seriously the Al results. The Al 2p spectrum (Fig. 22) is a doublet with a less intense higher binding energy peak. Deconvolution yields the intensity ratio lower $E_b$/higher $E_b$

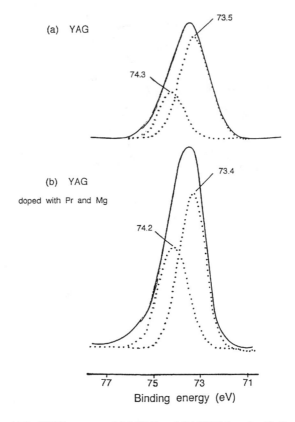

FIG. 22. The Al 2p ESCA spectra of (a) YAG and (b) YAG doped with Pr and Mg [136].

as ca. $3/2$. In line with our earlier arguments, we assign the lower binding energy (73.4 eV) to the three tetrahedral Al sites and the higher binding energy (74.2 eV) to the two octahedral Al sites. Other YAGs with varying amounts of dopants give in general an almost identical doublet Al 2p spectra with similar $3/2$ to $4/3$ intensity ratios and similar binding energies. The binding energy of 74.2 eV is consistent with [6]Al in the presence of cations and an adjacent octahedral group, as in glaucophane, while the peak at 73.4 eV is consistent with relatively negative tetrahedral groups in this case enhanced by the presence of the $Y^{3+}$, for example, as would be similar to the effect of tetrahedral Al in the presence of $Na^+$ in zeolite Na−A.

9.4.2 Amphiboles.  As a continued form of this type of analysis XPS or ESCA has been successfully utilized to investigate a number of the double-chained (amphibole) silicates with general formulas, $A_{0-1} B_2 M_5 T_8 O_2 (OH)_2$.

These systems have been reviewed in detail in the literature [153, 154]. In this case the designation "double chain" refers to the arrangement of the tetrahedrally bonded (silicates or T-containing) unit, where one should note the double set of T1 and $T_2$ sites. These tetrahedral units are thus only 4 T's across in the page, but stretch to near infinite chains in and out of the page. The T sites are typically filled by Si, but may contain other species, particularly Al. Just as in the case of the framework silicates, when the latter are present there are sometimes counter-cations (e.g., $Na^+$) in the A sites. The M sites are corresponding octahedral sites that can be seen to "cement" two double-chained tetrahedra together into a complex "sandwich." This sandwich, very resilient inside itself, is commonly referred to as the "double anvils" [153]. Thus, whereas elemental exchange is relatively common inside the double anvil, breakage and "sectional extraction," generally occur only on a zigzag course around these units [153, 154]. The $M_1$, $M_2$ and $M_3$ sides tend to be occupied by typical octahedral cations similar to those found for sheet silicates, for example, Al, Mg, $Fe^{++}$, and $Fe^{+3}$ [153]. The $M_4$ site is the fundamental coupling site tying the octahedral to the tetrahedra. It is occupied by typical countercations ranging from $Mg^{++}$ (the most covalent) to $Na^+$ (among the most ionic) [153, 154].

A number of typical amphibole examples have been examined with XPS or ESCA by our group. These are identified in Table IX along with the *approximate*

TABLE IX.    Chemical Formulae and Elemental Binding Energies of the Amphiboles as Obtained from ESCA Results [136]

| Mineral | Formula | Binding energy (eV) | | | | | | |
|---------|---------|---------------------|---|---|---|---|---|---|
| | | K $(2p_{3/2})$ | Ca $(2p_{3/2})$ | Mg $(2p)$ | Fe $(2p_{3/2})$ | Si $(2p)$ | Al $(2p)$ | O $(1s)$ |
| Anthophyllite | $Mg_2[Mg_5]$ $[Si_8O_{22}](OH)_2$ | – | – | 50.95 50.5 | – | 103.45 | – | 532.6 |
| Tremolite | $Ca_2[Mg_5]$ $[Si_8O_{22}](OH)_2$ | – | 347.5 | 49.7 | – | 102.7 | – | – |
| Hornblende | $(Ca,Mg)$ $[Mg_{2.2}Fe_{1.8}{}^{2+}Al_{1.0}]$ $[Si_6Al_2O_{22}][OH,Cl]_2$ | – | 347.9 | 50.8 | 711.9 | 102.6 | 73.5 74.5 | – |
| Cummingtonite | $(Na,Ca,Fe_{1.5})$ $[Mg_{1.6}Fe_{2.2}Al_{1.3}]$ $[Si_{6.4}Al_{1.6}O_{22}](OH)$ | – | – | 50.25 | 710.7 | 102.2 | 73.5 74.2 | – |
| Actinolite | $(K,Ca)_2$ $[Mg_{4.3}Fe_{0.7}]$ $[Si_{6.5}Al_{1.5}O_{22}][OH,F]_2$ | 293.3 | 347.55 | 50.1 | 710.9 | 102.6 | 73.85 | 531.6 |

*surface compositional formula* as determined by the relative quantitative analyses realized by the XPS. Also found, but omitted in the later formulations, are small concentrations of lesser elements [147].

Also presented in Table IX are the average binding energies realized in this study [147]. It should be noted that, as in the case of the framework and sheet silicates, there are no single values for these amphibole systems but there are (as described in what follows) repetitive, but explicable binding energy, shifting patterns exhibited by the amphiboles, that seems in particular to undergo many changes in composition, particularly for the Si/Al ratio. One feature, however, that should be apparent is that the actual shifting compositional patterns realized by the amphiboles are far too complex to be understood in terms of the simple $M-O-M'$ covalency/ionicity arguments employed for the framework silicates. However, variations of that principal argument still seem to apply to these double-chained silicate forms. In our view, the key to the current analysis lies in the relative covalency/ionicity of the species that occupies the $M_4$ position, and how this element couples through the oxygens to the lattice cations in the $M_1$, $M_2$, $M_3$, and $T_1$, $T_2$ sites [147].

### 9.4.2.1 Anthophyllite and Tremolite.

Anthophyllite is an amphibole that contains small amounts of Ca, Fe and Al impurities. The aluminum is present only in the tetrahedral lattice. The ESCA method reveals that the presence of $Mg^{+2}$, instead of the more ionic $Ca^{+2}$, in the M4 sites produces a significant increase in the Si (2p) and Mg (2p) binding energies (Table IX) [2]. In the amphibole tremolite, a sharp increase in the ionicity of $Ca^{+2}$ over $Mg^{+2}$ in the M4 sites results in an increase in the covalency of the $Si-O$ (tetrahedral) and $Mg-O$ (octahedral) bonds involved and a corresponding decrease in their key binding energies. The detailed argument is as follows. Consider the $Si-O-M$ bond, where M is a nonnegative species. When M is a metal cation, and the $M-O$ bond is relatively ionic, this forces the $Si-O$ bond to become even more covalent than in $SiO_2$. This results in a reduction in the binding energy of Si relative to silica, where Si (2p) $= 103.2$ eV. These features are revealed using ESCA through a substantial decrease in the Si (2p) and Mg (2p) binding energies (see Table IX) [2, 148, 152, 155–159].

### 9.4.2.2 Cummingtonite, Hornblende, and Actinolite.

The examination of the amphibole cummingtonite, hornblende, and several actinolites of different origins revealed significant amounts of Al, as well as large amounts of very ionic (typically $Ca^{+2}$ and $Na^+$ as opposed to $Mg^{+2}$) countercations. All these features make the $Si-O$ bonds in the silicate sub-lattice even more covalent, once again reducing the Si (2p) binding energy relative to that for $SiO_2$ [2, 27, 144] and anthophyllite (see preceding text).

The influence on the various binding energies of significant amounts of Fe in the octahedral sublattice and of aluminum in both sublattices is not well understood. The relative concentrations of these species are generally consistent with the trends established in bulk analysis except for the presence of significant amounts of Al in cummingtonite [153]. The resulting binding energies seem to be particularly low for the actinolite materials. This may be a result of the significant presence in the latter of $K^+$, a species which when doubled in amount, substantially exceeds the "field" influence of $Ca^{++}$. In the case of glaucophane we also find a particularly large relative amount of $Na^+$. This feature seems to contribute to the most substantial downfield shift (enhanced covalency) yet observed for the lattice Si Al, Mg and Fe species in these amphiboles [75]. It has been suggested that this is uncommon, but we note that cummingtonite is often found in hybrid distributions with such aluminum-rich materials as hornblende [153].

The broad Al (2p) manifold of most of the amphiboles studied contains two reproducible subpeaks, indicating the presence of two different aluminum-containing species (Fig. 23) similar to those already described here for the YAG materials. The Al (2p) peaks of cummingtonite at 73.5 and 74.2 eV and of hornblende at 73.5 and 74.5 eV suggest the presence of distinctive chemical species (structural). As a result, the Si (2p) and Al (2p) binding energies in the tetrahedral subsheet become more negative compared to those in the corresponding simple oxides, with Al (2p) at 73.5 eV. Similar bonding features are found in cummingtonite. The Mg and Fe in the octahedral sites also give more positive binding energies due to the relatively cationic nature of the octahedral subsheet. The relative covalency/ionicity of the different species in the octahedral sublattice is difficult to quantify [2]. These binding energy values are consistent with the binding energy of [4]Al to recognize the presence of a relatively positive octahedral layer and the interstitial cations.

The [6]Al 2p has a larger binding energy than in glaucophane, because the cations are multivalent and reduce the binding energy of Al in the octahedral sites by less than when the cation is monovalent, as in Na-glaucophane. In addition to the enhanced complexity of cummingtonite, hornblende, and actinolite as compared to tremolite and anthophyllite, the minerals in the latter group contain iron and are much more fibrous, which makes them of particular interest.

The deconvolution of the two Al sites indicates that in both hornblende and cummingtonite [6]Al dominates over [4]Al, with an [6]Al/[4]Al spectral intensity ratio of from 2/1 to 3/2. This has been incorporated into the ESCA-derived formula [143] for these amphiboles (Table IX), but one must remember that neither mineral is likely to be monophasic.

### 9.4.2.3 Importance of Linewidth Analysis in ESCA Studies of Silicates.

In some aluminates the structure of the Al 2p ESCA peak is rather symmetric (Gaussian—Lorentzian). This often occurs because only one type of Al is present (as in glaucophane), and sometimes because the amounts of [6]Al and [4]Al are

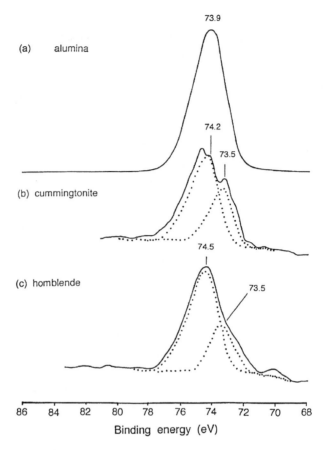

FIG. 23. The Al 2p ESCA spectra of (a) alumina, (b) cummingtonite, and (c) horn-blende [136].

approximately equal and the Al 2p binding energies are very similar, as in sil-limanite. The absence of a well-defined shoulder complicates the use of ESCA for structural identification. Fortunately, when this problem arises one may often resort to secondary means of identification. In all the cases considered so far, the energies of the [6]Al 2p and [4]Al 2p peaks differ by at least 0.5 eV. As a result, even when low resolution prevents the detection of two readily identifiable peaks, there is still evidence of broadening of the overall Al 2p manifold compared to compounds with only one type of aluminum. Thus, for example, when only one type of Al is present, the Al 2p/Si 2p linewidth ratio is always $ca.$ 1.0. On the other hand, when significant amounts of $Al_T$ and $Al_O$ coexist, this ratio is $>1.0$ (as shown in the mixed amphiboles in Table X).

TABLE X.    The ESCA linewidths ($\pm 0.2$ eV) for Aluminosilicates[a]

| Single Al (2p) structure | Si (2p) | Al (2p) |
|---|---|---|
| Kaolinite | 2.28 | 2.21 |
| Montmorillonite | 2.05 | 1.98 |
| Glaucophane | 2.2 | 1.95 |
| NaA | 1.72 | 1.56 |
| NaY | 1.68 | 1.55 |
| Analcime | 2.30 | 1.95 |
| Multiple Al (2p) structure | | |
| Muscovite | 1.89 | 2.02 |
| Hornblende | 2.32 | 2.56 |

[a] Note that some linewidths for a particular material may differ from others due to variations in macromorphology of the sample and differences in spectrometer resolution, charging effects and performance.

# 10  ESCA IN SELECTED ORGANIC SYSTEMS

Compared to its inorganic applications, ESCA has not been extensively employed in the organic materials area, but where it has been used the results have been rewarding. Several research groups have used XPS in the medical and dental areas. These include Andrade et al. [160], Gardella et al. [161], Cooper [162], Gassman [163], and, in particular, Castner and Ratner [164]. Once again, however, significant leadership in this area has sprung from the point of origin of ESCA in Upsala [165, 166]. The rapid growth of interest in the use of XPS in bioorganic materials research is indicated in the two well-attended symposia held under the sponsorship of the Physical Electronics division of the Perkin Elmer Co., in Minneapolis [167]. In most cases the use of ESCA in support of these studies has concentrated on investigations of the interfacial features of the point of interaction between various biological systems and a number of different inorganic substrates.

## 10.1 ESCA Studies of Cell–Silicate Interactions

Why do we study such cell-silicate interactions? Human exposure to certain mineral silicate dusts has been associated frequently with the development of pneumoconiosis and other similar lung diseases. This has led to alterations of practices, enactment of programs of removal and replacement, and the virtual creation of major US government agencies, for example, the EPA. In many instances fear and misunderstanding have combined with a lack of knowledge to create situations for all silicate minerals that border on "witch hunts" with both exposures and castigations of the culprits. In some circles it has been assumed that essentially any material containing Si is a poison. All of this is promulgated on a planet whose crust and mantle are, and, as far as man's existence is concerned, always have been more than 90% silicates. How man survived (and positively evolved) during several million years of living in the caves, gullies, savannas, and deserts of Africa is inexplicable, if everything from crushed sand to common clays are extremely pathogenic. Silicosis, a fibrotic reaction of the lung, is apparently induced by overt exposure to some configurations of silica particles. Other particulate or fibrous silicates, such as asbestos and the zeolite erionite, can also induce fibrotic and cancerous lung diseases [168, 169]. Of the silicate minerals, the amphiboles crocidolite, certain forms of anthophyllite, amosite asbestos, and some other fibrous minerals exhibit pathogenic activity towards living systems, evidently by processes that are related to their fibriform morphology and dimensions [170].

Chemical modifications by derivitizing the surfaces of mineral fiber silicates have revealed alterations in reactivity in *in vitro* cell toxicity studies, suggesting a multifactorial character of particle/cell interactions [171]. In addition, the adsorption of bovine serum albumin onto asbestos fibers has been shown by infrared spectroscopy and NMR to be mediated by $O-H-N$ hydrogen bonds [172]. **In view of the various hypotheses put forward in the bulk biochemical studies and the apparent surface-oriented origins of these reactions, it is important to examine the features of the silicate/cell interfaces and surfaces.** Herein we present a typical case of the interaction between a related alumino-silicate (i.e., cummingtonite) and bioorganic cell interaction.

10.1.1 Cummingtonite and Ehrlich Cell (rat tumor) Interaction. Living cells contain varying amounts of water, proteins, carbohydrates, lipids, nucleic acids, and other organic species, along with anions and inorganic cations. Despite the inertness of silicates it should be noted that contact between cells and select silicates induces rapid adsorption of water, ions and proteins onto silicate surfaces. Contact between the silicate and the cell membrane thus may trigger cellular reactions, which alter the complex nature of proteins both inside the cell and within the cell membrane [171–173]. A list of the results relevant to this case is presented in Table XI, for cummingtonite. It should be noted for both "parent" materials that the resulting binding energies vary a bit, but even more dramatic than these slight alterations is the general overriding similarity of all of the numbers. Thus, it seems reasonable to state that subjecting cummingtonite to the various forms of autoclaving, incubation, washing, drying, and even doing this while immersing these silicates in the MEM produces *little, if any, change in the silicate chemistry* as indicated in the resulting binding energies. The situation that results when the cells are present will be shown to be different.

TABLE XI.    Binding Energies in eV for Cummingtonite Subjected to Various Treatments [146]

| | Fresh | A, W, D[a] | A, I, W, D[a] | With cells (freeze-dried) | Separated (from cells) | With (medium) |
|---|---|---|---|---|---|---|
| Si (2p) | 102.2 | 102.2 | 102.3 | 101.4 (Si—C) | 102.1 | ? |
| | 102.1(2) | no Si—O—C | | 102.2 | | |
| | | | | 103.2 (Si—O—C) | | |
| Mg (2p) | 50.0 | – | – | – | 49.6 | ? |
| O (1s) | 531.4 | 531.2 | 531.3 | 531.3 | 531.5 | 531.4 |
| Fe (3p) | 55.4 | – | – | 55.8 | 55.35 | 54.8 |
| | 55.5(2) | | | | | |
| Fe (2p) | 709.7(2) | 710.8 | 710.8 | | 711.2 | 710.0 |
| Al (2p) | 74.45 | 73.6 | 74.2 | 74.? | 73.9 | ~75.0 |
| | 74.4(2) | | 73.4 | 73.? | | 73.8 |
| | 73.4 | | | | | |
| | 73.65(2) | | | | | |

[a] A = autoclave; W = wash; D = dry; I = incubation; (2) = second run.

The relative amount of surface Al wrt. Si increases very markedly when cummingtonite is treated with cells as compared to a fresh sample or one that is placed only in a medium. The Si/Al ratio can decrease in two ways—either $Al^{+3}$ ions are released from the tetrahedral and octahedral lattices and drawn to the cell/silicate interface in the presence of cells, or the reverse happens to the silicon. On the other hand, no aluminum was detected either in the medium or in the cells after contact with the aluminosilicate. During cell/silicate interaction there is also a definite increase in the concentration of Fe on the surface and a slight increase in the level of Mg.

Dramatic changes in the XPS results seem to occur following cell growth on these silicates (and subsequent cell removal) [147]. Both the binding energies and relative elemental distributions experience these changes. In Table XI we first note that the Al (2p) for the cummingtonite that readily revealed two peaks before and during the "preliminary" treatment now, in the presence of cells, exhibits for the part only one recognizable peak. The binding energy of this Al peak suggests that the Al is either located in the tetrahedral subsheet or results from $Al_2O_3$ formed during the interaction.

In Table XI there also seem to be finite changes in the binding energies of the resulting Mg and Fe species. In view of the possible movement of some of the lattice Al to the surface as $Al_2O_3$, it is not a surprising feature for the other octahedral cations also to shift in binding energy to compensate for the removal of Al. A distinct enhancement seems to be created, for example, in the surface visibility of Al, Fe, and Mg relative to the detected Si, with the rise in the Al particularly dramatic, suggesting that much of it is now on the surface of the silicate [147].

**10.1.2 Freeze-Dried Cells Contacted with Cummingtonite.** The ESCA method has also been used to study the chemistry of freeze-dried cells contacted with cummingtonite, where the cells were grown on a surface of the silicate wafer under controlled conditions and the sample was subsequently freeze-dried (see, e.g., Fig. 24). We have detected substantial quantities of C, N, P, and Cl (in various chemical forms) and lesser amounts of S, indicating that ESCA readily detects the elemental components of the cells themselves. Also found were significant amounts of Si and Fe, together with smaller amounts of Al, Mg, Na, and other elements, suggesting simultaneous ESCA sensitivity to the composition of the amphibole. More cells and less silicate were found in the upper part of the sample before freeze-drying, but a substantial presence of both species is found in the freeze-dried sample. The ESCA signal from 6-coordinate aluminum detected in cummingtonite prior to treatment with cells disappeared after treatment. When the cummingtonite-cell combined (freeze-dried) system is examined by ESCA or (XPS), as expected, all of the silicate peaks are muted in size due to the surface presence of the cellular material [147]. In addition, the system exhibits the appar-

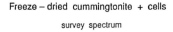

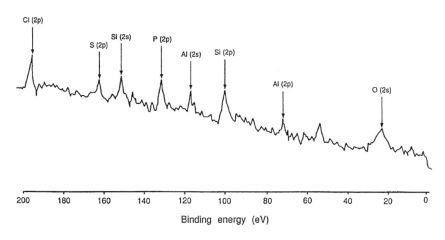

FIG. 24. The ESCA 0-200 eV spectra of freeze-dried cummingtonite with organic cells [146] and the Ph.D. dissertation of S. Seal (1996).

ent retention of the double Al peaks in approximately the same ratio as detected previously for the higher to lower binding energy forms. The binding energies for these two peaks are, however, both $\sim$0.4 eV lower than the corresponding peaks detected previously for the cummingtonite system.

The major differences experienced by the lattice cations for the cell-bound cummingtonite occur in the complexity of the Si spectra (see Fig. 25). In this case, several different samples repeated a similar three-peaked manifold. The linewidth of this manifold for cummingtonite ($\sim$2.07 eV) is indicative of the presence of several peaks grouped together. (Note that the corresponding O (1s) has a linewidth of 2.0 eV, while for all other cummingtonite systems, the linewidth for the O (1s) is at least 20% broader than that for Si (2p).) The largest part in the Si (2p) manifold of peaks occurs at $\sim$102.2 eV, that is, in the same area as "unperturbed" cummingtonite. The other two peaks are at $\sim$103.2 and $\sim$101.4 eV, respectively. In view of the fact that this reproducible structure seems to be reserved for mixed cases of silicates and cells we propose an explanation (presented in what follows) that is based on a chemical interaction involving the creation of Si$-$O$-$C (higher BE) and C$-$Si (lower BE) bonds. In the details of these systems, we were also able to discern some of the surface characteristics of the frozen cells and surface-retained medium materials. Much of this can be seen in the C (1s) spectra (Fig. 26), which suggest the presence of C=O (BE$_C$ = 288 eV), C$-$O$-$C (BE$_C$ = 285.8 eV), C$-$H and possibly C$-$H (BE$_C$ = 284.6 eV), C=N (BE$_C$ = 286 eV), and C$-$OH (BE$_C$ = 286.5 eV) bonds (BE$_C$ denotes the binding energy of C (1s)). The C$-$Si linkages should give a C (1s) peak at ca. 283 eV and a possible extra shoulder [174–176]. Structural units of this type are not uncommon in cases where a silicate system was in contact with materials containing Group IV elements under conditions of substantial physical strain.

Other related cell (rat tumor and lung cells)/silicate (Chrysotile-sheet silicate asbestos, silicon, silica, hornblende) interaction studies performed using ESCA have been reviewed in detail elsewhere by Seal and his coworkers [177–180]. The general feature inherent to all of these systems is that there seems to be a particular type of chemical interaction between the cells and the silicate involving direct coupling between the cell and the tetrahedral silicate subunit with a corresponding release of these components to the octahedral subunit (i.e., the Al, Mg, and particularly the Fe). The results also suggest that some of the Fe is inserted into the cell.

## 10.2 Key Examples in Polymers

Since the early 1970s, XPS has been the favored method for the surface chemical analysis of polymers [181, 182] but it is not without problems, not least of which is the small dynamic range of the chemical shift of the C 1s photoelectron line and other peaks of interest (e.g., O 1s, N 1s, Si 2p, S 2p, etc.). There have

SEAL AND BARR

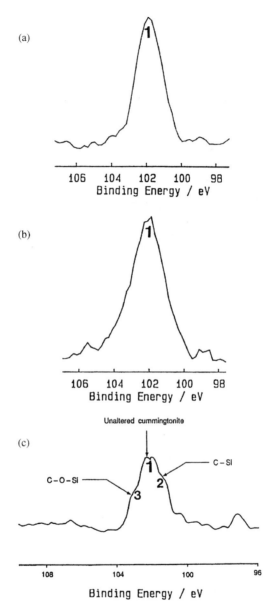

FIG. 25. Key deconvoluted ESCA Si (2p) spectra of cummingtonite following various treatments with Ehrlich cells: (a) as received; (b) treated with cell culture medium (MEM); and (c) freeze-dried with cells [146]. The intensity of the individual spectra is not the same. Possible peak identification: (1) Si—O; (2) Si—C; and (3) Si—O—C. S. Seal, Ph.D. dissertation (1996).

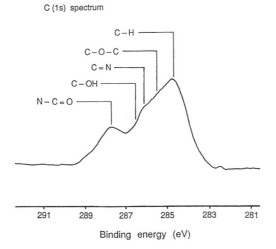

C (1s) spectrum

C–H

C–O–C

C=N

C–OH

N–C=O

291    289    287    285    283    281

Binding energy (eV)

FIG. 26. The ESCA C (1s) spectra of freeze-dried cummingtonite contacted with Ehrlich cells [146] and the Ph.D. dissertation of S. Seal (1996).

been various approaches to overcoming this shortcoming, both by sample preparation methods and improved instrumentation. Briggs [182] and Reilly et al. [183] have listed detailed ESCA studies of polymeric reactions. Application of high-resolution XPS has recently been successfully applied by Beamson and Briggs and reported on in a series of publications [184–186]. Most notable was the recognition of small (0.1 eV) shifts that are functional group specific and where the shake-up spectra of solid polymers is directly comparable with their gas phase analogs. Various publications list a detailed application of ESCA in polymeric adhesion, failure, and degradation behavior [187–191].

We present here a case of a polymeric membrane failure observed following long-term operation of a low-pressure reverse osmosis (RO) membrane pilot plant to treat highly organic surface water from the Hillsborough River in Tampa, Florida, using ESCA. The RO membranes are currently used worldwide to desalinate seawater and brackish waters for human consumption. Results from three different commercial RO membranes, made of cellulose acetate (industry brand name CALP) or polyamide (industry brand names LFC1 and ESNA), showed performance deterioration due presumably to membrane fouling and degradation under given experimental conditions. The cellulose acetate polymer consists of gluco-pyranose rings connected by $\beta$-glycoside (ether) linkages. Hydroxyl groups of this polymer are acetylated at various degrees. In the polyamide membrane, a permeate flux is improved by providing an ultrathin, cross-linked skin layer, approximately 200–500 nm in thickness. This thin layer is fabricated by interfacial polymerization and deposited on a microporous substrate, usually consisting of a polysulfone. The resulting composite membrane contains amide or urea linkages in various stages of substitution.

It is important to note that the chemical structures of the membranes are proprietary and local variations can occur due to the nature of membrane manufacturing processes. In order to avoid possible misinterpretation of membrane chemical structure due to local variations, ESCA analysis was repeated for selected membranes. The XPS analysis revealed that CALP was a cellulose acetate membrane represented for the most part by tri-acetate species (acetate $(O-C=O)_{BE} \sim 289.7$ eV) (Fig. 27). The LFC1 membrane showed a polyamide active layer apparently crosslinked to a polysulfone (S $(2p)_{BE} \sim 170-168$ eV) infrastructure as suggested by the XPS surface scan.

After use in the pilot study to purify surface water, the XPS surface scans of the membranes showed remarkably different results. The nitrogen, sulfur, chlorine and calcium peaks were the main focus of this research. New CALP had no nitrogen substituent as detected from surface spectrum, but a strong nitrogen presence was found in the used sample of CALP. This observation suggested that CALP had been under attack from a biological vector. The difference in the FWHM of the C (1s) peak before and after use indicated a change in amplitude of two chemical substances as well as the addition of a different species (carboxyl $(COOH)_{BE} \sim 289.4 \pm 0.15$ eV). Further investigation found the nitrogen to be linked in an amino acid structure (N $(1s)_{BE} \sim 400.5$ eV), very similar to the digestive juices of the adsorptive/absorptive feeder found on the membrane itself.

A series of detailed charts showing the elements and their chemical bonds (as derived from XPS or ESCA results) relating to these polymeric membranes are listed in Table XII. The amine group is part of the ESNA (brand name) structure as indicated by N (1s) $(N-C=O_{BE} \sim 287.5-287.2$ eV). Used ESNA showed evidence of chlorine uptake (Fig. 27) while treated LFC1 exhibited evidence of chlorine (Cl $(2p1, 2p3)_{BE} \sim 202.0-200.3$ eV) and calcium (Ca $(2p1, 2p3)_{BE} \sim 350.3-346.9$ eV) on the surface after use. The chlorine in LFC1 is present as $C-Cl$ bonds while the calcium is indicative of an oxide $(Ca-O)$.

The XPS analysis of unwashed membranes showed that ESCA could identify the organic fouling layers of the membranes as well as the surface structure. Peak fitting regimens determined whether the materials discovered were foulants deposited (as indicated by the N (1s) data) on the surface or materials that had become chemically bound to the surface. To be chemically bound to the surface, the foulant must show an oxidation state indicative of a bond with membrane substituents. If the foulant showed a chemical composition not associated with the membrane, it was determined to be lying on the surface (physisorbed) and not chemically attached. This was the case for calcium on LFC1 (Ca $(2p1, 2p3)_{BE} \sim 350.3-346.9$ eV). However, this finding does not preclude the possibility of calcium complexation with functional groups of the RO membranes used

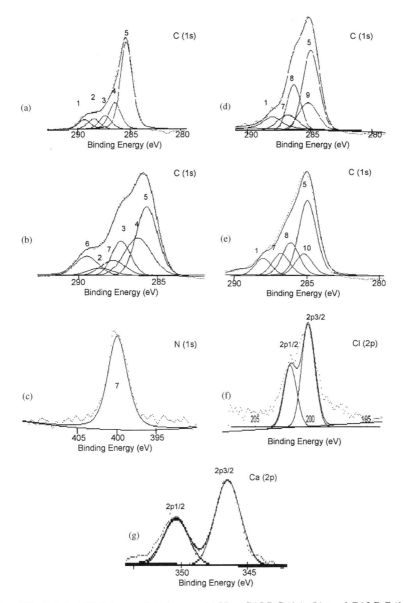

FIG. 27.  Detailed XPS deconvoluted scans: (a) New CALP C (1s); (b) used CALP C (1s); (c) used CALP N (1s); (d) new LFC1 C (1s); (e) used LFC1 C (1s); (f) used LFC1 Cl(2p); and (g) used LFC1 Ca (2p). Possible Peak Identification: 1) O−C=O; 2) O−C−O; 3) C−O−C; 4) C−C; 5) aromatic C−H; 6) COOH; 7) N−C=O; 8) C−N; 9) C−SO$_2$; and 10) C−Cl. Reprinted with permission from S. Beverly, S. Seal, and S. K. Hong, *J. Vac. Sci. Technol.* A18(4), 1107, 2000, in press. Copyright (2000), American Vacuum Society.

TABLE XII.   Detailed XPS Binding Energy Data of RO Membranes (For C (1s): ±0.1 eV; N (1s): ±0.15 eV; O (1s): ±0.1 eV)

| | C (1s) Binding energy | State | Total FWHM | O (1s) Binding energy | State | N (1s) Binding energy | State |
|---|---|---|---|---|---|---|---|
| New CALP | 289.7 | O−C=O | 2.4 | 533.1 | O−C−O | | |
| | 288.6 | O−C−O | | 533.8 | [a]O−C=O | | |
| | 287.2 | C−O−C | | 532.4 | O−C=O[a] | | |
| | 286.5 | C−C | | | | | |
| | 285.4 | CH | | | | | |
| Used CALP | 289.4 | COOH | 3.1 | 534.2 | COOH | 400.5 | N−C=O |
| | 288.6 | O−C−O | | 533.8 | [a]O−C=O | | |
| | 287.7 | N−C=O | | 533.3 | C−O−C | | |
| | 286.7 | C−O−C | | 532.4 | O−C=O[a] | | |
| | 286.5 | C−C | | 531.5 | N−C=O | | |
| | 285.4 | CH | | | | | |
| New ESNA | 288.9 | O−C=O[c] | 2.3 | 533.9 | [a]O−C=O | 400.4 | N−C=O |
| | 287.5 | N−C=O | | 532.6 | O−C=O[a] | | |
| | 287.0 | C−N | | 531.8 | N−C=O | | |
| | 285.0 | CH | | | | | |
| Used ESNA | 288.7 | O−C=O[c] | 2.6 | 533.8 | [a]O−C=O | 400.4 | N−C=O |
| | 287.2 | N−C=O | | 532.6 | O−C=O[a] | | |
| | 286.7 | C−N | | 531.6 | N−C=O | | |
| | 285.1 | C−Cl | | | | | |
| | 285.0 | CH | | | | | |
| New LFC1 | 288.4 | O−C=O[c] | 3 | 533.9 | [a]O−C=O | 400.4 | N−C=O |
| | 287.0 | N−C=O | | 532.6 | O−C=O[a] | | |
| | 286.6 | C−N | | 532.9 | O=S=O | | |
| | 285.5 | C−SO$_2$ | | 531.9 | N−C=O | | |
| | 285.0 | CH | | | | | |
| Used LFC1 | 288.0 | O−C=O[c] | 3 | 533.8 | [a]O−C=O | 400.1 | N−C=O |
| | 287.2 | N−C=O | | 532.9 | O−C=O[a] | | |
| | 286.5 | C−N | | 531.9 | N−C=O | | |
| | 285.2 | C−Cl | | | | | |
| | 285.0 | CH | | | | | |

Reprinted with permission from S. Beverly, S. Seal, and S. K. Hong, *J. Vac. Sci. Technol.* A18(4), 1107 (2000). Copyright American Vacuum Society (2000).

[a] Binding energies are reported relative to indicated oxygen atom.

[b] For final fit, the "additional adjust" function was used after placing the peaks; this may shift the final peak positions by an additional 0.2 to 0.3 eV.

[c] This bond is indicative of surface oxidation.

in water utility plants. Once again ESCA is being useful in identification of the chemical alterations in these industrial membranes.

## 10.3  ESCA Studies of Selected Biological Systems

In addition to the studies reported here for polymers and biocells/silicate inter-action, we have conducted studies using ESCA in the field of Botany. In this sec-tion, we exemplify this with a detailed surface study of leaves before, during, and after the process of senescence. Senescence of leaves of most deciduous trees is accompanied by a decline in photosynthesis [192, 193] and by alterations in struc-ture and function of chloroplasts. These lead to a decrease of the photosynthetic pigment content and photochemical activity [192, 194–198]. The mechanism by which the color of a leaf changes is fairly well understood. Thus in the autumn chlorophyll begins to disintegrate. Until this happens, to the naked eye the green of the chlorophyll dominates the yellow, orange and brown colors of the always-present carotenes and the red of the anthocyanins. It is known that senescence alters the accessibility of electron transport chains to the exogenous electron ac-ceptors and donors and causes a shift in the photochemical activity [199, 200], which reflects the changes in structural and functional integrity of the membrane systems [201]. Thus, the gradual decreases in chlorophyll content during senes-cence release first one color and then another. We anticipate that the outer surface of the leaves is experiencing concurrent changes caused by a combination of in-ternal chemical processes and external atmospheric effects. We report therefore the first detailed ESCA study of these surface processes and their relationship to internal senescence.

Green, yellow-green, yellow and brown leaves of the silver maple were re-moved from three groups of trees and from the surrounding areas of the Uni-versity of Wisconsin-Milwaukee. The trees were selected for their isolation from road and foot traffic and were first examined by optical microscopy to ensure that only those not adversely affected by the environment were studied.

10.3.1  **Green Leaf.**    The upper surface of a freshly picked green maple leaf was found by ESCA to contain significant amounts of oxygen and nitrogen. The N (1s) peak positions suggest the presence of epidermal amino acids, and the O (1s) peak positions suggest substantial air-induced passivation (C$-$O$-$C units covered by >C=O as shown in Table XIII). The C (1s) confirmed a variety of long-chain hydrocarbon waxes that probably originated in the cuticle (Fig. 28). Some of the conclusions were based on comparative ESCA studies of many car-bonaceous materials [2, 35]. The valence band spectrum from the lower sur-face of the green leaf confirms the presence of long-chain hydrocarbons, while a reduced amount of C$-$O$-$C and >C=O species suggest that oxidation is re-tarded on the underside in comparison to the upper surface. The small N (1s) peak indicates that a smaller amount of amino acids ejected by the outer sur-

TABLE XIII.    Relative Concentrations of Oxygen and Nitrogen in the
Surface Layer of Silver Maple Leaves—Three Samples of Leaves in Each
Category Were Examined

| Collected from the tree | Upper oxygen | Surface nitrogen | Lower oxygen | Surface nitrogen |
|---|---|---|---|---|
| Green | L | L | ML | S |
| Green-yellow | VL | L | L | ML |
| Yellow | L | ML | M | M |
| Brown | S | 0 | ML | 0 |
| Collected from the ground | | | | |
| Green | MS | 0 | | |
| Green (3 weeks) | ML | S | ML | M |
| Yellow-brown (3 weeks) | M | L | MS | M |
| Brown | S | 0 | ML | 0 |

Reprinted with permission from T. L. Barr, S. Seal, S. E. Hardcastle, M. A.
Maclauran, L. M. Chen, and J. Klinowski, *Bull. Pol. Acad. Sci.* 45, 1 (1997).

[a] Abbreviations: VL = very large, L = large, ML = moderately large, M moderate, MS = moderately small, S = small, 0 = absent.

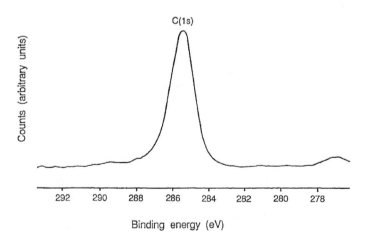

FIG. 28.  The C (1s) ESCA peak from the upper surface of a green leaf, representative
of spectra from leaves in all stages of senescence. The principal peak is due to C—C and
C—H bonds; carbon-oxygen bonds create the shoulders on the high binding energy side.
Reprinted with permission from T. L. Barr et al., *Bull. Pol. Acad. Sci.* 45, 1 (1997).

face of the epidermal tissue is present on this side. We suggest that there is more surface cuticle exposure at the lower surface of this leaf than on its upper surface, because cutin should be less susceptible to oxidation than are the epidermal species.

**10.3.2 Yellow-Green Leaf.** The valence band region of the upper surface of a freshly picked yellow-green leaf (Fig. 29a) is found by ESCA to be dominated by long-chain hydrocarbons. The nature of the oxygen species in these hydrocarbons is reflected in the peak structures in the range of 23–30 eV. The peak at ca. 28 eV shows the presence of significant amounts of $C-O-C$ bonds, while a shoulder peak at ca. 25 eV reflects the presence of $>C=O$. Although fairly small, the amount of oxygen is much larger than on the upper surface of the green leaf. Significant amounts of nitrogen are also detected, and the fine structure of the N (1s) spectrum in Figure 30a indicates the presence of at least three chemical types of nitrogen. The structure at 398.5 eV suggests the possibility of small amounts of nitride. Two other kinds of nitrogen give peaks at 400.5 and 399.5 eV. Both are in the energy range typical of organic amines, the latter being typical of peptide nitrogens. These probably come from the epidermal amino acids. The chemical composition of the lower surface of the yellow-green leaf is generally similar to the upper surface, but there may be several significant differences. The distribution of oxygen on the lower surface is reflected in the valence band spectrum in the range of 23–30 eV (Fig. 29b). There is less total oxygen than on the upper surface of the leaf, but more than on the lower surface of the green leaf. Once again, most of the carbon is in the form of long-chain hydrocarbons. There is now much more $C-O-C$ (binding energy 27.5 eV) type species than $>C=O$ (25 eV) bonds, another feature which is different than for the green leaf. The nitrogenous species on the lower surface are also substantially different—the N (1s) spectrum in Figure 30b indicates that there is a reduction in the amount of amine nitrogen and a change of its nature. This suggests a different distribution of epidermal residue structures, perhaps indicating the presence of transferase materials, while the upper surface of the yellow-green leaf may contain a mixture of outer epidermal and extracellular species.

**10.3.3 Yellow Leaf.** The upper surface of a typical freshly picked yellow leaf contains a significant amount of oxygen, primarily in $C-O-C$ bonds (Fig. 31). In fact, almost the same amount of O was found as detected on the upper surface of the green leaf, but less than on the yellow-green leaf. This seems to reflect a slight removal of air-induced passivation and an enhanced amount of cuticle material, with a corresponding reduction in ESCA-visible amines expelled from the epidermis (Table XIII). The surface also contains a small amount of silicon, as did the green leaf collected near the base of a tree. A very small amount of sodium chloride is also present. Sodium chloride is often found on the surfaces of many materials, which have experienced environmental exposure, but this does not influence our results.

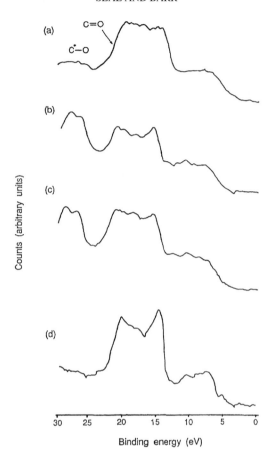

FIG. 29. The ESCA valence band spectra of: (a) upper surface of green leaf; (b) upper surface of yellow-green leaf; and (c) lower surface of yellow-green leaf. The heights of key spectral features reflect relative quantification. (d) Most of these spectral features are present in the spectrum of a representative hydrocarbon, low-density polyethylene [35]. Alteration of a hydrocarbon by adding extra links to the $(C-C)_n$ chain or introducing branching or multiple bonds "blunts" the peaks and valley features, as seen in (a)–(c). On the other hand, carbon-oxygen bonds produce O (2s)-dominated peaks at ca. 27.5 eV if $C-O-C$ species are present (first arrow) and/or add a shoulder at ca. 24 eV if $>C=O$ species are present (second arrow). Reprinted with permission from T. L. Barr, S. Seal, S. E. Hardcastle, M. A. Maclauran, L. M. Chen, and J. Klinowski, *Bull. Pol. Acad. Sci.* 45, 1 (1997).

### 10.3.4 Yellow Leaf Which Gradually Turned Brown.

*10.3.4.1 Upper Surface.* The N (1s) spectrum shown in Figure 30c indicates that the nitrogen is primarily of the amine type, pointing to a single type of surface-oriented extracellular species on the exterior of the epidermis.

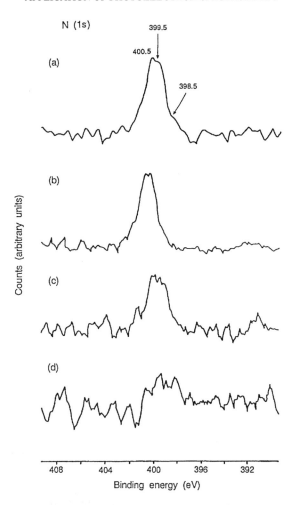

FIG. 30. N (1s) spectrum of (a) upper surface of yellow green leaf; (b) lower surface of yellow green leaf; (c) upper surface of yellow leaf; and (d) lower surface of yellow leaf. Note the relative heights and half-widths and shifts of peaks which reflect the quantity of species and the changes in their chemistry. Primary peak positions are marked by arrows. Reprinted with permission from T. L. Barr, S. Seal, S. E. Hardcastle, M. A. Maclauran, L. M. Chen, and J. Klinowski, *Bull. Pol. Acad. Sci.* 45, 1 (1997).

*10.3.4.2 Lower Surface.* The lower surface of this leaf contains no silicon to the depth ca. 5 nm. Figure 30d shows that the relative amount of nitrogen is now quite small and similar to that detected on other "off the tree" leaf surfaces. As in the case of the upper surface of the green leaf stored for 3 weeks after picking, there is now a mixture of at least two distinct types of amine species.

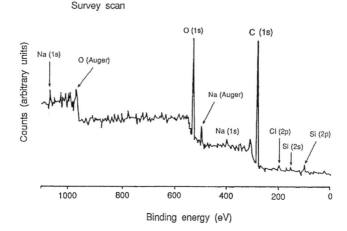

FIG. 31. Qualitative elemental survey scan of the surface of a yellow leaf. Reprinted with permission from T. L. Barr, S. Seal, S. E. Hardcastle, M. A. Maclauran, L. M. Chen, and J. Klinowski, *Bull. Pol. Acad. Sci.* 45, 1 (1997).

### 10.3.4.3 Brown Leaf.

**Upper Surface.**  All ESCA spectra for the upper surface of the brown leaves show that they are covered by a layer dominated by the hydrocarbon parts of the cuticle material, which is incapable of preventing significant amounts of air-induced oxidation. Although reduced in comparison to the other leaf surfaces, the limited amount of oxygen found by grazing incidence is predominantly on the outermost surface. The waxy esterified parts of the cuticle appear to be dissipated and thus, as previously mentioned, all surfaces exposed to air, even the most non-reactive, are subject to at least some $O_2$- and $H_2O$-induced oxidation. Thus, what we call natural passivation is indicative of extensive oxidation, which prevents further corrosive oxidation [33, 202]. Hence, in the case of the brown leaf we observe the botanical equivalent of metal corrosion.

**Lower Surface.**  There is no nitrogen, but more oxygen than on the upper surface. The leaves were picked up off the ground and although dried in storage and UHV, the results for the lower surface, with its more numerous stomata, may reflect its ability to retain water. Brown leaves show a definite cessation of these processes. It may be reasonable to compare the surface of the brown leaf to that of a corroded metal or alloy.

All of the analyses put together suggest that our brown leaf has a surface first dominated by naturally passivated hydrocarbon deposits, with more passivation on the bottom of the leaf than on its top. All this is of course interesting, but it is a very limited study. From this limited study we cannot be certain as to what is transpiring, but a general pattern may be emerging that we can speculate upon. First, as we have already documented, senescence results in a destruction of chlorophyll and assertion of other constituents. All of this is accompanied at the surface by a distinct presence of components with varying amounts of nitrogen, which are partially air oxidized. Simultaneously, hydrocarbon-dominated cuticular materials are being excreted to the outer surface and oxidized by these environmental oxidants. After life ceases the oxidative process turns to the equivalent of metal corrosion and the leaf browns, loses its structural integrity, and disintegrates.

# 11  CONCLUSION

In this chapter, the key principles of XPS or ESCA have been illustrated. The main advantages of ESCA are its simplicity, flexibility in sample handling, and high scientific information content. In addition, several key examples of ESCA in inorganic and organic materials systems have been described. In the last 30 years, ESCA has gone from a system primarily devoted to basic research to a practical, widely available, surface analytical tool in both academia and industry. For further experimental and theoretical details the reader is encouraged to consult the various texts and review articles cited throughout this chapter. The continued interest in materials science and engineering, biomedical sciences, and surface-related phenomena in general, make it probable that ESCA will remain one of the predominant surface analysis techniques in the coming future.

## Acknowledgment

The authors kindly acknowledge the support from their fellow colleagues A. Kale, S. Sundaram, V. Desai, L. Chow, D. Himenez, S. Hong, S. Beverly, K. Wozniak, J. Klinoswski, C. C. Kao, L. M. Chen, T. Dugall, E. Hoppe, S. Hardcastle, M. Maclauran, P. Shah, and E. Benko, support from AAS-UWM, Florida Space Grant Consortium (FSGC), Advanced Materials Processing Analysis Center (AMPAC), MCF and Mechanical, Materials and Aerospace Eng. (MMAE)—University of Central Florida.

## References

1. J. B. Hudson, "Surface Science." Butterworth and Heinemann, MA, 1991.
2. T. L. Barr, "Modern ESCA." CRC Press, Boca Raton, FL, 1994.
3. R. S. Swingle and W. M. Riggs, *CRC Crit. Rev. Anal. Chem.* 267 (1975).
4. K. Siegbahn, *Science* 217, 111 (1982).
5. P. K. Ghosh, "Introduction to Photoelectron Spectroscopy." John Wiley and Sons, New York, 1983.

184    SEAL AND BARR

6. T. A. Carlson, "Photoelectron and Auger Electron Spectroscopy." Plenum Press, New York, 1975.
7. B. D. Ratner and B. J. McElroy, in "Spectroscopy in the Biomedical Sciences" (R. M. Gendreau, Ed.), p. 107, CRC Press, Boca Raton, FL, 1986.
8. J. D. Andrade, in "Surface Chemistry and Physics" (J. D. Andrade, Ed.), Vol. 1, pp. 105–195, Plenum Press, New York, 1985.
9. A. D. Baker and D. Bitteridge, "Photoelectron Spectroscopy." Pergamon, New York, 1972.
10. T. L. Barr and I. E. Davis (Eds.), "Applied Surface Science." ASTM, STP 699, Philadelphia, 1980.
11. H. Hertz, *Wiedemannsche Ann. Phys.* 31, 982 (1887).
12. M. Planck, *Verh. Dtsch. Phys. Ges.* 2, 202 (1900).
13. H. Robinson, *Philos. Mag.* 50, 2441 (1925).
14. K. Siegbahn, *Royal Soc. London* (1985).
15. U. Gelius and K. Siegbahn, *Faraday Discuss. Chem. Soc.* 54, 257 (1972).
16. U. Gelius, E. Basilier, S. Svensson, T. Bergmann, and K. Siegbahn, *J. Electron Spectrosc. Rel. Phen.* 2, 405 (1974).
17. M. A. Kelly and C. E. Taylor, *Hewlett-Packard J.* 24, 2 (1972).
18. D. Briggs and M. P. Seah (Eds.), "Practical Surface Analysis." Wiley, Chichester, 1990.
19. J. Q. Broughton and P. S. Bagus, *J. Elec. Spec.* 20, 261 (1980).
20. K. M. Minachev and E. S. Shapiro, "Catalyst Surface: Physical Methods of Studying." CRC Press, Boca Raton, FL, 1990.
21. T. Koopman, *Physica* 1, 104 (1934).
22. J. Q. Broughton and D. L. Perry, *J. Elec. Spec. Rel. Phen.* 16, 45 (1979).
23. L. Hedin and A. Johansson, *J. Phys.* B2, 1336 (1969).
24. C. D. Wagner, *Anal. Chem.* 47, 1201 (1975).
25. R. T. Lewis and M. A. Kelly, *J. Electron Spectrosc. Rel. Phenomenon* 20, 105 (1980).
26. P. Swift, D. Shuttleworth, and M. P. Seah, "Practical Surface Analysis" (D. Briggs and M. P. Seah, Eds.), App. 2, John Wiley and Sons, Chichester, England, 1983.
27. T. L. Barr, *Appl. Surface Sci.* 15, 1 (1983).
28. G. Barth, R. Linder, and C. Bryson, *Surface Interface Analysis* 11, 307 (1986).
29. C. C. Kao, T. L. Barr, and R. P. Merill (to be published); C. C. Kao, Ph.D. Dissertation (Thesis), Cornell University, Ithaca, NY, 1988.
30. This widely employed term is generally attributed to C. D. Wagner, see, for example, C. D. Wagner, *J. Electron Spectrosc. Rel. Phenomena* 18, 345 (1980).
31. P. Swift, D. Shuttleworth, and M. P. Seah, in "Practical Surface Analysis" (D. Briggs and M. P. Seah, Eds.), p. 437, J. Wiley & Sons, Chichester, England, 1983; P. Swift, *Surface and Interface Anal.* 4, 47 (1982); R. J. Bird and P. Swift, *J. Electron Spec.* 21, 227 (1980); C. D. Wagner and A. Joshi, *Surface and Interface Anal.* 6, 215 (1984); S. Kohiki, *Appl. Surface Sci.* 17, 497 (1984).
32a. M. G. Mason and R. C. Baetzold, *J. Chem. Phys.* 64, 271 (1976).
32b. P. H. Citrin and G. K. Wertheim, *Phys. Rev.* B27, 3176 (1983).
32c. P. S. Bagus, C. J. Nelin, and C. W. Bauschlicher, *Surface Sci.* 156, 615 (1985).
33. T. L. Barr, *J. Phys. Chem.* 82, 1801 (1978).

34. T. L. Barr, "NACE." Corrosion 50, p. 296, Houston, 1990.
35. T. L. Barr and M. P. Yin, *J. Vac. Sci. and Tech.* A10, 2788 (1992).
36. T. L. Barr, "Modern ESCA." Chap. 7, CRC Press, Boca Raton, FL, 1994.
37. T. L. Barr, *Critical Rev. Anal. Chem.* 22, 113 (1991).
38. G. Barth, R. Linden, and C. Bryson, *Surface and Interface Anal.* 11, 307 (1988).
39. T. L. Barr, in "Practical Surface Analysis" (D. Briggs and M. P. Seah, Eds.), 2nd ed., Chap. 8, John Wiley & Sons Inc., Chichester, England, 1990.
40. T. L. Barr, in "Preparation and Characterization of Catalysts" (S. A. Bradley, R. J. Bertolacini, and M. J. Gattuso, Eds.), ACS Book Series, Vol. 411, p. 203, 1984.
41. K. M. Minachev and E. S. Shapiro, "Catalyst Surface: Physical Methods of Studying." CRC Press, Boca Raton, FL, 1990.
42. T. L. Barr, *J. Vac. Sci. and Tech.* 47, 1677 (1989).
43. K. Siegbahn and C. Nordling, *Nova Acta Regiae Soc. Sci. Ups.* 4, 20 (1967).
44. D. H. Stephenson and H. J. Binkowski, *J. Noncrystalline Solids* 22, 399 (1976).
45. S. Tougaard, *J. Vac. Sci. Tech.* A8, 2197 (1990).
46. C. D. Wagner, in "Quantitative Surface Analysis of Materials" (N. S. McIntyre, Ed.), Chap. 2, ASTM STP 643, American Society for Testing and Materials, Philadelphia, 1978.
47. C. D. Wagner, *Anal. Chem.* 44, 1050 (1972).
    C. D. Wagner, *Anal. Chem.* 49, 1282 (1977).
48. C. D. Wagner, L. E. Davis, M. V. Zeller, J. A. Taylor, R. H. Raymond, and L. H. Gale, *Surface Interface Anal.* 3, 211 (1981).
49. V. Nefedov, N. P. Serguskin, I. N. Band, and M. B. Trzhakovskaya, *J. Electron Spectrosc. Rel. Phenomena* 2, 383 (1973).
50. M. P. Seah, *Surface Interface Anal.* 2, 222 (1980).
51. C. J. Powell, in "Quantitative Surface Analysis" (N. S. McIntyre, Ed.), Chap. 1, ASTM STP 643, American Society of Testing and Materials, Philadelphia, 1978.
52. M. P. Seah, *Surface Interface Anal.* 9, 85 (1986).
53. S. J. Mroczkowski, *J. Vac. Sci. Tech.* A7, 1529 (1989).
54. M. P. Seah, in "Practical Surface Analysis" (D. Briggs, and M. P. Seah, Eds.), Chap. 5, John Wiley & Sons, Chichester, England, 1983.
55. C. J. Fadley, R. J. Baird, W. Siekhaus, T. Novakov, and S. A. Bergstrom, *J. Elec. Spectrosc. Rel. Phenomena* 4, 93 (1974).
56. C. J. Fadley, *J. Electron Spectrosc. Rel. Phenomena,* 5, 725 (1974).
57. H. D. Eland, "Photoelectron Spectroscopy." Butterworths, London, 1974.
58. H. Ibach, "Electron Spectroscopy for Surface Analysis." Springer Verlag, Berlin, 1977.
59. A. W. Czanderna, "Methods of Surface Analysis." Elesevier, New York, 1975.
60. W. S. McIntyre (Ed.), "Quantitative Surface Analysis of Materials." ASTM STP 643, Philadelphia, 1978.
61. H. Siegbahn and L. Karlsson, "Handbook of Physics." Vol. 31, p. 215, Springer Verlag, Berlin, 1982.
62. See, for example, A. Nilsson, Ph.D. Dissertation (Thesis), No. 182, Faculty of Science, University of Uppsala, Uppsala, Sweden, 1989.
63. T. L. Barr, S. Seal, L. M. Chen, and C. C. Kao, *Thin Solid Films* 253, 277, (1994).

64. See, for example, S. A. Bradley, M. J. Gattuso, and R. J. Bertolacini (Eds.), "Characterization and Catalyst Development." American Chemical Society, Washington, DC, ACS Symposium Series 411, 1989.
65. D. L. Cocke, E. D. Johnson, and R. P. Merrill, *Catal. Rev. Sci. Engg.* 26, 163 (1984).
66. C. F. McConville, D. L. Seymour, D. P. Woodruff, and S. Bao, *Surface Sci.* 188, 1 (1987).
67. A. Bradshaw, in "Interactions of Atoms and Molecules with Solid Surfaces" (V. Botolani, N. March, and M. Tosi, Eds.), Chap. 15, Plenum, London, 1990.
68. E. D. Johnson, Ph.D. Dissertation (Thesis), Cornell University, 1984.
69. C. C. Kao, Ph.D. Dissertation (Thesis), Cornell University, 1987.
70. See, for example, S. A. Flodström, B. Z. Bachrach, R. S. Bauer, and S. B. M. Hagstrom, *Phys. Rev. Letters* 37, 1282 (1976).
71. D. Norman and D. P. Woodruff, *J. Vac. Sci. and Technol.* 15, 1580 (1978).
72. S. A. Flodstrom, C. W. B. Martinsson, R. Z. Bachrach, S. B. M. Hagstrom, and R. S. Bauer, *Phys. Rev. Letters* 40, 907 (1978).
73. K. Siegbalm et al., "ESCA: Atomic, Molecular and Solid Structure Studies by Means of Electron Spectroscopy." Nova Acta Regial Soc. Sci., Upsaliensis, Sweden, 1967.
74. D. Briggs in "Practical Surface Analysis" (D. Briggs and M. P. Seah, Eds.), 2nd edn., Chap. 9, Wiley, Chichester, est Sussex, 1990.
75. T. L. Barr, *J. Vac. Sci. and Technol.* A9, 1793 (1991).
76. F. P. Fehlner and N. F. Mott, *Oxidation of Metals* 2, 59 (1970).
77. N. Cabrera and N. F. Mott, *Reports in Progress in Phys.* 12, 163 (1948).
78. F. P. Fehlner, "Low Temperature Oxidation." John Wiley & Sons, New York, 1986.
79. T. L. Barr, M. Mohsenian, and L. M. Chein, *Applied Surface Sci.* 51, 71 (1991).
80. T. L. Barr and C. R. Brundle, *Phys. Rev.* B46, 9199 (1992).
81. Y. I. Chen and J. G. Duh, *Surf. Coat Technol.* 48, 163 (1991).
82. J. G. Duh and J. C. Doong, *Surf. Coat Technol.* 56, 257 (1993).
83. V. R. Parameswaran, J. P. Immarigeon, and D. Nagy, *Surf. Coat Technol.* 52, 251 (1992).
84. F. Hohl, H. R. Stock, and P. Mayr, *Surf. Coat Technol.* 54/55, 160 (1992).
85. J. L. Peytvi, A. Lebugle, G. Montel, and H. Pastor, *High Temperatures-High Pressures* 10, 341 (1978).
86. H. Karner, J. Laimer, H. Stori, and P. Rodhammer, "12th Plansee Seminar, 1989." Vol. 3, Tirol, Austria, 8–12 May (1989).
87. J. Aromma, H. Ronkainen, A. Mahiout, S. P. Hannula, A. Leyland, A. Matthews, B. Matthews, and E. Brozeit, *Mater. Sci. Eng.* A140, 722 (1991).
88. C. Mitterer, M. Reuter, and P. Rodhammer, *Surf. Coat. Technol.* 41, 351 (1990).
89. W. Herr, B. Matthews, E. Broszeit, and K. H. Kloss, *Mater. Sci Eng.* A140, 616 (1991).
90. O. Knmotek, R. Breidenbach, F. Jungblut, and F. Loffler, *Surf. Coatings Technol.* 43/44, 107 (1990).
91. G. Dearnley and A. T. Peacock, UK Patent GB 2.197.346, A and B, 1988.
92. T. Friesen, J. Haupt, W. Gissler, A. Barna, and P. B. Barna, *Vacuum* 43, 657 (1992).
93. H. Novotny, F. Benesovsky, C. Brukl, and O. Schob, *Mh. Chem.* 92, 403 (1961).
94. A. Erdemir and C. C. Cheng, *Ultramicrosc.* 29, 266 (1989).

95.  A. Erdemir and C. C. Cheng, *J. Vac. Sci. Technol.* A7, 2486 (1989).
96.  A. Erdemir and C. C. Cheng, *Surf. Coat. Technol.* 41, 285 (1990).
97.  C. C. Cheng, A. Erdemir, and G. R. Fenske, *Surf. Coat. Technol.* 39/40, 365 (1989).
98.  U. Helmersson, B. O. Johansson, J. E. Sundgren, H. T. G. Hentzell, and P. Billgren, *J. Vac. Sci. Technol.* A3, 308 (1985).
99.  D. S. Rickerby and R. B. Newbery, in "Proceedings of IPAT 87" (CEP Consultants, Brighton, Edinburgh, UK, 1987), 224, 1987.
100. R. Vardiman, *Defect and Diffusion Forum* 57–58, 135 (1988).
101. J. Halbritter, H. Leiste, H. J. Mathes, and P. Walk, *J. Anal. Chem.* 341, 320 (1991).
102. K. S. Robinson and P. M. A. Sherwood, *Surf. Interface. Anal.* 6, 261 (1984).
103. T. L. Barr, *JVST* (1991). *Crit. Rev. Anal. Chem.* 22, 220 (1991).
104. J. S. Colligon, H. Kheyrandish, L. N. Lesnervsky, A. Naumkin, A. Rogozin, I. I. Shkarban, L. Vasilyev, and V. E. Yurasova, *Surf. Coat. Technol.* 70, 9 (1994).
105. R. D. Arnell, J. S. Colligen, K. F. Minnebaev, and V. E. Yurasova, *Vacuum* 47, 425 (1996).
106. A. Ermolieff, M. Girard, C. Raow, C. Bertrand, and T. M. Duci, *Appl. Sur. Sci.* 21, 65 (1985).
107. C. Ernsberger, J. Nickerson, A. E. Miller, and J. Moulder, *J. Vac. Sci. Technol.* A3, 2415 (1985).
108. A. R. Gonzalez-Elipe, G. Munnera, J. P. Espinos, and J. M. Sanz, *Surface Sci.* 220, 368 (1989).
109. F. Elstner, A. Ehrlich, H. Gregenback, H. Kupfer, and F. Richter, *J. Vac. Sci. Technol.* A12, 476 (1994).
110. C. T. Huang and J. G. Duh, *Vacuum* 46, 1465 (1995).
111. O. Knotek, W. Bosch, and T. Leyendecker, "Proc. 7th Int. Conf. Vac. Met." Linz, Austria, 1985.
112. W. D. Müntz, *J. Vac. Sci. Technol.* A4, 2717 (1986).
113. H. A. Jehn, S. Hofmann, V. A. Rückborn, and W. D. Müntz, *J. Vac. Sci. Technol.* A4, 2701 (1986).
114. O. Knotek, W. D. Müntz, and T. Leyendecker, *J. Vac. Sci. Tech.* A4(6), 2695 (1986).
115. S. Seal, J. H. Underwood, M. Uda, H. Osawa, A. Kanai, T. L. Barr, E. Benko, and R. C. C. Perera, *J. Vac. Sci. Technol.* A16(3) (1998).
116. S. Seal, T. L. Barr, N. Sobczak, and J. Morgiel, in "Polycrystalline Thin Films Structure, Texture, Properties and Applications III" (J. Im, Ed.), MRS Proc., 472, 269, Pittsburgh, PA, 1997.
117. N. Sobczak, E. Benko, S. Seal, and T. L. Barr, "Third International Conference on Composites Engineering" (D. Hui, Ed.), ICCE/3, 111, New Orleans, LA, 1996.
118. I. E. Campbell, High Temperature Technology, John Wiley and Sons, New York, 1st Edition, 1956.
119. P. Sommerkamp, Application of E. B. Coating in the Metal Industries, 2nd Electron Beam Sem. Frankfurt, 1972.
120. S. Schiller, H. Förster, and G. Jäsch, *J. Vac. Sci. Technol.* 12, 800 (1975).
121. W. D. Müntz, "CEI Course, Nitride and Carbide Coatings." LSRH-Neuchâtel, Switzerland, 1985.
122. M. Ohring, "The Materials Science of Thin Films." Academic Press, San Diego, CA, 1991.

123. W. Jeitschko, H. Nowotny, and F. Benesovsky, *Monatsh. Chem.* 94, 1198 (1963).
124. J. C. Schuster and J. Bauer, *J. Solid State Chem.* 53, 260 (1984).
125. D. McIntyre, J. E. Greene, G. Håkansson, J. E. Sundgren, and W. D. Münz, *J. Appl. Phys.* 67, 1542 (1990).
126. S. Seal, T. L. Barr, N. Sobczak, and S. J. Kerber, *J. Mat. Sci.* 33, 4147 (1998).
127. H. A. Jehn, S. Hofmann, and W. D. Müntz, *Thin Solid Films* 153, 45 (1987).
128. S. Hofmann, *Surf. Int. Anal.* 12, 329 (1988).
129. H. H. Madder, *Surf. Sci.* 150, 39 (1985).
130. T. L. Barr, S. Seal, L. M. Chen, and C. C. Kao, *Thin Solid Films* 253, 277 (1994).
131. E. D. Johnson, Ph.D. Dissertation (Thesis), Cornell University, 1984.
132. S. Seal and S. Shukla, "Surface Interface Analysis" (to be published).
133. A. Biaconi, R. F. Bachrach, S. B. M. Hagström, and S. A. Flodström, *Phys. Rev.* B 19, 2847 (1979).
134. K. S. Chari and B. Mathur, *Thin Solid Films* 81, 271 (1978).
135. K. Shimizu, S. Tajima, G. E. Thompson, and G. C. Wood, *Electrochem. Acta.* 25, 1481 (1980).
136. T. L. Barr, S. Seal, K. Wozniak, and J. Klinowski, *J. Chem. Soc. Far. Trans.* 93, 181 (1997).
137. S. Seal, T. L. Barr, N. Sobczak, and E. Benko, *J. Vac. Sci. Technol.* A15, 505 (1997).
138. A. Ermolieff, M. Girard, C. Raoul, C. Bertrand, and T. H. Dre, *Appl. Surf. Sci.* 21, 65 (1985).
139. K. S. Robinson and P. M. A. Sherwood, *Surf. Int. Anal.* 6, 261 (1984).
140. C. Ernsberger, J. Nickerson, A. E. Miller, and J. Moulder, *J. Vac. Sci. Technol.* A3, 2415 (1985).
141. T. L. Barr and M. A. Lishka, *J. Am. Chem. Soc.* 108, 3178 (1986).
142. C. D. Wagner, D. E. Passoja, H. E. Hillary, I. G. Kinisky, H. A. Six, W. I. Jansen, and J. A. Taylor, *J. Vac. Sci. Technol.* 21, 933 (1982).
143. B. Herreros, H. He, T. L. Barr, and J. Klmowski, *J. Phys. Chem.* 98, 1302 (1994).
144. T. L. Barr, *Zeolites* 10, 760 (1990).
145. A. Corma, V. Fornês, D. Paletta, 1. H. Cruz, and A. Ayerba, *J. Chem. Soc., Chem. Commun.* 10, 333 (1986).
146. S. Seal, S. Krezoski, D. Petering, T. L. Barr, J. Klinowski, and P. Evans, *Proc. R. Soc. London* B, 263, 943 (1996).
147. S. Seal, S. Krezoski, S. Hardcastle, T. L. Barr, D. Petering, C. F. Cheng, J. Klinowski, and P. Evans, *J. Vac. Sci. Technol.* A13, 1260 (1995).
148. T. L. Barr, S. Seal, H. He, and J. Klinowski, *Vacuum* 46, 1391 (1995).
149. T. L. Barr, L. M. Chen, M. Mohsenian, and M. A. Lishka, *J. Am. Chem. Soc.* 110, 7962 (1988).
150. T. L. Barr, *Microporous Mater.* 3, 557 (1995).
151. B. Herreros, H. He, T. L. Barr, and J. Klinowski, *J. Phys. Chem.* 98, 1302 (1994).
152. H. He, T. L. Barr, and J. Klinowski, *Clay Miner.* 30, 201 (1995).
153. W. A. Deer, R. A. Howie, and J. Zussman, "An Introduction to the Rock Forming Minerals." 2nd edn. Longman Scientific and Technical Press, London, 1992.
154. W. E. Ernst, "Amphiboles." Springer-Verlag, New York (1968).
155. H. He, T.L. Barr, and J. Klinowski, *J. Phys. Chem.* 98, 8124 (1994).

*156.* B. Herreros, H. He, T. L. Barr, and J. Klinowski, *J. Phys. Chem.* 98, 1302 (1994).

*157.* B. Herreros, T. L. Barr, P. J. Barrie, and J. Klinowski, *J. Phys. Chem.* 98, 4570 (1994).

*158.* B. Herreros, T. L. Barr, and J. Klinowski, *J. Phys. Chem.* 98, 738 (1994).

*159.* H. He, C. F. Cheng, S. Seal, T. L. Barr, and J. Klinowski, *J. Phys. Chem.* 99, 3235 (1995).

*160.* J. D. Andrade, V. Hlady, J. Herron, and J. N. Lin, Symposium on Surfaces in Biomaterials, Minneapolis, MN, October, 1991.

*161.* J. A. Gardella, G. L. Grobe, and L. Salvatti, Symposium on Surfaces in Biomaterials, Minneapolis, MN, October, 1991.

*162.* S. L. Cooper, Symposium on Surfaces in Biomaterials, Minneapolis, MN, October, 1991.

*163.* P. G. Gassman, Symposium on Surfaces in Biomaterials, Minneapolis, MN, October, 1991.

*164.* D. G. Castner and B. D. Ratner, Symposium on Surfaces in Biomaterials, Minneapolis, MN, October, 1991.

*165.* N. Larsson, P. Steiner, J. C. Eriksson, R. Maripuu, and B. J. Lindberg, *J. Colloid Interface Sci.* 90, 127 (1982).

*166.* R. Maripuu, Ph.D. Dissertation (Thesis), University of Uppsala, 1983.

*167.* First Symposium on Surfaces in Biomaterials, Minneapolis, MN, 1991.

*168.* K. R. Spurney, *Sci. Tot. Environ.* 30, 147 (1983).

*169.* B. T. Mossman, *Br. J. Industr. Med.* 50, 673 (1993).

*170.* M. F. Stanton, M. Layard, A. Tegeris, E. Miller, M. May, E. Morgan, and A. Smith, *J. Nat. Cancer Inst.* 67, 965 (1981).

*171.* P. H. Evans, R. C. Brown, and A. Poole, *J. Toxicol. Environ. Health* 11, 535 (1983).

*172.* R. Dumitru-Stanescu, C. Mandravel, and C. Bercu, *Analyst* 119, 689 (1994).

*173.* J. D. Andrade and V. Hlady, *Advances in Polymer Science* 79, 1 (1986).

*174.* P. K. Weathersby, T. A. Horbett, and A. S. Hoffman, *Trans. Am. Soc. Artif. Internal Organs* 22, 242 (1976).

*175.* T. A. Horbett and J. L. Brash, ACS Symp. Ser. 343, American Chemical Society, Washington, DC, 1987.

*176.* D. Briggs and M. P. Seah, in "Practical Surface Analysis" (D. Briggs and M. P. Seah, Eds.), 2nd ed., Chap. 9, Wiley, Chichester, 1990.

*177.* S. Seal, S. Krezoski, T. L. Barr, and D. Petering, *J. Vac. Sci. Tech.* A15, 1235 (1997).

*178.* T. L. Barr, S. Seal, S. Krezoski, and D. H. Petering, *Surface and Interface Analysis* 24, 99 (1996).

*179.* S. Seal, S. Krezoski, D. H. Petering, T. L. Barr, J. Klinowski, and P. Evans, *Journal of Hazardous Materials* 53, 55 (1997).

*180.* S. Seal, S. Krezoski, D. Petering, and T. L. Barr, in "Thin Films and Surfaces for Bioactivity and Biomedical Applications" (C. M. Cotell, S. M. Gorbatkin, G. Grobe, and A. E. Meyer, Eds.), p. 183, Vol. 414, MRS Proceedings, Pittsburgh, PA, 1996.

*181.* D. T. Clark, in "Handbook of X-Ray and Ultraviolet Photoelectron Spectroscopy" (D. Briggs, Ed.), pp. 211–247, Heyden, London, 1977.

*182.* D. Briggs, in Reference [10, pp. 437–483]. "Practical Surface Analysis." Vol. 1, Wiley, Chichester, 1990.

183. C. N. Reilley, D. S. Everhart, and F. F.-L Ho, in "Applied Electron Spectroscopy for Chemical Analysis" (A. Windawi and F. F.-L. Ho, Eds.), pp. 105–133, Wiley, New York, 1982.

184. G. Beamson, A. Bunn, and D. Briggs, *Surface Interface Anal.* 17, 105 (1991).

185. G. Beamson, D. T. Clark, J. Kendrick, and D. Briggs, *J. Electron Spectrosc. Rel. Phenom.* 57, 79 (1991).

186. D. Briggs and G. Beamson, *Anal. Chem.* 65, 1517 (1993).

187. J. F. Watts and J. F. Castle, *J. Mater. Sci.* 18, 2987 (1983).

188. F. Castle and I. F. Watts, *Ind. Eng. Chem. Prod. Rec. Dev.* 24, 361 (1985).

189. J. F. Watts, *Surface Interface Anal.* 12, 497 (1988).

190. J. F. Watts and F. M. Gibson, *Int. J. Adhes.* II, 105 (1991).

191. F. Watts, M. M. Chehimi, and E. M. Gibson, *J. Adhes.* 39, 145 (1992).

192. S. Gepstein, in "Senescence and Aging in Plants" (L. D. Nooden and A. C. Leopold, Eds.), pp. 85–109, Academic Press, San Diego, 1988.

193. H. Thomas and J. L. Stoddart, *Annual Reviews of Plant Physiology* 31, 83 (1980).

194. R. D. Butler and E. W. Simmon, *Advances in Gerontological Research* 3, 73 (1970).

195. P. J. Camp, S. C. Huber, J. J. Burke, and D. E. Moreland, *Plant Physiology* 70, 1641 (1982).

196. M. T. Greening, F. I. Butterfield, and N. Hans, *New Phytologist* 92, 279 (1982).

197. J. Hudak, *Photosynthetica* 15, 174 (1981).

198. H. W. Woodhouse, *Canadian Journal of Botany* 62, 2934 (1984).

199. S. C. Sabat and S. Saha, *Indian Journal of Experimental Biology* 23, 711 (1985).

200. P. A. Siengenthaler and P. J. Rawler, *Photochem. Bi-Biol.* 3, 175 (1989).

201. U. P. Alldley and C. S. Sillghal, *Photobiochemistry and Photobiophysics* 6, 135 (1983).

202. T. L. Barr, *Journal of Vacuum Science and Technology* 14, 660 (1977).

# 3. SECONDARY ELECTRON FINE STRUCTURE—A METHOD OF LOCAL ATOMIC STRUCTURE CHARACTERIZATION

Yu. V. Ruts, D. E. Guy, D. V. Surnin

Physical Technical Institute, Ural Branch, Russian Academy of Sciences,
426000 Izhevsk, Russia

V. I. Grebennikov

Institute of Metal Physics, Ural Branch, Russian Academy of Sciences,
620219 Ekaterinburg, Russia

## 1 INTRODUCTION

The use of an electron probe for analyzing the surface of a substance is one of the conventional experimental methods extensively used in the physics of condensed matter and, in particular, in the physics of thin films. This is due to the extremely strong interaction of electrons with the substance, which ensures a surface sensitivity of the electron methods of investigation from several angstroms to tens of angstroms. Electron optics allows high spatial resolution, which sometimes becomes a necessity. Availability and compactness of the necessary experimental equipment play a great role in the extent of application of electron methods.

In parallel with traditional techniques of study of the surface crystalline structure (low- and high-energy electron diffraction and photo- and Auger-electron diffraction) and surface chemical and electron structure (Auger, X-ray photoelectron, and characteristic electron energy loss spectroscopies), in the last two decades, methods of analysis of the local atomic structure of thin films, surfaces, and subsurface layers have been developed. The local atomic structure of an atom of a given chemical element is characterized by the lengths and angles of the bonds with the surrounding atoms. The experimental methods of analysis of the local atomic structure of surface layers are based on obtaining and processing the extended fine structures of the electron energy loss spectra and the secondary electron extended fine structure spectra. Extended fine structures of electron spectra— extended energy loss fine structures (EELFS) and secondary electron fine structures (SEFS), consist in oscillations of the electron emission intensities (Fig. 1), whose extent is several hundred electron volts and whose period amounts to tens of electron volts. In the electron energy loss spectra the extended fine structure is located on the low-energy side (on the scale of kinetic energies of detected electrons) of the loss edge, which is due to ionization of the core level (C-level) of an atom by an incident electron. In the secondary electron spectrum the extended fine structure is located on the high-energy side of the CVV Auger line (the CVV Auger transition involves two valence electrons and a hole on the core C-level).

EXPERIMENTAL METHODS IN THE PHYSICAL SCIENCES
Vol. 38
ISBN 0-12-475985-8

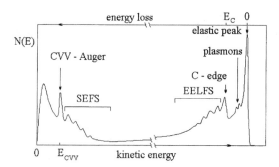

FIG. 1. Schematic picture of the electron spectrum.

The intensity of the extended fine structure, as a rule, is an order of magnitude less than the intensity of the ionization loss edge for EELFS, and than the intensity of the CVV Auger line for SEFS. The analyzed layer depth amounts to about 5–7 Å in the SEFS method and about 10–30 Å in the EELFS method, and is determined by the energy of the electrons detected in experiment. So, in the EELFS method, by changing the experimental conditions (namely, the excitation energy or/and angle geometry) one can change the analyzed layer depth.

The origin of the extended fine structures of electron spectra is closely connected with the effect of coherent elastic scattering of secondary electrons by atoms on the intensity of the atom core level ionization by the electron impact. More precisely, the final state is defined as the state of an electron wave scattered by the nearest atomic environment of the ionized atom. The extended fine structures of electron spectra are similar in their physical nature to those of X-ray absorption spectra (EXAFS) [1, 2]. In this connection SEFS and EELFS structures are classified as EXAFS-like phenomena [3]. Since the extended fine structure is formed by the secondary electron coherent elastic scattering occurring during ionization of the atom core level, EELFS and SEFS spectra are determined, as in the case of EXAFS, by the local atomic environment of the ionized atom. In addition the different scattering ability of different chemical elements allows a chemically selective analysis of the local atomic structure of the substance under study. One of the most significant advantages of EXAFS-like methods is their ability to formalize the inverse problem for finding the local atomic structure parameters. In this case, the kind of the inverse problem and the way of its solution do not depend on the atomic structure of the substance studied and allow the matching of the method of solution to a test object.

Among the experimental methods of analysis of the surface atomic structure based on electron spectral data, the EELFS technique is the most developed. This is explained in part by the rather simple mechanism of the EELFS formation: an atom goes into an excited state (characterized by a hole on the core level and a

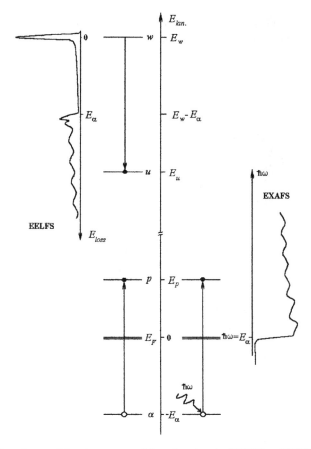

FIG. 2. The scheme of the electron transitions forming the EELFS and EXAFS spectra.

secondary electron) as a result of its interaction with an incident electron. The incident electron, having lost energy, is deteceted in the experiment and produces the EELFS. Since the mechanism of EELFS spectrum formation is similar to that of EXAFS (Fig. 2), both theoretical approaches and methods of treatment of the EXAFS spectroscopy have been used in the EELFS spectroscopy. This has made it possible to perform a number of interesting studies of the local atomic structure of both surface layers of pure samples and thin films on their surfaces with the help of the EELFS method (see reviews [4, 5]). In contrast to EELFS, the mechanism of formation of the secondary electron extended fine structures has proved to be much more complicated. This is reflected in a number of characteristic features of SEFS spectra and makes the extraction of local atomic structure parameters from the experimental data more difficult, although information obtained by the SEFS method is unique in characterizing superthin atomic surface layers of a substance.

## 1.1 The History of SEFS: Experimental Results and Theoretical Approaches

Since 1971, nonmonotonic (oscillating) behavior of the 3d transition metal secondary electron spectrum above the MVV Auger line has been observed [6, 7]. The first investigations of these spectral peculiarities were performed by McDonnell et al. [8] and Becker and Hagstrum [9]. They obtained extended fine structure spectra above the MVV Auger line from Cu and Co single-crystal and polycrystal samples under different experimental conditions: different sample temperatures, different orientations of the studied single-crystal plane, and different angular geometries. The experimental evidence pointed to the fact that nonmonotonic behavior of the secondary electron spectrum above the MVV Auger line is a result of the coherent scattering of secondary electrons by the atomic structure of the sample surface layers. McDonnell et al. [8, 10, 11] suggested considering the diffraction of electrons by atomic planes of the sample surface as the mechanism that forms the oscillating structure of the secondary electron spectrum that corresponds to low-energy electron diffraction (LEED).

An alternative approach to the description of oscillations in the secondary electron spectrum was proposed by DeCrescenzi et al. [12, 13]. They suggested considering the SEFS structure as the result of EXAFS-like scattering of an intermediate state electron in the second-order process, i.e., an autoionization-type process. This made it possible to offer an explanation of the location of the extended fine structure above the CVV Auger line, since the autoionization process is of a threshold character, the threshold coinciding with the position of the Auger line in the spectrum. Unlike classical resonance (autoionization) processes, the SEFS mechanism implies the presence of two-step transitions over a wide range of secondary electron energies, i.e., it does not imply a resonance condition. The EXAFS-like nature of the secondary electron extended fine structure, viz., the dependence of the SEFS spectrum on the local atomic structure, has been shown in studies of crystalline and amorphous surfaces.

Investigations of experimental SEFS spectra have been aimed at obtaining structures from samples with different surface chemical composition, different degree of surface atomic structure ordering, and different temperatures of samples under study. The investigations performed have revealed a number of characteristic features of SEFS spectra as compared to other EXAFS-like structures [8–24]:

1. Studies of SEFS spectra obtained from the surfaces with different degrees of atomic ordering (a single- crystal plane, a polycrystal surface, and an

amorphous surface) have shown that the secondary electron extended fine structure is formed as a result of the EXAFS-like scattering of secondary electrons by the nearest atomic environment of the ionized atom.

2. The extended fine structures are well observed in 3d- and 4d-transition metal spectra. MVV SEFS spectra from Cr to Cu and Ag NVV SEFS spectra, as well as Fe and Ni LVV SEFS spectra and Ag MVV SEFS spectra, have been obtained.

3. Sample temperature variation over a relatively small range (300–900 K) causes quite considerable changes in secondary electron extended fine structures.

4. Investigations of the dependence of SEFS spectra on the valence band structure of an oxidic film at a Ni surface have shown the interrelationship between the extended fine structure and the valence band structure of the substance under study.

5. A strong dependence of the extended fine structure on the angular geometry of the experiment, namely, on the value of the collection space angle of the energy analyzer and the angular orientation of the studied single crystal with respect to the energy analyzer, is observed in experiments on single-crystal samples.

Developing an approach to the study of SEFS (suggested in [8–11]), McRae [25] described the process of scattering of the electrons emitted by a sample in the Bragg approximation. The signal obtained on the basis of this description contained oscillations whose period corresponded essentially to the period of the main (the most intense) oscillation in the observed SEFS spectrum. All the other theoretical approaches and calculations are based on the EXAFS-like nature of SEFS.

The most complete calculation in the context of the first-order theory (the secondary electron direct emission from core and valence states) was made for Cu (Fig. 3) by Aebi et al. [26]. Partial densities of states (of s, p, d, and f types) were obtained on a Cu atom in the framework of the cluster calculation. A comparison between the calculated results and the oscillating part of the Cu MVV SEFS spectrum has shown that the main contribution to the experimental signal is made by the partial density of f-states. The experimental spectrum was in good agreement with the calculated total density of states (Fig. 3) in the region of secondary electron energies from 150 to 200 eV, but the calculated result proved to be shifted to higher energies. In the region of secondary electron energies greater than 200 eV the calculation did not reproduce the experimental result at all. The extended fine structure of the Ni MVV spectrum has been also calculated by Vedrinsky et al. [27] within the cluster formalism. The result of calculations was compared with the Fourier-transformed experimental spectrum. The calculated and the experi-

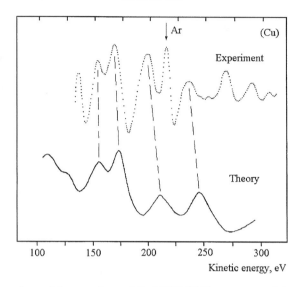

FIG. 3. Comparison of the experimental Cu MVV SEFS spectrum and the calculated Cu atomic density of states.

mental oscillating part were in good agreement with respect to the oscillation period, but they differed in their amplitude behavior.

Estimation of the relative intensity of the autoionization process [27] has shown that the second-order process intensity constitutes 10–15% of the first-order process intensity. Based on the results obtained and the dipole approximation for electron transitions [28], the authors of [27] draw the conclusion that the SEFS structure is formed in the process of coherent scattering of a secondary electron emitted from the valence band as a result of excitation by an incident electron (the first-order process). By contrast, the intensity of autoionization, i.e., the second-order process, was estimated [29–31] with hydrogenlike wave functions. The autoionization intensity in the region of the existence of the SEFS spectrum was shown to be comparable to the intensity of the first-order processes.

An ingenious approach to the autoionization process was suggested by Tomelini and Fanfoni [32]. They examined diffraction processes in the framework of autoionization, i.e., a second-order process. More exactly, they studied only the second stage of autoionization, namely, the secondary electron emission from the intermediate state where the core hole is coherently distributed over the crystal. With such an approach it was shown that in crystals this diffraction contribution enhances fine structure as compared to the amorphous substance. However, the applicability of such a model is determined by the probability of the occurrence of the core hole distributed coherently over the crystal. More conventional is the picture of generation of mutually independent holes on different atoms upon interaction of incident electrons with the substance.

A qualitatively new approach to the extended fine structure in the secondary electron spectrum was proposed in [33–35]. Bearing in mind the autoionization mechanism, the authors suggested allowing for the coherent elastic scattering of electrons of both intermediate state of autoionization and the final state, i.e., the electrons detected in an experiment. In the context of this approach, the secondary electron extended fine structure results from superposition of two interference terms, both of which are determined by the local atomic environment of the ionized atom. One of them is determined by the wave number of the intermediate-state electron of autoionization, and the other by the wave number of the final-state electron detected in the experiment.

## 1.2 Mechanisms of the SEFS Signal Formation

Consider the processes of secondary electron creation in the inelastic scattering of high-energy electrons (3–10 keV). Figure 4b and c show direct transitions of an electron from the core level $\alpha$ ($E_\alpha$ is the binding energy) and the valence band $\beta$ into a certain final state $p$ with energy ranging from zero to 1 keV, which is just registered in experiment. These standard transitions are described by first-order perturbation theory.

Transitions in the second order of the theory to the same final state $p$ are illustrated in Figure 4d and e, where Figure 4e shows the direct transition to the final state and Figure 4d shows the exchange transition. These transitions proceed in two steps: first, the core electron goes into a certain intermediate state $q$, and then the hole arising in the core level $\alpha$ annihilates with the participation of the $\beta$ valence electron and the $q$ intermediate-state electron. In this case, in the final state there is a hole in the valence band and an electron $p$ registered in experiment. Transitions presented in Figure 4d and e resemble the known autoionization process, with the difference that its second step, i.e., the autoionization itself, in the present case originates not from the localized low-energy resonance state but from the state $q$ that belongs to the continuous spectrum.

The second-order process amplitude contains an integration over all possible intermediate states $q$. However, due to the known resonance denominator, the contribution from the excited states, whose energy—accurate to within their width (in our case, several electron volts)—satisfies the energy conservation law, becomes predominant, so we can put $E_p - E_q \approx |E_\beta - E_\alpha|$. Hence it follows, in particular, that the second-order process has the energy threshold corresponding to the Auger line given in Figure 4a, d, and e, while the direct transitions represented in Figure 4b and c start at zero kinetic energy of the secondary electrons.

The experimental spectrum is determined by the sum of contributions of all processes shown in Figure 4 as well as of the transitions from other levels that

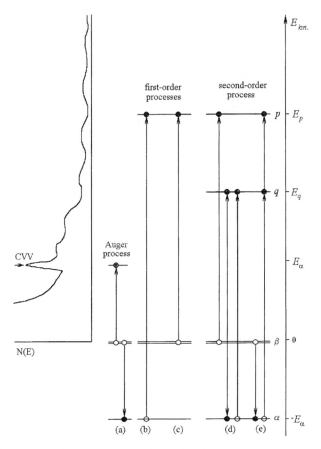

FIG. 4. The scheme of the electron transitions forming the secondary electron spectra: (a) Auger process, (b) emission from the core level, (c) emission from the valence state [(b) and (c) are the first-order processes], (d) exchange transition, and (e) direct transition of the second-order process (autoionization). The process of the incident-electron energy loss is not shown.

are omitted in the figure. There are two ways of clarifying which of them are most important. The first is to make a theoretical estimation of probabilities of all transitions in a solid (the fine structure is formed by electrons scattered by surrounding atoms). The second is to make an effort to establish the qualitative differences between the intermediate $q$ and final $p$ states and, in the case they exist, to determine the role of the second-order processes by comparing the characteristic features of the experimental signal with the theoretically predicted one. Consider this problem within the second approach.

Since both the intermediate $q$ and final $p$ states belong to the continuous spectrum, it may appear at first glance that there is no fundamental difference between them. Such is indeed the case in atoms. However, in condensed matter, wave functions are distorted because of the elastic scattering of waves by the nearest neighbors of the ionized atom that gives rise to SEFS, as in EXAFS spectroscopy. This scattering is caused by wave interference. In angle averaging [36, 37] (according to the experimental condition—for example, in polycrystal, in an amorphous sample, etc.) the main contribution to the oscillating signal comes from single backscattering by neighboring atoms with the maximum difference in wave path of $2R$ ($R$ is the interatomic distance). The wave interference takes place in both final $p$ and intermediate $q$ states; therefore the oscillating part of a signal will contain two terms. The former can be approximated by $\sin(2pR_j)$, the latter by $\sin(2qR_j)$, where summation over all atoms $j$ of the ionized atoms environment is implied and $p$ and $q$ are the wave numbers of the electrons in the final and intermediate states, respectively. The relation between wave numbers is determined by the energy conservation law, which, for the simplest case, can be given by the expression $E_p - E_q = E_\alpha$. The amplitude of oscillations of the intermediate state $q$ is determined by the intensity of the second-order atomic process, and the amplitude of the oscillations of the $p$-state is determined by intensities of both the first- and second-order processes.

Thus, with allowance for both the first-order and second-order processes, the secondary electron spectrum fine structure is formed by oscillations of two types, which are determined by the same local atomic structure but different wave numbers; this is the main difference of SEFS spectra from EXAFS and EELFS spectra. It is just this qualitative difference that must determine the characteristic features of SEFS spectra, and it must be taken into account in obtaining parameters of the local atomic structure from experimental data. However, it should be pointed out that a signal from two final states can be observed also in EXAFS and EELFS spectra in the case of the excitation of two closely spaced levels. And though the mechanism of appearance of these signals differs from that in the case of SEFS, nevertheless conceivably the analogous problem must be solved also for these traditional methods.

## 1.3 Analysis of the Local Atomic Structure of Surface Layers from SEFS Experimental Data

There are two approaches to obtaining parameters of the local atomic structure in EXAFS and EXAFS-like experimental methods.

The first approach is based on a calculation of the oscillating signal of the *a priori* prescribed local atomic configuration and a comparison between calculated and experimental results. By varying the local atomic configuration, the calculated result is then fitted to the experimental spectrum, and thereby the local atomic structure of the sample under study is simulated. Methods of direct calculation of the extended fine structure of a preset atomic configuration are mainly based on a cluster calculation of the electron structure. In the simplest cases the partial density of states determining the oscillating part of the spectrum and other scattering parameters are recalculated by analytical formulae from the found electron structure and the preset atomic cluster structure.

The second approach to obtaining parameters of the local atomic structure of a solid under study is based on the fact that the oscillating structure is mainly (80–90%) formed as a result of single backscattering by nearest neighbors of the ionized atom. In the context of this process the oscillating part of the extended fine structure is determined by the atoms within twice the nearest-neighbor distance. Thus the experimental signal turns out to be connected with the atomic pair correlation function, which can be found by solving the inverse problem. In methods based on solving the inverse problem, the atomic pair correlation functions are obtained from the experimental result by solution of an integral equation. In the framework of this approach the most straightforward and widespread technique for obtaining the structural information is the Fourier transform method. Since the oscillating part of the EXAFS spectrum and of most EXAFS-like structures is a sine Fourier transform of the radial atom distribution function that is accurate in amplitude, the required parameters of the local atomic structure can be obtained by Fourier transformation of the experimental oscillating part of a spectrum [38]. Though extremely simple, the Fourier transform method has a number of essential disadvantages: the restricted interval of integration of the experimental signal, neglect of the parameters of decay, and so on. All of them usually produce errors in determination of the local atomic structure parameters. The method based on the direct solution of the integral equations expressing the relation between the oscillating part of the spectrum and the atomic pair correlation functions [39–41] is significantly more precise but more complicated. The problem reduces in essence to the solution of a system of Fredholm integral equations of the first kind in the case of a multicomponent substance, or to the solution of one equation in the case of a one-component substance [41–44].

Both the methods of the direct calculation of extended fine structure and those of the solution of the inverse problem have their advantages and disadvantages. The main disadvantage of direct calculation methods is that the determination of local atomic structure parameters is in essence performed by fitting the calculated result to the experiment; at the same time, in the framework of this method the possibility exists of taking into account the majority of physical processes that produce the extended fine structure of a spectrum. The main disadvantage of methods based on the inverse problem solution is incomplete allowance for the physical processes producing the extended fine structure. The main advantage of these methods is that the form of the inverse problem and the method of its solution are independent of the atomic structure of the material under study. It should be noted that to obtain parameters of the local atomic structure from experimental data by either direct calculation or solving the inverse problem, it is essential to obtain detailed theoretical description of the processes that produce the extended fine structure of the spectra.

At the present time, of all EXAFS-like methods of analysis of local atomic structure, the SEFS method is the least used. The reason is that the theory of the SEFS process is not sufficiently developed. However the standard EXAFS procedure of the Fourier transformation has been applied also to SEFS spectra. The Fourier transforms of MVV SEFS spectra of a number of pure 3d metals have been compared with the corresponding Fourier transforms of EELFS and EXAFS spectra. Besides the EXAFS-like nature of SEFS oscillations shown by this comparison, parameters of the local atomic structure of studied surfaces (the interatomic distances and the mean squared atomic deviations from the equilibrium positions [12, 13, 15–17, 21, 23, 24]) have been obtained from an analysis of Fourier transforms of SEFS spectra. The results obtained have, at best, a semi-quantitative character, since the Fourier transforms of SEFS spectra differ qualitatively from both the bulk crystallographic atomic pair correlation functions and the relevant Fourier transforms of EXAFS and EELFS spectra.

The direct methods of solving inverse problems in SEFS spectroscopy have been used, taking into account oscillations of two types, for Cu MVV SEFS spectra [45–47]. Satisfactory agreement of the results obtained with the bulk atomic pair correlation functions and with the results of the solution of the inverse problem in EXAFS and EELFS spectroscopy makes it possible to conclude that to obtain correct structural information from the SEFS experimental data it is essential to take into account oscillations of two types, which rules out the application of the Fourier transformation and necessitates the direct solution of the inverse problem. The use of Tikhonov regularization as a method for solving the inverse problem in SEFS spectroscopy is the subject of Section 6. Up to now, extraction of atomic structure parameters through direct calculation of the SEFS spectrum has not been performed.

## 1.4 Characteristic Features of SEFS; Advantages and Disadvantages

Compared to the diffraction methods, the main advantages of the spectral methods of structural analysis, in particular SEFS spectroscopy, are due to their physical nature, namely, their sensitivity to the local atomic structure only. That is, no matter what the structural state of the matter (crystalline or amorphous), the oscillating signal in the spectrum is determined by the nearest atomic environment of the ionized atom of the particular chemical element. Owing to this, the spectral methods of structural analysis make it possible to use the results obtained from a sample with known atomic structure as standards in the study of the structure of unknown objects by determining all parameters of the extended fine structure of the test object. The sensitivity of spectroscopic methods to chemical elements arises because the secondary electron scattering amplitude is different for various chemical elements. In the case of a multicomponent sample there is a possibility of determining the atomic arrangement according to the chemical species of the atoms.

In the framework of spectroscopic methods the possibility exists of formalizing the inverse problem, whose form and solution technique do not depend on the atomic structure of the studied sample. At the same time, the spectral methods of structure analysis require a detailed theoretical description of the scattering processes that lead to the oscillatory structure of spectra, including inelastic electron scattering. The mentioned peculiarities of spectroscopic methods of atomic structural analysis as compared to diffraction techniques are also true for the SEFS as compared to the LEED method. On the one hand, SEFS as a method of structural analysis calls for detailed theoretical description, and on the other hand, contrary to the LEED method, it does not require *a priori* modeling of the atomic surface structure. Besides, the SEFS method can be used with the same success for analysis of both crystalline and amorphous surfaces.

Among the spectroscopic (EXAFS-like) methods of analysis of local atomic structure, the SEFS method is a purely surface technique. The analyzed-layer depth in the SEFS method is determined by the mean free path of the secondary electrons and amounts to about 5–7 Å. It is not feasible to obtain such a small depth of the analyzed layer by other EXAFS-like methods of surface structure analysis. It should be mentioned that from the standpoint of the physics of surfaces, information on the atomic structure of superthin surface layers is of most interest, since it is precisely in these surface layers that one encounters the most considerable changes due to the presence of a free surface in a solid. The use of electron optics in the SEFS method makes it possible to obtain extrahigh resolution for the analyzed area (down to several nanometers), which presently is not attainable with X-ray optics used in the EXAFS method.

Compared to the EELFS method, which is also based on the data of electron spectroscopy, the SEFS method has a number of advantages. Namely, the absence of strong angular dependence of the intensity of the secondary electron emission makes it possible to obtain SEFS spectra of high signal-to-noise ratio, irrespective of the angular geometry of the experiment, including the standard angular geometry of electron spectrometers (backscattering geometry). Electrons scattered directly by the nearest neighbors are detected in the SEFS method; quite different electrons, containing only integrated (through the energy conservation law) information, are detected in the EELFS method. The SEFS angular dependence gives information on the local atomic structure that EELFS does not. Thus, in the case of crystal surfaces the SEFS method provides a possibility of obtaining information not only on spherically averaged parameters, but also on the angular atom distribution of the local atomic structure.

But in spite of all the advantages of the SEFS method as compared to both diffraction and spectroscopic methods of structure analysis, this technique has not yet been applied to analyze the local atomic structure of surfaces and thin films. This is explained by the diversity and complexity of the processes forming the secondary electron spectrum and the corresponding fine structures, and the resulting difficulties in their theoretical description and in the mathematical formalization of the problem of determining local atomic structure parameters from the experimental data.

In the present work we use the abbreviation SEFS to designate the extended fine structures of the secondary electron spectra. In a number of publications the abbreviation EXFAS, suggested by DeCrescenzi et al. [12, 13], is used to designate these structures. We do not use the abbreviation EXFAS here, because of its similarity to the abbreviation EXAFS, which often leads to confusion.

The structure of the present review is as follows: In Section 2 the experimental procedure for obtaining the SEFS spectra and extraction of their oscillating parts is described. In Section 3, in the framework of the one-electron approach, a theoretical description of processes forming the SEFS is given, with allowance for the first- and second-order processes, and a simple analytical SEFS formula is deduced. In Section 4, estimates are made of the amplitudes and intensities of electron transitions that determine the SEFS oscillating terms. In Section 5, the SEFSs are calculated on the basis of the expressions obtained, and an assessment of their correctness is made by comparing the calculated results with the experimental data. In Section 6, the formalization of the inverse problem of finding the atomic pair correlation functions from SEFS spectra is conducted, and its solution by the Tikhonov regularization method is presented for Fe, Ni, and Cu MVV SEFS spectra.

## 2 THE METHOD OF OBTAINING THE SEFS EXPERIMENTAL DATA

As mentioned above, by now secondary electron spectroscopy is one of the main methods of studying superthin surface layers of condensed matter. This situation is partly the result of the successful development of the necessary ultrahigh vacuum facilities, electron optics, and methods for the recording of electron emission spectra. In this chapter the procedure for obtaining the secondary electron spectra is not treated, since it is universally accepted; we will focus only on those peculiarities that are typical of the method of obtaining the SEFS experimental data.

To obtain the experimental data in electron spectroscopy and, in particular, to obtain SEFS spectra, a standard electron spectrometer will suffice. Such a device includes the following units: an ultrahigh-vacuum analyzing chamber (vacuum

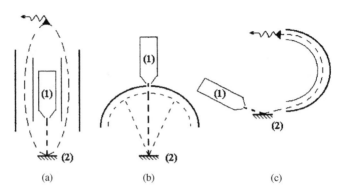

FIG. 5. Electronic equipment used to detect electron spectra: (a) the cylindrical mirror analyzer (CMA); (b) the retarding potential analyzer; (c) the concentric hemispherical analyzer (CHA). The electron gun is labeled as (1), the specimen (2).

not worse than $10^{-7}$ Pa), a kinetic energy electron analyzer (energy analyzer), a primary electron source, and a manipulator with a sample holder. The kinetic energy of electrons emitted from the surface is mainly measured by energy analyzers of the following types (Fig. 5): the cylinder mirror type, the grid analyzer, and the hemispherical analyzer [48–50]. Each of these energy analyzers has its advantages and disadvantages. The most-used type in electron spectroscopy is the cylindrical mirror. Its resolution is characterized by the ratio $\Delta E/E$, where $\Delta E$ is the energy analyzer, dispersion and $E$ is the actual value of the kinetic energy of the observed electrons. The energy resolution of a standard electron spectrometer is usually $\Delta E/E = 0.1$–$0.4\%$. The source of primary electrons is an electron gun with a monoenergetic electron beam. The primary beam energy can be established, as a rule, in the region from several tens of electron volts to 10 keV. The current density of the electron beam is the same as in Auger electron spectroscopy and is several microamperes per square millimeter.

   The method of carrying out the experiment for obtaining the SEFS spectra does not differ essentially from the methods used in electron spectroscopy for the study of a surface. The intensity of the oscillations in the secondary electron spectra is, as a rule, an order of magnitude lower than the intensities of the main Auger lines. In this connection, the acquisition of SEFS spectra may require a long time (up to several hours), because of the need to obtain a structure intensity at least 10 times the noise intensity. Otherwise, one cannot reasonably process the SEFS spectra to obtain information on the local atomic structure of the object under study. It is worth noting that view of the long duration of the SEFS spectrum acquisition it is essential to control the state of the studied surface, which can be accomplished by standard methods (Auger electron spectroscopy, the LEED method, etc.) and should preserve the surface state for the whole time of the experiment.

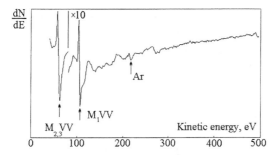

FIG. 6. The Cu secondary electron spectrum.

The experimental SEFS spectra are obtained by the direct counting of electrons (the integral mode) or by synchronous detection (the differential mode) [48, 49, 51]. The use of the differential mode of obtaining the SEFS spectra will now be considered in more detail. Along with a number of problems (such as distortion of the spectrum background component, and uneven blurring of spectral features by the apparatus function) the differential mode has a number of advantages. The signal contrast is inherently better than in the integral mode, and there is the further possibility of improving the signal-to-noise ratio for certain spectral features by choosing the modulation amplitude. The latter is sometimes the most important factor in obtaining the low-intensity extended fine structure.

The secondary electron spectrum of a copper sample in the region of the Cu MVV Auger line is shown in Fig. 6 as an example. The spectrum is given for clarity in the differential mode. This spectrum is a typical one: nonmonotonic behavior above the low-energy Auger lines exists for all 3d and 4d transition metals (above MVV in 3d and above NVV in 4d transition metals) [51]. This nonmonotonic behavior is seen also in high-energy Auger spectra (above LVV in 3d and above MVV in 4d transition metals) [51].

## 2.1 Extraction of the Oscillating Part of the Secondary Electron Spectrum (Mathematical Pretreatment)

The secondary electron extended fine structure consists in oscillations of the intensity of the secondary electron emission. The characteristic period of these oscillations is several tens of electron volts, and the structure extends over several hundred electron volts. To extract local atomic structure parameters from the spectrum and to analyze them it is essential to separate the oscillation part from the background experimental spectrum. This procedure is called *background subtraction*, and it is needed for all the EXAFS-like methods. It is worth noting that of all the stages of treating EXAFS and EXAFS-like spectra, the procedure of separating out the oscillating part (subtracting the background) is the least formalized, i.e., there are no rigorous criteria for performing this procedure. This problem becomes somewhat simpler if the nonoscillating spectrum component is a simple function; in that case, whatever the procedure of the background subtraction, the obtained result (the oscillating part) remains practically unchanged.

In comparison with the traditional EXAFS, in the SEFS spectrum the nonoscillating part (background) has a much more complicated form, so that the procedure of extracting the oscillating part has to be more sophisticated. The nonoscillating part of the SEFS spectrum, as a rule, is approximated either by polynomials or by splines. To assure that the obtained result is correct, the separation procedure is carried out repeatedly with the use of different functions approximating the background; then, by comparison of the results obtained, a conclusion regarding the correctness of the performed procedure is drawn.

In the case of the differential mode of obtaining the experimental secondary electron spectrum, a spectrum integration must be performed in addition to the procedure of separating out the oscillating part.

It is well known that an experimentally obtained spectrum is the convolution of the actual electron distribution with the apparatus function of the electron spectrometer [52–57]. The presence of the apparatus broadening decreases the oscillation amplitude, thereby worsening the signal-to-noise ratio and decreasing the extent of the oscillating structure. At present, as a rule, the apparatus broadening in the SEFS method is not taken into account.

## 3 THEORETICAL DESCRIPTION OF THE SEFS PROCESS

Before examining the processes forming extended fine structures of the secondary electron spectrum, we dwell on some characteristic experimental peculiarities of obtaining SEFS spectra, on the strength of which a number of essential approximations have been made in the present work. The fine structure spectra of the secondary electrons are obtained with both the standard Auger and LEED experimental equipment. The geometry of electron backscattering from the sample surface is measured with the standard geometry of Auger analyzers coaxial with the electron gun. In the LEED geometry of the energy analyzer, the collection solid angle is $2\pi$. Monochromatic beams of incident electrons with kinetic energy of 5–10 keV are used, as a rule, in obtaining the SEFS spectra. Thus, the energies of incident electrons, both for MVV and LVV SEFS spectra of 3d metals and for NVV and MVV SEFS spectra of 4d metals, far exceed the binding energy of the corresponding C core level participating in the formation of the CVV Auger line. The actual extent of the experimental SEFS is several hundred (from 200 to 400) electron volts. For low-energy CVV SEFS spectra (MVV for 3d and NVV for 4d metals), the energy region where SEFS is detected covers from one to several binding energies of the C level. For the high-energy CVV SEFS spectra the extent of the fine structure does not exceed the value of binding energy of the corresponding core C level.

The main part of this section is presented in [20, 22, 30, 33, 35, 37, 45].

In this section we shall use the atomic (Hartree–Fock) system of units where $\hbar = e = m = 1$, the energy unit is $e^2/a_0 = 2$ Ry $\approx 27.2$ eV, and the unit of length is the Bohr radius $a_0 \approx 0.52$ Å.

Consider the processes that are induced by an electron beam incident on a sample and give rise to a current of secondary electrons whose kinetic energies are comparable to or higher than the energy of a certain CVV Auger line. The secondary electron emission current is formed by the true secondary electron current and the current of electrons from atoms. The true secondary electron current is determined by multiple incoherent energy losses of the incident electron beam upon its passage through the matter. It forms the most intense, but structureless, part of the spectrum, the so-called *diffusive* part; in other words, it forms a background of true secondary electrons. The structural parts of the spectrum, namely Auger lines and fine structure, are formed by the secondary electron current from atoms, which is mainly determined by the atomic spectrum of the secondary electron emission. The extended fine structure of the secondary electron spectrum (SEFS) is formed by coherent scattering of secondary electrons emitted from an atom by its nearest neighbors.

The emission current, $dJ(E_p, \hat{\boldsymbol{p}})$, of secondary electrons with energies ranging from $E_p$ to $E_p + dE_p$ and a momentum direction from $\hat{\boldsymbol{p}} = \boldsymbol{p}/p$ to $\hat{\boldsymbol{p}} + d\hat{\boldsymbol{p}}$, is written as

$$dJ(E_p, \hat{\boldsymbol{p}}) = \langle j(E_w, \hat{\boldsymbol{w}}) \, d\sigma(E_p, \hat{\boldsymbol{p}}) \rangle \tag{1}$$

where $j(E_w, \hat{\boldsymbol{w}})$ is the flux density of the electron beam incident on the atom, with electron energy $E_w = w^2/2$, and momentum direction $\hat{\boldsymbol{w}} = \boldsymbol{w}/w$. In Eq. (1) $\langle \ldots \rangle$ denotes averaging over all initial states. The cross section for secondary electron creation, $d\sigma(E_p, \hat{\boldsymbol{p}})$, is given by the following expression:

$$d\sigma(E_p, \hat{\boldsymbol{p}}) = \frac{2\pi}{w} \sum |T_{fi}|^2 \delta(E_i - E_f) \tag{2}$$

In the process under study, the energy conservation law is given by $\delta(E_i - E_f)$, where $E_i$ is the energy of the system in its initial state $|i\rangle$, and $E_f$ is the energy of the system in its final state $|f\rangle$. The initial state of the system, $|i\rangle$, is characterized by the nonexcited electron subsystem of the sample and an incident electron $|w\rangle$. The final state, $|f\rangle$, is characterized by the incident electron that has lost energy, $|u\rangle$, and a secondary electron ($E_p, \hat{\boldsymbol{p}}$), which is just detected in the experiment, as well as by a hole either on the sample atom core level or in the valence band. Summation over all final states that are not detected in the experiment is implied in Eq. (2).

Consider the process of secondary electron emission from an atom whose electron subsystem consists of a core level, $|\alpha\rangle$, with $2n_\alpha$ electrons, and the valence states, $|\beta\rangle$, involving $2n_\beta$ electrons. In the case of the secondary electron emission from the core level $|\alpha\rangle$ the energy conservation law gives $E_w - E_u = E_p + E_\alpha$, where $E_w - E_u$ is the energy loss of the incident elecron, $E_\alpha$ is the binding energy of the core level $|\alpha\rangle$, and $E_p$ is the energy of the secondary electron detected in the experiment. Then, after the summation in Eq. (2) over the states $|w\rangle$ and $|u\rangle$ that are not detected in the experiment, the cross section of the secondary electron emission from the core level $|\alpha\rangle$ takes the form

$$d\sigma\left(E_p, \hat{\boldsymbol{p}}\right) = 2n_\alpha \, d\sigma_0 \left\langle |T_{fi}|^2 \right\rangle_{\hat{\boldsymbol{u}}} \tag{3}$$

where the following quantity has been used:

$$d\sigma_0 = 64\pi^5 a_0^2 \frac{pu}{w} \, dE_p \, d\hat{\boldsymbol{p}} \tag{4}$$

In the case of the secondary electron emission from the valence state $|\beta\rangle$ its cross section is also defined by Eq. (3) with $|\beta\rangle$ instead of $|\alpha\rangle$. The matrix element $T_{fi}$ determines the amplitude of the transition from the initial state $|i\rangle$ into the final state $|f\rangle$.

## 3.1 Emission from the Core Level: The First-Order Process

Consider the amplitude for the creation of secondary electrons upon atom excitation by electron impact. As a result of the Coulomb interaction with the atom, the incident electron loses a part of its energy and goes into an inelastically scattered, state and the atom goes into an excited state characterized by a core hole and a secondary electron. In the context of the single-electron approach, the initial state of the system is characterized by $|i\rangle = |w, \alpha\rangle$ and the final states are characterized by $|f\rangle = |p, u\rangle$ where $|w\rangle$ and $|u\rangle$ are single-electron wave functions of the incident and inelastically scattered electrons, and $|p\rangle$ and $|\alpha\rangle$ are single-electron wave functions of the secondary electron and the core level electron, respectively. Then the amplitude for creation of the secondary electron is defined by the matrix element

$$T^{(1)}(\boldsymbol{p}) = \left\langle u, p | V | w, \alpha \right\rangle \tag{5}$$

where $V$ is the operator of the Coulomb electron–electron interaction. In Eq. (5) we have neglected the exchange matrix element, which is a good approximation in the case when the incident electron energy far exceeds the binding energy of the core electron, $E_w \gg E_\alpha$. By the energy conservation law, in the process under study, $E_w - E_u = E_p + E_\alpha$, where $E_w - E_u$ is the energy loss of the incident electron, $E_w = w^2/2$ the incident electron energy, $E_u = u^2/2$ the inelastically scattered electron energy, $E_p = p^2/2$ the kinetic energy of the secondary electron, and $E_\alpha = Z^*\alpha^2/2$ the binding energy of the core electron ($Z^*$ is an effective charge, and $\alpha$ is the inverse radius of the core-electron wave function localization). Then the energy conservation law gives the relation $w^2 - u^2 = p^2 + Z^*\alpha^2$.

The matrix element (5) describes the amplitude for the direct emission of the core electrons into the final state. But in the case of ionization of an atom of the matter, the secondary electron may go into the final state in a more complicated manner, namely through intermediate elastic scattering of the secondary electron by neighboring atoms.

### 3.1.1 Single Scattering.
Consider the amplitude of emission from the ionized atom when the secondary electron has undergone single elastic scattering by a neighboring atom. As a result of the interaction of the atom with the incident electron, the atom goes into the excited state characterized by a hole on the core level $|\alpha\rangle$ and a secondary electron in an intermediate state $|p'\rangle$. The secondary electron goes into the final state $|p\rangle$ as a result of elastic scattering by the $j$th neighboring atom. The amplitude of this process is given by the matrix element

$$T_j^{(1)}(p) = \sum_{|p'\rangle} \langle p|t_j|p'\rangle \frac{1}{E_p - E_{p'} + i\eta} \langle p', u|V|\alpha, w\rangle \qquad (6)$$

The process of the transition into the intermediate excited state is described by the matrix element $\langle p', u|V|\alpha, w\rangle$, i.e., the matrix element of the Coulomb electron-electron interaction. The evolution of the system in the intermediate state is described by the resolvent $(E_p - E_{p'} + i\eta)^{-1}$, where $E_p = p^2/2$ and $E_{p'} = p'2/2$ are the kinetic energies of the secondary electron in the final and intermediate states, respectively, and $i\eta$ is a nonzero imaginary addition taking into account effects of decay of the many-electron excited subsystem of the sample. The transition of the secondary electron into the final state is described by the matrix element $\langle p|t_j|p'\rangle$, where $t_j$ is the operator of elastic scattering of the secondary electron by the $j$th neighboring atom. The amplitude of this transition may be presented graphically as a diagram (Fig. 7b).

Let us turn to the MT coordinate system in which the coordinates of an electron are determined by the coordinates of the center of one of a set of nonoverlapping atomic MT spheres and the coordinates within the appropriate sphere. We also assume that the coordinates of the MT sphere of the ionized atom are referred to the origin of coordinates. Then the free Green's function, which in Eq. (6) describes the propagation of an electron from the ionized (0th) atom to the neighboring ($j$th) atom, can be presented in the following form:

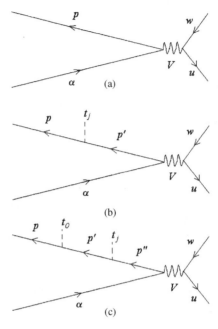

FIG. 7. Diagrams of the amplitudes of the secondary electron emission from the core level: (a) direct emission into the final state; (b) single scattering process; (c) double scattering process.

$$\sum_{|p'\rangle} |p'\rangle \frac{1}{E_p - E_{p'} + i\eta} \langle p'| = \frac{\exp(ip^+|r_1 - r_0|)}{-2\pi|r_1 - r_0|}$$

$$\approx \exp(i\,p_j \cdot \rho_1) \frac{\exp(ip^+ R_j)}{-2\pi R_j} \exp(-i\,p_j \cdot \rho_0) \tag{7}$$

where $R_j$ is the center of the MT sphere of the $j$th atom, and $\rho_0$ and $\rho_j$ are locations within the 0th and $j$th MT spheres. The approximation (7) is the well-known plane-wave approximation for a free two-site Green's function, and is a standard approximation of the EXAFS theory [1, 58, 59]. Within the framework of this approximation, the secondary electron wavefront in the region of the neighboring atom is supposed to be plane [60]. However, this approximation is valid only for atoms far away from the ionized one, whereas the most intense contributions to SEFS come from the electron scattering by the nearest atoms, where one must take into account the curvature of the electron wavefront. Curvature of the divergent electron wave can be taken into account in Eq. (7) by including an additional dependence on $R_j$ [61, 62], as in EXAFS, which finally results in the dependence of the amplitudes of secondary electron scattering by neighboring atoms on interatomic distances [63–68].

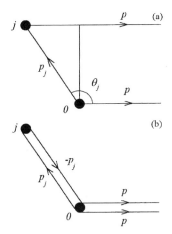

FIG. 8. The scheme of the secondary electron emission: (a) single scattering process; (b) double scattering process.

The following new designations are entered into Eq. (7): the momentum of the secondary electron, $p_j = p\hat{R}_j$, is directed toward the $j$th atom; $(p^+)^2 = (p + i\gamma)^2 \approx p^2 + 2i\eta$ is the complex wave number of the secondary electron, which takes into account attenuation of the excited state, where $p^2 + Z^*\alpha^2 = w^2 - u^2$ follows from the energy conservation law. With this approximation the amplitude of the transition [Eq. (6)] can be presented in the following form:

$$T_j^{(1)}(\boldsymbol{p}) = f_j\left(\widehat{\boldsymbol{p}, \boldsymbol{p}_j}\right)\frac{\exp(ip^+L_j)}{R_j}T^{(1)}(\boldsymbol{p}_j) \qquad (8)$$

where the matrix element $T^{(1)}(\boldsymbol{p}_j)$ describes the secondary electron creation amplitude with the momentum $p_j = p\hat{R}_j$ for the ionized (0th) atom, and $L_j = R_j(1 - \cos(\widehat{\boldsymbol{p}, \boldsymbol{p}_j}))$ is the path difference of the scattered and nonscattered electron waves. The amplitude of the secondary electron elastic scattering by the $j$th atom through an angle $\theta_j = \widehat{\boldsymbol{p}, \boldsymbol{p}_j}$ is determined by the standard expression

$$f_j(\theta_j) = \int \frac{d^3\rho}{-2\pi} \exp(-i\boldsymbol{p} \cdot \boldsymbol{\rho})t_j(\boldsymbol{\rho}) \exp(i\boldsymbol{p}_j \cdot \boldsymbol{\rho}) \qquad (9)$$

The plane-wave approximation allows the emission of the secondary electron to the final state, taking into account its single scattering by a neighboring atom, to be presented in the form of a simple scheme (Fig. 8a).

Taking into account the curvature of the wavefront of the electron emitted by the ionized atom leads to the effective dependence of the scattering amplitude on the interatomic distance. In this review we will not discuss this question in more detail, because it has already been successfully solved in EXAFS spectroscopy [63–68].

Taking into consideration the processes of secondary electron scattering, we restrict ourselves to single scattering by neighboring atoms, which is a standard approximation in obtaining the EXAFS formula [1, 2]. The applicability of such an approximation is due to a high kinetic energy (from several tens to several hundreds of electron volts) of the secondary electrons forming the extended fine structure, which makes the scattering amplitude in Eq. (9) a small parameter. The contributions to SEFS from the higher-order scattering processes constitute about 10–20% of the intensity of the process discussed above. They are more significant in the nearest fine structure region, i.e., the region of small kinetic energy of the secondary electrons.

However, the process we have considered is not sufficient to allow for all the processes of secondary electron single scattering by neighboring atoms. It is necessary besides to take into account one of the processes of the pair scattering of the secondary electron.

3.1.2 Pair Scattering.    Let us consider the scattering by the neighboring ($j$th) and ionized (0th) atoms, where the amplitude of this process is determined by the matrix element

$$T_{0j}^{(1)}(\boldsymbol{p}) = \sum_{|\boldsymbol{p}''\rangle,|\boldsymbol{p}'\rangle} \langle \boldsymbol{p}|t_0|\boldsymbol{p}''\rangle \frac{1}{E_p - E_{p''} + i\eta}\langle \boldsymbol{p}''|t_j|\boldsymbol{p}'\rangle$$
$$\times \frac{1}{E_p - E_{p'} + i\eta}\langle \boldsymbol{p}', \boldsymbol{u}|V|\alpha, \boldsymbol{w}\rangle \tag{10}$$

where all symbols of Eq. (6) are retained. A diagram of this transition is shown in Figure 7(c). Applying the plane-wave approximation to the free Green's function [Eq. (7)], we can present the matrix element (10) in the form

$$T_{0j}^{(1)}(\boldsymbol{\rho}) = ipf_0(\widehat{\boldsymbol{\rho}, -\boldsymbol{\rho}_j})\chi(p, R_j)T^{(1)}(\boldsymbol{\rho}_j) \tag{11}$$

where

$$\chi(p, R_j) = f_j(\widehat{-\boldsymbol{\rho}_j, \boldsymbol{\rho}_j})\frac{\exp(i2p^+ R_j)}{ipR_j^2} \tag{12}$$

That is, taking into account the pair scattering of the secondary electron by the neighboring and ionized atoms naturally results in the transition amplitude determined by the secondary electron backscattering by the neighboring atom. A scheme of the secondary electron emission with consideration of pair scattering by the neighboring and ionized atoms is shown in Figure 8b in comparison with the secondary electron direct emission into the final state.

Thus, the amplitude of the secondary electron emission from the core level within the secondary electron single-scattering approximation can be presented as the sum $T^{(1)}(\boldsymbol{\rho}) + T_j^{(1)}(\boldsymbol{\rho}) + T_{0j}^{(1)}(\boldsymbol{\rho})$, where the direct emission into the final state is described by the matrix element $T^{(1)}(\rho)$ [Eq. (5)] and the processes with the secondary electron scattering are described by the matrix elements $T_j^{(1)}(\rho)$ [Eq. (8)] and $T_{0j}^{(1)}(\rho)$ [Eq. (11)].

## 3.2 Emission of the Secondary Electrons from the Valence Band

The process we discussed above describes emission of the secondary electron from a core level of an atom with regard to the secondary electron single scattering by the nearest atomic neighbors of the ionized atom. Consider now the emission of the secondary electrons from the valence state of an atom, with particular emphasis on the processes forming the secondary electron spectrum above a certain CVV Auger line. The amplitude of the transition from the valence state $|\beta\rangle$ into the final state $|p\rangle$ as a consequence of its interaction with the incident electron is determined by Eq. (5) with $|\beta\rangle$ in place of $|\alpha\rangle$ in the matrix element of the transition and with $E_\beta$ in place of $E_\alpha$ in the energy conservation law. In the same manner as in Eqs. (8) and (11), the scattering of the secondary electron during its emission from the valence band can be taken into account in this process. By energy conservation, $E_w - E_u = E_p + E_\beta$, where we can put $E_w - E_u = E_p$ while considering the processes forming the secondary electron spectrum above the CVV Auger line.

However, besides the emission of the secondary electron from the valence band as a consequence of its direct excitation by an incident electron, the atom can also eject secondary electrons from the valence band upon excitation of the atomic core level.

The Auger transitions are the best example of such a process. The Auger process is a radiationless channel for annihilation of the core hole created due to ionization of the atom by the incident electron. The CVV Auger process of annihilation of the hole on the C core level occurs with the participation of two V valence electrons as a result of the Coulomb electron–electron interaction. In the final state there remain two holes in the valence band. Note that the CVV Auger process is connected with the primary process of the core level ionization through the product of the probabilities of the CVV Auger transition and of the ionization of the corresponding core level. Leaving aside the classical Auger process, in which the creation of the core hole is not of fundamental importance, we shall consider alternative processes of electron emission from the valence band.

The creation of the core hole should be taken into account in studying the process of its annihilation if the electron generated by the core level ionization is involved in the annihilation of the created hole. That is to say, the excited atom goes into an intermediate state with a core hole and a secondary electron. Next, the radiationless annihilation of the hole created occurs with the participation of the same secondary electron and an electron of the valence state. The amplitude of these two-step processes is determined by the second-order term of perturbation theory. The spectral features caused by transitions of this type are well known in atomic spectroscopy. It is precisely these transitions that are responsible for the rather intense (up to 10–20%) satellite peaks and are connected with autoioniza- tion transitions or Fano effects [69–71]. In condensed-matter spectroscopy, the Fano effects manifest themselves in a similar manner, with allowance made for the structural difference between the valence band and the atomic valence state. The Fano processes result in the appearance of low-intensity satellite peaks in the secondary electron emission spectrum that are due to the presence of an un- filled resonance (energetically localized) level in the valence band of the studied substance.

But, in addition to localized satellite peaks, the second-order autoionization processes may also form the extended features of the secondary electron spec- trum.

### 3.2.1 Emission in the Second-Order Process (Autoionization Emission).

The secondary electron spectrum on the high-energy side of the CVV Auger line is influenced, apart from the processes discussed above, by a more complicated second-order autoionization process. As a result of the Coulomb interaction with an incident electron, the atom goes into an excited intermediate state characterized by a hole on the core level $|\alpha\rangle$ and an electron in the continuous spectrum $|q'\rangle$. The transition into the final state with creation of the electron $|p\rangle$ occurs as the result of radiationless annihilation of the core hole $|\alpha\rangle$ with the participation of the valence electron $|\beta\rangle$ and the electron $|q'\rangle$ of the intermediate state. Unlike the first-order processes discussed in the previous section, the second-order process is of a threshold character, its threshold coinciding with the CVV Auger line in the secondary electron spectrum. The threshold character of the second-order process is due to the fact that the intermediate-state electron energy $E_{q'} = q'^2/2$ should not be less than the Fermi energy. The amplitude of the final electron $|p\rangle$ creation in the second-order process is determined by the matrix element

$$T^{(2)}(\rho) = \sum M(\rho, q') \frac{1}{E_q - E_{q'} + i\eta} T^{(1)}(q') \tag{13}$$

where the summation is over all intermediate states. The transition of the atom into the intermediate state is governed by the matrix element $T^{(1)}(q')$ from Eq. (8), with $|p\rangle$ replaced by $|q'\rangle$. In this process, according to energy conserva- tion, $E_p = E_q + E_\alpha = E_w - E_u$ (or $p^2 = q^2 + Z_\alpha^* \alpha^2 = w^2 - u^2$). As previously, we have neglected the binding energy of the valence electron in comparison with that of the core electron. The evolution of the system in the intermediate state is described by the resolvent $(E_q - E_{q'} + i\eta)^{-1}$, where $E_q = q^2/2$ by energy conservation $E_{q'} = q'^2/2$ is the kinetic energy of the secondary electron in the in- termediate state, and $i\eta$ describes decay of the intermediate many-electron state. As in the case of Eq. (7), after inserting $(q^+)^2 = (q + i\gamma)^2 = q^2 + 2i\eta$, the

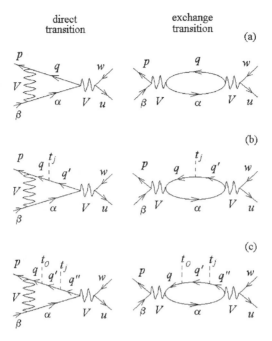

FIG. 9. Diagrams of the second-order process amplitudes: (a) without scattering in the intermediate state; with single and double scattering in the intermediate state, respectively.

resolvent in Eq. (13) can be presented in the form $2/(q^+)^2 - q'^2$. The transition of the system from the intermediate into the final state is described by the matrix element of the Coulomb electron–electron interaction:

$$M(p, q') = \langle p, \alpha | V | q', \beta \rangle - 2 \langle p, \alpha | V | \beta, q' \rangle \qquad (14)$$

where factor 2 multiplying in the exchange matrix element arises from the spin summation. The amplitude diagram of secondary electron creation in the second-order process is shown in Fig. 9a for the direct and exchange transitions.

The scattering of the final electron created due to the second-order process is taken into account in essentially the same way as in the first-order process. Thus, with $T^{(1)}(p_j)$ replaced by $T^{(2)}(p_j)$, Eq. (8) describes the amplitude of the final-electron creation in the second-order process in the approximation of final-electron single scattering by neighboring atoms.

But allowance for the second-order process gives rise to one more (low-energy) secondary electron, an intermediate-state electron, whose scattering by the local atomic structure must be taken into account. As in the case of the final electron, we restrict our consideration to single scattering by neighboring atoms.

(a) *Single scattering in the intermediate state.* On interaction with an incident electron, the atom goes into an intermediate state, and the incident electron goes into an inelastically scattered state. The intermediate-state electron that was scattered by a neighboring atom is involved, together with the valence electron, in the process of radiationless annihilation of the core hole, as a result of which the atom ejects the final electron. The amplitude of this process is expressed by the matrix element

$$T_j^{(2)}(p) = n_\alpha \int d^3q'\, M(p, q') \frac{2}{(q^+)^2 - q'^2} T_j^{(1)}(q') \qquad (15)$$

where $2n_\alpha$ is the number of electrons on the core level $|\alpha\rangle$. The relevant diagrams are shown in Figure 9b. The emission of the intermediate-state secondary electron and its subsequent scattering by a neighboring $j$th atom are determined by the matrix element $T_j^{(1)}(p)$ [Eq. (8)] with $p$ replaced by $q'$. In the plane-wave approximation for the free Green's function the amplitude of this process is given by the matrix element

$$T_j^{(1)}(q') = f_j(\widehat{q', q'_j}) \frac{\exp\{i R_j[q^+ - q' \cos(\widehat{q', q'_j})]\}}{R_j}$$
$$\times T^{(1)}(q_j) \qquad (16)$$

Then the amplitude of the second-order process with consideration for the single scattering of the intermediate-state electron in the range $qR \gg 1$ is written as

$$T^{(2)}(p) = -n_\alpha 4\pi^2 i q \chi(q, R_j) M(p, -q_j) T^{(1)}(q_j) \qquad (17)$$

where the structure-dependent term $\chi(q, R_j)$ is defined by Eq. (12) with $q$ in place of $p$.

(b) *Pair scattering in the intermediate state.* To take into complete account the processes of the single scattering of the secondary electron by neighboring atoms, in addition to the process discussed above, allowance should be made for the process of the secondary electron pair scattering by the neighboring ($j$th) and ionized (0th) atoms. The amplitude for final-electron creation in the second-order process with consideration for pair scattering in the intermediate state (the diagrams are presented in Fig. 9c) is determined by the matrix element

$$T_{0j}^{(2)}(p) = n_\alpha \int d^3q'\, M(p, q') \frac{2}{(q^+)^2 - q'^2} T_{0j}^{(1)}(q') \qquad (18)$$

where the creation of the intermediate-state electron on the ionized atom and its subsequent scattering by the neighboring and ionized atoms are described by the matrix element $T_{0j}^{(1)}(q')$, which is given by Eq. (11) with the replacement of $p$ by $q'$. Then the asymptotic behavior of Eq. (18) at $qR \gg 1$ has the form

$$T_{0j}^{(2)}(p) = -n_\alpha f_0(\widehat{q, -q_j}) 4\pi^2 iq\chi(q, R_j)$$
$$\times M(p, q)T^{(1)}(q_j) \quad (19)$$

where the amplitude of scattering by the ionized atom, $f_0(\widehat{q, -q_j})$, is determined by Eq. (9), the structure-dependent part $\chi(q, R_j)$ by Eq. (12), and the amplitudes of radiationless transitions by Eq. (15).

The validity of the estimates made for the second-order process amplitudes is determined by the condition $qR \gg 1$. For the experimental SEFS of 3d and 4d transition metals, the $qR$ value is no less than 10, i.e., the condition $qR \gg 1$ is fulfilled in a quite satisfactory way.

## 3.3 The SEFS Formula

The extended fine structure of the secondary electron spectrum is a result of secondary electron emission from an atom when the direct secondary electron emission interferes with the emission of secondary electrons scattered by the nearest neighbors of the ionized atom. In the previous section we have considered the amplitudes of these processes in the single-scattering approximation.

### 3.3.1 Intensity of the Secondary Electron Emission from the Atomic Core Level.
The intensity of the secondary electron emission from the core level of an atom is determined by the average of the squared matrix element that describes the amplitude of the first-order process. With allowance for the processes of the secondary electron single scattering by the nearest atomic neighbors of the ionized atom, the amplitude of the secondary electron emission is determined by the sum of three matrix elements, $T^{(1)}(p) + T_j^{(1)}(p) + T_{0j}^{(1)}(p)$, of which the first describes the amplitude of the secondary electron direct emission into the final state [Eq. (5)], and the second [Eq. (8)] and third [Eq. (11)] describe the amplitude of the emission into the final state through secondary electron single scattering by a neighboring atom. Then, taking into account the interference of these processes, the ionization cross section of the atom core level [Eq. (3)] becomes

$$d\sigma^{(1)}(E_p, \hat{p}) = d\sigma_{at}^{(1)}\left\{1 + 2\,\mathrm{Re}\left[\mu^{(1)}(\widehat{p, p_j})\left(f_j(\widehat{p, p_j})\right.\right.\right.$$
$$\times \frac{\exp(ip^+ L_j)}{R_j} + ipf_0(\widehat{p, -p_j})\chi(p, R_j)\left.\left.\left.\right)\right]\right\} \quad (20)$$

where the summation is over all atoms $j$, i.e., all neighbors of the ionized atom, and $d\sigma_{at}^{(1)}$ is the atomic cross section for ionization expressed by

$$d\sigma_{at}^{(1)}(E_p, \hat{p}) = 2n_\alpha \, d\sigma_0 \langle |T^{(1)}(p)|^2 \rangle_{\hat{u}} \tag{21}$$

with $d\sigma_0$ determined in Eq. (4). The effect of the nearest atomic neighbors on the emission intensity reduces to the appearance of an interference term. The intensity of the interference term is determined by the angular correlation function, written as

$$\mu^{(1)}\left(\widehat{p, p_j}\right) = \frac{\langle T^{(1)*}(p)T^{(1)}(p_j) \rangle_{\hat{p}}}{\langle T^{(1)*}(p)T^{(1)}(p) \rangle_{\hat{p}}} \tag{22}$$

The interference term in Eq. (20) determines the oscillating part of the intensity of the secondary electron emission current from the ionized atom and depends on the spatial arrangement of neighboring atoms, i.e., on their chemical species and the bond length and direction. The dependence on the bond direction leads to the sensitivity of SEFS to the angular orientation of single-crystal samples. When the studied sample is not a single crystal and/or the spectrometer energy analyzer has a large spatial collection angle, Eq. (20) can be significantly simplified by averaging over directions to neighboring atoms (averaging over a *polycrystal*).

Averaging over a polycrystal, i.e., over $\theta_j = \widehat{p, p_j}$, brings the expression for the secondary electron emission intensity [Eq. (20)] to the form:

$$d\sigma^{(1)}(E_p, \hat{p}) = d\sigma_{at}^{(1)}\left[1 + \text{Re}\left(a^{(1)}\chi(p, R_j)\right)\right] \tag{23}$$

where the following symbol is used:

$$a^{(1)} = \mu^{(1)}\left(\widehat{p_j, -p_j}\right) + 2ip\langle \mu^{(1)}\left(\widehat{p, p_j}\right) f_0\left(\widehat{p, -p_j}\right)\rangle_{\widehat{p, p_j}} \tag{24}$$

After the expansion of the angular correlation function and the scattering amplitude in terms of the Legendre polynomials

$$\mu^{(1)}\left(\widehat{p, p_j}\right) = \sum_l \mu_l^{(1)} P_l\left(\cos \widehat{p, p_j}\right) \tag{25}$$

$$f_0\left(\widehat{p, -p_j}\right) = \sum_l \frac{\exp(i2\delta_l^0)}{2ip}(-1)^l(2l+1)P_l\left(\cos \widehat{p, p_j}\right) \tag{26}$$

Eq. (24) may be presented in the form

$$a^{(1)} = \sum_l \mu_l^{(1)}(-1)^l \exp\left(i2\delta_l^0\right) \tag{27}$$

After averaging over a polycrystal, the dependence on the direction to neighboring atoms disappears from the expression for the secondary electron emission intensity. As a result, the interference term becomes EXAFS-like, i.e., is governed by backscattering of the secondary electron by neighboring atoms. Hereinafter, for simplicity, we shall use polycrystal expressions for the intensity of the secondary electron emission.

### 3.3.2 Intensity of Emission from the Valence Band.

In principle, the intensity of the electron emission from the valence state of the atom in the first-order process is determined by equations identical to those for the intensity of the emission from the core level [Eq. (23)]. The distinction lies in the matrix elements describing the atomic amplitude of this process. As mentioned above, the electron emission from the valence band may result from both the first- and the second-order processes. If the final state of the system formed as a result of these transitions is the same, these two processes must interfere. This interference is ignored in the present work. Such an approximation is justified by the fact that the final state of the system is determined by the secondary electron and the many-electron subsystem of the sample with a hole in the valence band. Neglect of the interference of the first- and second-order processes corresponds to the assumption that those processes give rise to different final states of the many-electron subsystem of the sample. Moreover, the contribution from the first-order processes of emission from the valence band is neglected in this work. The reason for that approximation is discussed in detail in Section 4. Thus, of all processes forming the spectrum of the secondary electron emission from the valence band of an atom, we shall consider only the second-order process.

The intensity of emission of the secondary electrons created in the second-order process differs fundamentally from that in the first-order process in the fact that in this case the secondary electron spectrum is determined by the scattering of both the final- and intermediate-state electrons. Then the amplitude of the second-order process in the approximation of single scattering by neighboring atoms is determined by the sum of five matrix elements, namely, the matrix element of the atomic process, two matrix elements taking account of the final-electron scattering, and two matrix elements allowing for the intermediate-state electron scattering. Then, in the polycrystal approximation the intensity of the secondary electron emission from the atom is given by the expression

$$d\sigma^{(2)}(E_p, \hat{p}) = d\sigma_{\text{at}}^{(2)}\big\{1 + \text{Re}\big[a^{(2)}\chi(p, R_j)$$
$$+ 2b\cos\phi\, e^{i\phi}\chi(q, R_j)\big]\big\} \tag{28}$$

where $d\sigma_{\text{at}}^{(2)}$ is determined by Eq. (21) with $2n_\alpha$ replaced by $2n_\beta$ and $T^{(1)}(p)$ replaced by $T^{(2)}(p)$. The relation $q^2 = p^2 - Z^*\alpha^2$ results from the energy conservation law $E_q = E_p - E_\alpha$. The intensity of the interference term deriving from scattering of the intermediate-state electron is determined by the quantity

$$b = v\big(\widehat{q_j, -q_j}\big) + 2iq\big\langle v\big(\widehat{q, q_j}\big)f_0\big(\widehat{q, -q_j}\big)\big\rangle_{\widehat{q_j, -q_j}} \tag{29}$$

depending on the angular correlation function of the intermediate state. As in the case of $a^{(l)}$ [Eq. (24)], after the expansion of $v(\widehat{\boldsymbol{q}, \boldsymbol{q}_j})$ and $f_0(\widehat{\boldsymbol{q}, \boldsymbol{q}_j})$ in terms of the Legendre polynomials [Eq. (29)], $b$ is presented in the form

$$b = \sum_l v_l (-1)^l \exp(i2\delta_l^0) \tag{30}$$

where coefficients $v_l$ are defined by the expansion of the angular correlation function

$$v(\widehat{\boldsymbol{q}, \boldsymbol{q}_j}) = \frac{\langle T^{(2)*}(\boldsymbol{p}) M(\boldsymbol{p}, \boldsymbol{q}) T^{(1)}(\boldsymbol{q}_j) \rangle_{\hat{q}}}{\langle T^{(2)*}(\boldsymbol{p}) M(\boldsymbol{p}, \boldsymbol{q}) T^{(1)}(\boldsymbol{q}) \rangle_{\hat{q}}} \tag{31}$$

in terms of the Legendre polynomials $P_l(\cos \widehat{\boldsymbol{q}, \boldsymbol{q}_j})$. Furthermore, the intensity and the phase of the interference term of the intermediate state are determined by the phase of the matrix element of the second-order process,

$$T^{(2)}(\boldsymbol{p}) = |T^{(2)}(\boldsymbol{p})| e^{i\phi(\boldsymbol{p})} \tag{32}$$

The interference term of the final state is the same as defined above with the replacement of $a^{(1)}$ by $a^{(2)}$, where $a^{(2)}$ is determined by Eq. (27) in which $\mu_l^{(2)}$ are coefficients of the expansion of the angular correlation function of the final state of the second-order process,

$$\mu^{(2)}(\widehat{\boldsymbol{p}, \boldsymbol{p}_j}) = \frac{\langle T^{(2)*}(\boldsymbol{p}) T^{(2)}(\boldsymbol{p}_j) \rangle_{\hat{p}}}{\langle T^{(2)*}(\boldsymbol{p}) T^{(2)}(\boldsymbol{p}) \rangle_{\hat{p}}} \tag{33}$$

in terms of Legendre polynomials [Eq. (25)].

### 3.3.3 The SEFS Formula.

In the region above the CVV Auger line, the current of the electron emission from the atom is determined by the flux density of the electron beam incident on the atom and the obtained cross sections for secondary electron generation. Considering the fact that in polycrystal the interference terms in Eqs. (23) and (28) do not depend on the energy or the momentum direction of the electron incident on the atom, the intensity of the electron emission current from the atom is determined by Eqs. (23) and (28) with $d\sigma_{at}^{(1)}(E_p, \hat{\boldsymbol{p}})$ replaced by $dJ_{at}^{(1)}(E_p, \hat{\boldsymbol{p}})$ and with $d\sigma_{at}^{(2)} E_p, \hat{\boldsymbol{p}})$ replaced by $dJ_{at}^{(2)}(E_p, \hat{\boldsymbol{p}})$, where according to Eq. (1),

$$\begin{aligned}
dJ_{at}^{(1)}(E_p, \hat{\boldsymbol{p}}) &= \langle j(E_w, \hat{\boldsymbol{w}}) \, d\sigma_{at}^{(1)}(E_p, \hat{\boldsymbol{p}}) \rangle \\
&= \langle j(E_w, \hat{\boldsymbol{w}}) \, 2n_\alpha \, d\sigma_0 (|T^{(1)}(\boldsymbol{p})|^2)_{\hat{u}} \rangle \\
dJ_{at}^{(2)}(E_p, \hat{\boldsymbol{p}}) &= \langle j(E_w, \hat{\boldsymbol{w}}) \, d\sigma_{at}^{(2)}(E_p, \hat{\boldsymbol{p}}) \rangle \\
&= \langle j(E_w, \hat{\boldsymbol{w}}) \, 2n_\beta \, d\sigma_0 (|T^{(2)}(\boldsymbol{p})|^2)_{\hat{u}} \rangle
\end{aligned} \tag{34}$$

In the approximation of the secondary electron single scattering by neighboring atoms with neglect of the interference of the first- and second-order processes, the intensity of the secondary electron emission current, which forms the secondary electron spectrum above the CVV Auger line, is determined by superposition of the intensities of the first- and second-order processes:

$$dJ(E_p, \hat{p}) = dJ_{\text{at}}\left[1 + \text{Re}\left(A\chi(p, R_j) + 2B\cos\phi\, e^{i\phi}\chi(q, R_j)\right)\right] \quad (35)$$

where the intensity of the atomic process of emission is given by the sum $dJ_{\text{at}} = dJ_{\text{at}}^{(1)}(E_p, \hat{p}) + dJ_{\text{at}}^{(2)}(E_p, \hat{p})$. The weight functions $A$ and $B$, which determine the intensity of the interference terms of the final and intermediate states, are written as

$$A = \frac{a^{(1)}\, dJ_{\text{at}}^{(1)}(E_p, \hat{p}) + a^{(2)}\, dJ_{\text{at}}^{(2)}(E_p, \hat{p})}{dJ_{\text{at}}^{(1)}(E_p, \hat{p}) + dJ_{\text{at}}^{(2)}(E_p, \hat{p})} \quad \text{and}$$

$$B = \frac{b\, dJ_{\text{at}}^{(2)}(E_p, \hat{p})}{dJ_{\text{at}}^{(1)}(E_p, \hat{p}) + dJ_{\text{at}}^{(2)}(E_p, \hat{p})} \quad (36)$$

where $a^{(1)}$, $a^{(2)}$, and $b$ are given by Eqs. (27) and (34) and depend on the coefficients of expansion of the angular correlation functions $\mu_l^{(1)}$, $\mu_l^{(2)}$ and $\nu_l$ and on the partial phase shifts on the ionized atom $\delta_l^0$. In the considered approximation of the polycrystalline sample, the extended fine structure of the secondary electron spectrum is governed by two EXAFS-like interference terms:

$$\chi(p, R_j) = f_j\left(\widehat{-p_j, p_j}\right)\exp(-2\gamma_p R_j)\frac{\exp(i2pR_j)}{ipR_j^2}$$

$$\chi(q, R_j) = f_j\left(\widehat{-q_j, q_j}\right)\exp(-2\gamma_q R_j)\frac{\exp(i2qR_j)}{ipR_j^2} \quad (37)$$

both of which are determined by processes of the backscattering by neighboring atoms, but differ in the wave number: $p$ is the wave number of the final electron, and $q$ is the wave number of the electron of the intermediate state. The relation between $p$ and $q$ is determined by the energy conservation law: $E_q = E_p - E_\alpha$ where $E_\alpha$ is the binding energy of the core level electron; $E_p = p^2/2$ is the kinetic energy of the final electron, which is measured in the experiment; and $E_q = q^2/2$ is the kinetic energy of the electron of the intermediate state. In Eqs. (37) $\gamma$ is the inverse mean free path of the secondary electron produced by the decay of the corresponding state. In addition, the interference term of the intermediate state depends on the phase $\phi$ of the matrix element of the second-order process [Eq. (35)].

### 3.3.4 Allowance for the Thermal Vibrations of Atoms.

Up to now, in studies of secondary electron scattering we have assumed that atom positions in the local atomic structure are fixed. In reality, atoms are constantly changing their positions relative to each other. But at room temperatures of the sample the rate of the change of the interatomic distance is much less than the interference term formation rate, which is determined by the lifetime of the many-electron excited state of the sample electron subsystem. For room temperatures of the sample the interatomic distance variation with temperature is well described by the normal (Gaussian) distribution. Then the intensity of the emission current of the secondary electrons from a substance at a finite temperature can be presented in the form

$$
\begin{aligned}
dJ(E_p, \hat{p}) = dJ_{\text{at}}\big[1 &+ \text{Re}(AW(p, T)\chi(p, R_j) \\
&+ 2B\cos\phi\, e^{i\phi} W(q, T)\chi(q, R_j))\big]
\end{aligned}
\tag{38}
$$

where $R_j$ denotes the equilibrium interatomic distances. Taking into account temperature variations gives rise to the attenuation factors $W(p, T)$ and $W(q, T)$ in the interference terms, which are analogous to the Debye–Waller diffraction factor. In the framework of the approximations made, the temperature factors are determined in the form

$$
\begin{aligned}
W(p, T) &= \exp\left(-\frac{2}{3}\langle\Delta R_j^2\rangle p^2\right), \\
W(q, T) &= \exp\left(-\frac{2}{3}\langle\Delta R_j^2\rangle q^2\right)
\end{aligned}
\tag{39}
$$

where $\langle\Delta R_j^2\rangle$ is the temperature-dependent mean squared variation of the interatomic distance.

## 4  ESTIMATION OF AMPLITUDES AND INTENSITIES OF FIRST- AND SECOND-ORDER PROCESSES

The oscillating part of the secondary electron spectrum fine structure in the expression obtained is determined by two interference terms resulting from scattering of secondary electrons of final and intermediate states (the latter are due to the second-order process only). Here intensities of oscillating terms are determined by the amplitudes and intensities of electron transitions in the atom ionized. In this section we make estimations of these values within the framework of the simple hydrogen like model using the atomic unit system as in the preceding section. This section's content is based on papers [20, 22, 29–31, 33, 35, 37, 45–47].

## 4.1 Atomic Electron Emission from a Core Level: The First-Order Process

To make an evaluation of contributions of interference terms resulting from the electron scattering in final and intermediate states it is necessary to estimate amplitudes and intensities of atomic processes leading to emission of the final electron. Considering MVV and LVV SEFS spectra of 3d transition metals that are observed on exciting a sample with an electron beam of energy of 5–10 keV, we will choose the wave functions of incident and inelastically scattered electrons as plane waves, and those of secondary electrons as plane waves orthogonalized to the appropriate core level. Let us consider the wave function of the core level electron in the following form:

$$|1s\rangle = \psi_{1s}(r) = N^{-1}\exp(-\alpha r), \qquad \text{where} \quad N^{-1} = \frac{\alpha^{3/2}}{\sqrt{\pi}} \tag{40}$$

i.e., in the form of hydrogenlike wave function of the 1s type. Then the amplitude of the secondary electron emission from the atom core level [Eq. (5)] is determined by the expression

$$T^{(1)}(\boldsymbol{p}) = \frac{\sqrt{2}\alpha^{5/2}}{\pi^3}\frac{1}{\aleph^2}\left[\frac{1}{(\alpha^2+(\boldsymbol{p}-\aleph)^2)^2} - \frac{\xi}{(\alpha^2+p^2)^2}\right] \tag{41}$$

where $\aleph = \boldsymbol{w} - \boldsymbol{u}$ is the momentum transfer. Taking into account the orthogonality of the secondary electron and the core wave function [72], the correlation value is

$$\xi = \langle\alpha|\exp(i\aleph\cdot\boldsymbol{r})|\alpha\rangle \tag{42}$$

The approximation applied is an analog of the pseudopotential one and presents all the difficulties connected with its use.

### 4.1.1 Angular Dependence of the Intensity of the Secondary Electron Emission.
Using the obtained estimate of the amplitude of the secondary electron emission from the core level excited by the incident electron, we shall estimate the dependence of the intensity of the secondary electron emission on the angle between the incident and the secondary electron momenta. Then Eq. (21) for the ionization cross section of the core level may be presented as

$$d\sigma^{(1)}(E_p, \hat{\boldsymbol{p}}) = 2n_\alpha \, d\sigma_0 \frac{1}{(\alpha^2+p^2)^6}F(\theta_p) \tag{43}$$

where $\theta_p = (\widehat{\boldsymbol{p},\boldsymbol{w}})$. The angular dependence of the emission intensity is determined by the following expression:

$$F(\theta_p) = \xi + f(\theta_p)\left[5\lambda\eta\left\langle\frac{1+t}{tX^{7/2}}\right\rangle_t + \left\langle\frac{1+t}{t^2X^{5/2}}\right\rangle_t - 2R\xi\left\langle\frac{1+t}{t^2X^{3/2}}\right\rangle_t\right] \tag{44}$$

where $\theta_p = (\widehat{p, w})$ the following quantities independent of the angle are introduced:

$$t = \frac{\aleph^2}{\alpha^2 + p^2}, \qquad \eta = \frac{2p^2}{\alpha^2 + p^2}, \qquad \xi = \left(\frac{4 - 2\eta}{4 - 2\eta + t}\right)^2 \qquad (45)$$

as well as the following quantities dependent on the angle $\theta_p$:

$$f(\theta_p) = \left(1 - \frac{p}{w}\cos\theta_p\right)R^{-5/2},$$

$$R = 1 - 2\frac{p}{w}\cos\theta_p + \frac{p^2}{w^2} \qquad (46)$$

$$X = t^2 - 2(\lambda\eta - 1)t + 1, \qquad \lambda = \frac{\sin^2\theta_p}{R} \qquad (47)$$

The function $F(\theta_p)$ reaches its maximum value at the angle

$$\theta_{max} = \arccos\frac{p}{w} \approx \frac{\pi}{2} - \frac{p}{w} \qquad (48)$$

which corresponds to the direction required by the momentum conservation law $p = \aleph = w - u$. As the kinetic energy of the secondary electron increases in comparison with the binding energy of the core level, the fulfillment of the momentum conservation law becomes more restrictive, and in the case of $p^2/\alpha^2 \to \infty$ the intensity of the secondary electron emission is given by the expression

$$d\sigma^{(1)}(E_p, \hat{p}) = 2n_\alpha \frac{4a_0^2}{w^2 p^4}\delta(\cos\theta_p - \cos\theta_{max})\,dE_p\,d\hat{p} \qquad (49)$$

Thus, in the far asymptotics ($p^2/\alpha^2 \to \infty$) the secondary electron emission is governed by the energy conservation law $p^2 + Z^*\alpha^2/2 = w^2 - u^2$, that is, in this case $p^2 = w^2 - u^2$, and by the momentum conservation law $p = w - u$. Hence $\widehat{p, w} = \pi/2$.

For the case of the secondary electron kinetic energies close to the magnitude of the core-level binding energy, the angular dependence of the intensity of the secondary electron emission [$F(\theta_p)$] is represented in Figure 10. For secondary electron kinetic energies comparable with the core-level binding energy the angular dependence of the intensity of emission is not strong, and it will be considered hereafter as isotropic. This approximation is all the more true because the momenta of electrons incident on the atom concerned are not collinear, even though the electron beam incident on the sample surface is collimated. The scattering of incident electrons when passing through condensed matter results in unfocusing the electron beam.

4.1.2 The Angular Correlation Function.    The intensity of the final-state interference term is determined by the angular correlation function $\mu^{(1)}(\widehat{p, p_j})$ of

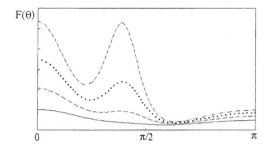

FIG. 10. Angular dependence of the secondary electron intensity with $E_p/E_\alpha$ equal to 1 (solid); 2 (dashed); 3 (dotted); 4 (dot-dashed). Here $E_w/E_\alpha = 50$ is used.

Eq. (22), or, to be more exact, it is determined by the coefficients of its expansion in terms of Legendre polynomials. Consider the angular correlation function as

$$\mu^{(1)}(\widehat{p,-p}) = \frac{\langle T^{(1)*}(p)T(-p)\rangle_{\hat{p}}}{\langle T^{(1)*}(p)T(p)\rangle_{\hat{p}}} = \sum_l \mu_l^{(1)}(-1)^l \tag{50}$$

Using an estimate of the emission amplitude [Eq. (41)], we find

$$\langle T^{(1)*}(p)T(\pm p)\rangle_{\hat{p},\aleph} = \frac{2}{\pi^6}\frac{1}{\alpha^7}\left(\frac{\alpha^2}{\alpha^2+p^2}\right)^2 F^\pm \tag{51}$$

where

$$F^+ = \left\langle \frac{1}{t^2}\left[\frac{8}{3}\frac{\eta t}{Y^3} + \left(\frac{1}{Y}-\xi\right)^2\right]\right\rangle_t \tag{52}$$

$$F^- = \left\langle \frac{1}{t^2}\left[\frac{1}{2(1+t)^2 Y} + \frac{1}{4(1+t)^3\sqrt{2\eta t}}\ln\frac{1+t+\sqrt{2\eta t}}{1+t-\sqrt{2\eta t}} - \frac{2\xi}{Y}+\xi^2\right]\right\rangle_t \tag{53}$$

Here we have made use of the variables introduced in Eq. (45) as well as

$$Y = 1 - 2(\eta - 1)t + t^2 \tag{54}$$

Thus, the angular correlation function of the backscattering is defined by the relation $\mu^{(1)}(\widehat{p,-p}) = F^-/F^+$.

To determine the coefficients of the expansion of $\mu^{(1)}(\widehat{p,p_j})$ in terms of Legendre polynomials $P_l(\cos\widehat{p,p_j})$ [Eq. (25)], we carry out the expansion of $T^{(1)}(p)$ [Eq. (41)] in terms of $2p\cdot\aleph/(\alpha^2+p^2+\aleph^2)$, resulting in the multiplet representation of the ionization amplitude. Note that in the region of the secondary electron kinetic energies that are less than or comparable to the core-level binding energy, the expansion proposed has no divergences at any momentum transfer value $\aleph$ as opposed to the expansion of the $\exp(i\,p\cdot r)$ plane wave traditionally applied in this case [4, 5, 73–77]. We restrict ourselves to the first three terms of the expansion for the ionization amplitude; then the averages of the matrix elements squared [Eq. (51)] that determine the angular correlation function $\mu^{(1)}(\widehat{p, p_j})$ may be presented in the following form:

$$\langle T^{(1)*}(p)T^{(1)}(p_j)\rangle_{\hat{p},\aleph} = \frac{2}{\pi^6}\frac{1}{\alpha^7}\left(\frac{\alpha^2}{\alpha^2+p^2}\right)^2 \tilde{F} \tag{55}$$

where

$$\tilde{F} \approx \tilde{F}_0 P_0(\cos\widehat{p, p_j}) + \tilde{F}_1 P_1(\cos\widehat{p, p_j}) + \tilde{F}_2 P_2(\cos\widehat{p, p_j}) \tag{56}$$

and the coefficients of the expansion are given by the following expressions:

$$\tilde{F}_0 = \left\langle \frac{1}{t^2}\left[ \frac{2\eta t}{(1+t)^4} + \frac{1}{(1+t)^2} - \xi \right]^2 \right\rangle_t \tag{57}$$

$$\tilde{F}_1 = \left\langle \frac{1}{t^2}\frac{8}{3}\frac{\eta t}{(1+t)^6}\right\rangle_t, \qquad \tilde{F}_2 = \left\langle \frac{1}{t^2}\frac{16}{5}\frac{\eta^2 t^2}{(1+t)^8}\right\rangle_t \tag{58}$$

In the multiplet representation the angular correlation function is determined by the expression

$$\mu^{(1)}(\widehat{p, p_j}) = \sum_l \mu_l^{(1)} P_l(\cos\widehat{p, p_j}) = \frac{\sum_l \tilde{F}_l P_l(\cos\widehat{p, p_j})}{\sum_l \tilde{F}_l} \tag{59}$$

The angular correlation functions of the backscattering obtained, provided that the kinetic energy of incident electrons is much higher than the binding energy of the core level, are given in Figure 11, both in the quadrupolar approximation

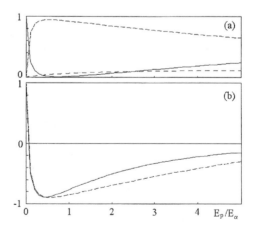

FIG. 11. (a) intensities of monopole (solid line), dipole (dashed line) and quadrupole (dot-dashed line) processes; (b) the angular correlation function: the "exact" calculation (solid line) and the quadrupole approximation (dashed line). Here $E_w/E_\alpha = 50$ is used.

[Eq. (56)] and "exactly" calculated [Eq. (52)], together with the quadrupole coefficients [Eqs. (57), (58)].

In the region of kinetic energies of the secondary electron that are much higher than the core-level binding energy, the suggested quadrupolar approximation turns out to be inadequate to describe the behavior of $\mu^{(1)}(\widehat{\boldsymbol{p}, -\boldsymbol{p}})$; because the $\delta$-function in Eq. (49) is expanded in an infinite series. This is explained by the fact that in the far range $(p^2/\alpha^2 \gg 1)$ it is necessary to take into account all the multiplet terms of the expansion. In the region of secondary electron kinetic energies comparable to the core binding energies the quadrupolar approximation closely matches the "exactly" calculated result, i.e., the dipole term $\tilde{F}_1$ in this region is dominant (Fig. 11a). In the case of low kinetic energies of the secondary electrons $(p^2/\alpha^2 \to 0)$, the obtained result is not quite correct, because of the roughness of the orthogonalization used [Eq. (42)], i.e., because the secondary electron wave function is orthogonalized to only one wave function—the wave function of the ionized atom.

For the 3d- and 4d-metal SEFS spectra the region of the experimental SEFS corresponds to the region of secondary electron kinetic energy given by $p^2/\alpha^2 \geq 1$. Then according to the preceding paragraph we can restrict ourselves to taking into account the dipole process of excitation of the atom core level by the incident electron. In this approximation the angular correlation function is determined by the following expression:

$$\mu^{(1)}(\widehat{\boldsymbol{p}, \boldsymbol{p}_j}) = P_1(\cos \widehat{\boldsymbol{p}, \boldsymbol{p}_j}) \tag{60}$$

i.e., we neglect all the coefficients of the expansion of the angular correlation function except for the dipole coefficient $\mu_1^{(1)} = 1$. In the range $p^2/\alpha^2 \geq 1$, where $\xi \approx 1$ in the dipole approximation, the matrix element of ionization of the core level is written

$$T^{(1)}(\boldsymbol{p}) = \frac{\sqrt{2}\alpha^{5/2}}{\pi^3} \frac{1}{\aleph^2} \frac{4\boldsymbol{p} \cdot \aleph}{(\alpha^2 + p^2 + \aleph^2)^3} \left[1 + O\left(\frac{2\boldsymbol{p} \cdot \aleph}{\alpha^2 + p^2 + \aleph^2}\right)\right] \tag{61}$$

Hereinafter, considering the amplitude of the secondary electron emission from the core level, we restrict ourselves to the dipole approximation [Eq. (61)] for the matrix element $T^{(1)}(\boldsymbol{p})$.

### 4.1.3 Intensity of the Atomic Emission Process.
The expressions obtained determine the atomic intensity of the secondary electron emission from the core level. But these expressions are rather cumbersome, which makes their use difficult. To describe the intensity of the emission process the following approximating expression is more conveniently used:

$$dJ_{1s}^{(1)}(E_p, \hat{\boldsymbol{p}}) = dJ(E_\alpha)\left[\frac{16p^2\alpha^6}{(\alpha^2 + p^2)^4}\right]^m \tag{62}$$

At $p^2/\alpha^2 = 1$ the intensity of the ionization process is determined by

$$dJ(E_\alpha) = \left\langle j\left(E_w, \hat{\boldsymbol{w}}\right) a_0^2 \frac{16}{3\pi} \ln\left(\frac{2E_w}{E_\alpha}\right) \frac{dE_p\, d\hat{\boldsymbol{p}}}{E_w (E_\alpha a_0^2)^2} \right\rangle \tag{63}$$

where $\langle \cdot \rangle$ denotes averaging over the incident-electron states. Equation (62) is an approximation of the intensity of the ionization process. Here the value $m = 3/2$ corresponds to the best approximation of the intensity of ionization in the range of $p^2/\alpha^2 < 1$, and this value fits the dipole approximation for the transition amplitude [Eq. (61)]. The value $m = 1$ corresponds to the best approximation in the region of secondary electron kinetic energies comparable to the core-level binding energy ($p^2/\alpha^2 \geq 1$). Furthermore, this approximation gives the correct (up to a constant factor) asymptotic behavior of the intensity of ionization at $p^2/\alpha^2 \to \infty$ [Eq. (49)].

4.1.4  Secondary Electron Emission from the $L_{2,3}$ and $M_{2,3}$ Core Levels. In calculating the emission of the secondary electrons from the 1s hydrogenlike core level, we succeeded in making analytical estimates and obtaining simple approximating expressions for the amplitude [Eq. (61)], the intensity of emission [Eq. (62)], and the angular correlation function [Eqs. (52), (53), (56)]. However in the study of the 3d metal SEFS it is essential to describe the emission of the secondary electrons from $L_{2,3}$ and $M_{2,3}$ core levels. Consider the ionization process of the $L_{2,3}$ and $M_{2,3}$ core level of the atom with the wave functions of the core level electron taken as

$$|2\mathrm{p}\rangle = \psi_{2\mathrm{p}}(\boldsymbol{r}) = \frac{\alpha \boldsymbol{r}}{\sqrt{3}} \psi_{1\mathrm{s}}(\boldsymbol{r}) \tag{64}$$

$$|3\mathrm{p}\rangle = \psi_{3\mathrm{p}}(\boldsymbol{r}) = \sqrt{\frac{8}{3}}\left(1 - \frac{\alpha \boldsymbol{r}}{2}\right) \psi_{2\mathrm{p}}(\boldsymbol{r}) \tag{65}$$

i.e., consider the ionization process within the frame of the simple hydrogenlike model. Then in the range $p^2/\alpha^2 \geq 1$ of the secondary electron kinetic energies comparable to the core-level binding energy, the amplitude of the secondary electron emission is determined by the following expressions:

$$T_{2\mathrm{p}}^{(1)}(\boldsymbol{p}) = \frac{4\alpha}{i\sqrt{3}} \frac{\boldsymbol{p}}{\alpha^2 + p^2} T_{1\mathrm{s}}^{(1)}(\boldsymbol{p}) \tag{66}$$

$$T_{3\mathrm{p}}^{(1)}(\boldsymbol{p}) = \sqrt{6}\frac{p^2 - \alpha^2}{\alpha^2 + p^2} T_{2\mathrm{p}}^{(1)}(\boldsymbol{p}) \tag{67}$$

Making an estimate of the intensity of the secondary electron emission from $L_{2,3}$ and $M_{2,3}$ levels, we use these results, namely, we neglect the angular dependence of the secondary electron emission and consider the ionization of the core level by an incident electron in the dipole approximation. Both of these approximations are quite good in the region where $p^2/\alpha^2 \geq 1$, which is just the region where experimental 3d-metal SEFS spectra are available.

The intensity of the interference term in Eq. (23) is determined by the angular correlation function $\mu^{(1)}(\widehat{\boldsymbol{p}, \boldsymbol{p}_j})$. The use of the dipole approximation for the ionization amplitude of $L_{2,3}$ and $M_{2,3}$ core levels reduces Eq. (59) to

$$\mu_{2p}^{(1)}(\widehat{\boldsymbol{p}, \boldsymbol{p}_j}) = \mu_{3p}^{(1)}(\widehat{\boldsymbol{p}, \boldsymbol{p}_j}) = \frac{1}{3}P_0(\cos\widehat{\boldsymbol{p}, \boldsymbol{p}_j}) + \frac{2}{3}P_2(\cos\widehat{\boldsymbol{p}, \boldsymbol{p}_j}) \qquad (68)$$

Then in the series expansion of the angular correlation function of the ionization of $L_{2,3}$ or $M_{2,3}$ core levels, only the coefficients $\mu_0^{(1)}$ and $\mu_2^{(1)}$ are not equal to zero. In this approximation $\mu_0^{(1)} = \frac{1}{3}$ and $\mu_2^{(1)} = \frac{2}{3}$.

The intensity of emission current from $L_{2,3}$ and $M_{2,3}$ core levels according to Eqs. (66), (67) is subject to the following expressions:

$$dJ_{2p}^{(1)}(E_p, \hat{p}) = dJ_{1s}^{(1)}(E_p, \hat{p})\left[\frac{4p\alpha}{\alpha^2 + p^2}\right]^2 \qquad (69)$$

$$dJ_{3p}^{(1)}(E_p, \hat{p}) = dJ_{2p}^{(1)}(E_p, \hat{p})\left[\frac{\alpha^2 - p^2}{\alpha^2 + p^2}\right]^2 \qquad (70)$$

The obtained expressions allow us to estimate the probability of the ionization of the atom core level by an incident electron, which is determined by the integrated intensity of the emission current of the secondary electrons [Eqs. (62), (69), (70)]. Then the probabilities of ionization of the 1s, 2p, and 3p core levels are proportional to $1.3/(\alpha a_0)^7$, $1.4/(\alpha a_0)^7$, and $1.2/(\alpha a_0)^7$, respectively, i.e., they are practically independent of the type of the core-level electron wave function. The dependences of intensities of the secondary electron emission on their relative energies [Eqs. (62), (69), (70)] for the case of the ionization of K, $L_{2,3}$, and $M_{2,3}$ core levels at $dJ(E_\alpha) = 1$ are given in Figure 12. Maxima of intensi-

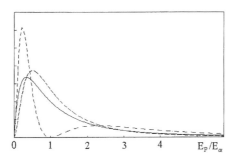

FIG. 12. The intensities of the secondary electron emission from K (solid line), $L_{2,3}$ (dashed line), and $M_{2,3}$ core levels (dot-dashed line).

ties of the electron emission from core levels are in the range of $p^2/\alpha^2 \leq 1$. The principal distinction between given dependences of intensities of the ionization of different (K, $L_{2,3}$, and $M_{2,3}$) core levels consists in the appearance for $M_{2,3}$ levels, of the sharp minimum in $dJ_{3p}^{(1)}(E_p, \hat{\boldsymbol{p}})$ at the value $p^2/\alpha^2 = 1$. This result is a consequence of the presence of a node in the radial part of the 3p wave function of the core electron [Eq.(65)]. However, in spite of the roughness of the hydrogenlike model used [78], the results obtained are qualitatively correct, i.e., they represent properly the functional behavior of atomic ionization intensities. This conclusion may be supported by the results of calculations of the generalized oscillator strengths of transitions when ionizing the Cu $L_{2,3}$ and $M_{2,3}$ levels [28], where the maxima of the ionization intensity are in good agreement with the obtained estimates and at $p^2/\alpha^2 \approx 1$ the ionization intensity of the $M_{2,3}$ core level has a sharp, well-localized minimum. Besides, the behavior of the nonoscillating parts of the 3d-metal $L_{2,3}$ and $M_{2,3}$ EELFS determined by atomic intensities of the ionization of the appropriate core level has a qualitative analogy to that given in Fig. 10; in particular, there is a minimum of the intensity of ionization in the $M_{2,3}$ EELFS nonoscillating part at $p^2/\alpha^2 \approx 1$ [54].

## 4.2 Atomic Emission of Secondary Electrons from the Valence State

The majority of secondary electrons forming the atomic spectrum are emitted from the valence band (see the estimates of the probability of ionization). However, according to the previous estimates the maximum of the intensity of emission is localized in the region of kinetic energies of the secondary electrons comparable to the binding energy of the appropriate atomic level. The intensity of the secondary electron emission is rapidly damped out in the region of kinetic energies that far exceed the binding energy of the ionized level [Eqs. (49), (62)]. Thus, although the integrated intensity of the secondary electron emission from the valence state is the most significant one, the contribution of this process to the atomic intensity of the secondary electron emission above the CVV Auger line is insignificant. The reason is that the maximum of the intensity of the secondary electron emission from the valence band falls in a region of kinetic energies much less than the energy of the CVV Auger line, and in the region of the CVV Auger line the intensity of the secondary electron current from the valence state is determined by the asymptotic form

$$dJ_{\beta}^{(1)}(E_p, \hat{\boldsymbol{p}}) = \frac{2n_\beta \, dE_p \, d\hat{\boldsymbol{p}}}{(E_p/\mathrm{Ry})^2} \left\langle \frac{j(E_w, \hat{\boldsymbol{w}})}{E_w} \right\rangle \tag{71}$$

where averaging is over the incident-electron states.

According to the estimates made, in the region of the CVV Auger line we neglect the contribution from the emission of the secondary electrons from the valence state as a consequence of its direct excitation by the electron impact (the first-order process). Thus we restrict ourselves to taking into account only the contribution of the second-order processes describing the two-step process of the secondary electron emission from the valence band.

**4.2.1 Atomic Emission of Electrons from the Valence State in the Second-Order Process.** Consider the atomic process of electron emission from the valence band on exciting the atom core level by an incident electron. As previously, the angular dependence of the intensity of emission of the final electron will be considered isotropic. To describe the emission of electrons from the valence band in the second-order process we will use hydrogenlike wave functions [Eqs. (40), (64), (65)] of an electron of the core level $|\alpha\rangle$, and the wave function of the valence-state electron will be given by

$$|\beta\rangle = \psi_\beta(\rho) = N_\beta^{-1} \exp(-\beta \cdot \rho) \tag{72}$$

where $\rho$ is the radius vector localized in the MT (atomic) sphere. The normalization of the wave function will be made with respect to one electron in the volume of the Wigner–Seitz atomic sphere with radius $\rho_0$. Under this condition the normalizing constant is determined from the following expression:

$$\begin{aligned} N_\beta^2 &= \int_{\rho \leq \rho_0} d^3\rho \exp(-2\beta \cdot \rho) \\ &= \frac{\pi}{\beta^3} \left[ 1 - \exp(-2\beta\rho_0)\left(12\beta\rho_0 + 2(\beta\rho_0)^2\right) \right] \end{aligned} \tag{73}$$

The normalization chosen corresponds to the correct normalization [Eq. (2)] of the ionization cross section of the atom. In the case of a slightly localized valence state, i.e., $\beta \to 0$, the normalizing constant is simply determined by the atomic volume: $N_\beta^2 = (4\pi/3)\rho_0^3$, i.e., by the volume of the Wigner–Seitz sphere. In the case of strong localization of the valence state the chosen wave function is merely a 1s hydrogenlike wave function with $N_\beta^2 = \pi/\beta^3$.

**4.2.1.1 Amplitude of the Core-Hole Annihilation.** The amplitude of the second-order process is determined by the amplitude of the ionization of the core level by the incident electron, $T^{(1)}(q')$, and by the amplitude of the annihilation of the core hole with the participation of an intermediate-state electron and a valence electron $|\beta\rangle$, that is, $M(p, q')$. Using the wave function of the valence electron [Eq. (72)] and the wave function of the core level [Eq. (40)], the amplitude of annihilation [Eq. (14)] is written as

$$M_{1s}(p, q') = \frac{4\alpha^{5/2}\beta^{3/2}}{\pi^2} \left[ \frac{1}{(p - q')^2} - \frac{2}{p^2} \right] \frac{1}{(\alpha^2 + (p - q')^2)^2} \tag{74}$$

Making use of Eqs. (64), (65) as the wave functions of the $L_{2,3}$ and $M_{2,3}$ core levels, the amplitude of the hole annihilation is determined by the following expressions:

$$M_{2p}(p, q') = \frac{4\alpha}{i\sqrt{3}} \frac{p - q\hat{p}}{\alpha^2 + (p-q)^2} M_{1s}(p, q') \qquad (75)$$

$$M_{3p}(p, q') = \sqrt{6} \frac{(p-q)^2 - \alpha^2}{\alpha^2 + (p-q)^2} M_{2p}(p, q') \qquad (76)$$

These expressions and those for the amplitude of ionization of the $L_{2,3}$ and $M_{2,3}$ core levels [Eqs. (66), (67)] are the approximations valid in the region $q \approx p$, i.e., for the intermediate-state electron energies comparable to or higher than the core-level binding energy ($q^2/\alpha^2 \geq 1$). Considering the process of ionization of the core level, we take into account the orthogonality of the wave function of the secondary electron to the wave function of the core level [Eq. (41), (42)]. In the study of the secondary electron emission from the valence state upon excitation of a 2p or 3p core level (the second-order process), we may restrict our consideration to the orthogonalization of the ionization amplitude only, the amplitude of the second-order process being automatically orthogonalized.

### 4.2.1.2 Amplitude of the Second-Order Atomic Process.

With the expressions obtained for the amplitudes of ionization [Eq. (61)] and annihilation [Eq. (74)] of the 1s core level, estimates of the second-order process amplitude are made. To do this we introduce a function analytical in the half semiplane:

$$F(z) = \ln \frac{p+z}{p-z} - \ln \frac{p+z+i\alpha}{p-z-i\alpha}$$
$$- \left(1 + \frac{2\alpha^2}{p^2}\right) \frac{\alpha}{2i} \left(\frac{1}{p+z+i\alpha} + \frac{1}{p-z-i\alpha}\right) \qquad (77)$$

The matrix element $T^{(2)}(p)$ is determined by the integral (13), the function $F(z)$ being an analytical continuation of the part of the integrand that does not contain any poles. Then, making use of analytical properties of the energy denominator in Eq. (13) and of the denominator in $T^{(1)}(q')$ [Eq. (61)], the matrix element of the second order is calculated analytically:

$$T_{1s}^{(2)}(p) = n_\alpha C L(p) T_{1s}^{(1)}(q\hat{p}) \qquad (78)$$

i.e., the matrix element of the second-order process is expressed through the matrix element of the first-order process (with $p$ replaced by $q\hat{p}$) and additional factors, where the constant $C$ is determined by the inverse radii of the localization of the valence and core states ($C^2 = [16/(a_0\alpha)^2]\beta^3/\alpha^3$) and the function $L(p)$ is expressed in terms of $F(z)$ as

$$L(p) = i\frac{2\alpha}{p}\left[F(q^+) - F(i\alpha) + \frac{\alpha^2 + q^2}{2\alpha}\frac{dF(i\alpha)}{d\alpha}\right] \qquad (79)$$

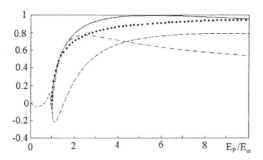

FIG. 13. The function $L(p)$: solid line for the modulus, dashed line for the imaginary part, dot-dashed line for the real part, and dotted line for the approximating function $q/p$.

The form of the real and imaginary parts of the function $L(p)$ and its modulus at $\gamma = 0$ are given in Figure 13. In the range $q^2/\alpha^2 \geq 1$ this function can be approximated to sufficient accuracy by the expression $L(p) \approx [(1+i)/\sqrt{2}]q/p = (q/p)\exp(i\pi/4)$ (see Fig. 13). The estimate of the amplitude of the second-order process is correct provided the kinetic energy of the incident electron is much greater than that of the final electron (or $w^2/\alpha^2 \gg 1$), which is fulfilled as a rule in actual SEFS experiments.

In the range of values $q^2/\alpha^2 \geq 1$ the matrix element of the second-order process, when the hole in the intermediate state is in the $L_{2,3}$ or $M_{2,3}$ level, is determined from Eqs. (66), (67) and Eqs. (75), (76) by the following expressions:

$$T_{2p}^{(2)}(\boldsymbol{p}) = n_\alpha C L(p) \frac{2\alpha}{\sqrt{3}q} T_{2p}^{(1)}(q\,\hat{\boldsymbol{p}}) \tag{80}$$

$$T_{3p}^{(2)}(\boldsymbol{p}) = n_\alpha C L(p) \frac{2\sqrt{2}\alpha}{p} T_{3p}^{(1)}(q\,\hat{\boldsymbol{p}}) \tag{81}$$

The distinction in amplitudes between two second-order processes with a hole in either intermediate state or in 1s [Eq. (78)], 2p [Eq. (80)], and 3p [Eq. (81)] core levels reduces to the appearance of insignificant additional factors in Eqs. (80), (81).

### 4.2.1.3 Estimates of the Angular Correlation Functions of the Second-Order Process.

The use of the dipole approximation for describing the core level ionization, i.e., the use of matrix elements [Eqs. (61), (66), (67)], leads to the dipole selection rules for the second-order processes too. Thus the angular correlation function of the final state [Eq. (33)] determined by $T_{2p}^{(2)}(\boldsymbol{p})$ and $T_{3p}^{(2)}(\boldsymbol{p})$ matrix elements is written as $\mu_{2p}^{(2)}(\widehat{\boldsymbol{p},\boldsymbol{p}_j}) = \mu_{3p}^{(2)}(\widehat{\boldsymbol{p},\boldsymbol{p}_j}) = P_1(\cos\widehat{\boldsymbol{p},\boldsymbol{p}_j})$, i.e., it is determined by the dipole selection rules for the final electron emission from the valence state. When making estimates of the amplitudes of the second-order processes, for simplicity we regard the wave function of the valence electron as spherically symmetric. As to the process of emission from the valence state of 3d metals, the angular correlation function of the final state in the dipole approxima- tion for the ionization of the core level is given by

$$\mu_{2p}^{(2)}\left(\widehat{\boldsymbol{p},\boldsymbol{p}_j}\right) = \mu_{3p}^{(2)}\left(\widehat{\boldsymbol{p},\boldsymbol{p}_j}\right) = \frac{2}{5}P_1\left(\cos\widehat{\boldsymbol{p},\boldsymbol{p}_j}\right) + \frac{3}{5}P_3\left(\cos\widehat{\boldsymbol{p},\boldsymbol{p}_j}\right) \qquad (82)$$

i.e., only the coefficients $\mu_1^{(2)}$ and $\mu_3^{(2)}$ of the angular correlation function expan- sion are not equal to zero. They are $\mu_1^{(2)} = \frac{2}{5}$ and $\mu_3^{(2)} = \frac{3}{5}$.

Taking into account the scattering in the intermediate state in the second-order process gives rise to an interference term whose intensity is determined by the an- gular correlation function of the intermediate state [Eq. (32)]. With the obtained amplitudes of the core level ionization and core-hole annihilation and the ampli- tude of the second-order process, we make an estimate of the angular correlation function of the intermediate state. The dipole approximation for the amplitude of core level ionization defines the angular correlation function by the following:

$$\nu_{2p}\left(\widehat{\boldsymbol{p},\boldsymbol{p}_j}\right) = \nu_{3p}\left(\widehat{\boldsymbol{p},\boldsymbol{p}_j}\right)$$
$$= \frac{1}{3}P_0\left(\cos[\widehat{\boldsymbol{p},\boldsymbol{p}_j}]\right) + \frac{2}{3}P_2\left(\cos[\widehat{\boldsymbol{p},\boldsymbol{p}_j}]\right) \qquad (83)$$

where only the coefficients $\nu_0$ and $\nu_2$ of the expansion [Eq. (32)] are not equal to zero ($\nu_0 = \frac{1}{3}$ and $\nu_2 = \frac{2}{3}$). As was to be expected, the estimate (32) is independent of the angular symmetry of the wave function of the valence state.

### 4.2.1.4 Intensity of the Second-Order Atomic Processes.

Using the am- plitudes of second-order processes [Eqs. (78), (80), (81)], we make estimates of atomic intensities of the secondary electron emission current from the valence state under excitation of the atom core level by an incident electron. In the case of the intermediate state characterized by the hole in K core level, the emission current of the secondary electrons is given by

$$dJ_{1s}^{(2)}\left(E_p, \hat{\boldsymbol{p}}\right) = n_\alpha n_\beta C^2 \left|L(p)\right|^2 \frac{q}{p} dJ_{1s}^{(1)}\left(E_q, \hat{\boldsymbol{p}}\right) \qquad (84)$$

When the intermediate state of the second-order process is defined by the holes on $L_{2,3}$ and $M_{2,3}$ levels, the intensity of the emission current of the secondary electrons in the range of $q^2/\alpha^2 \geq 1$ is determined by the following expressions:

$$dJ_{2p}^{(2)}\left(E_p, \hat{\boldsymbol{p}}\right) = n_\alpha n_\beta C^2 \left|L(p)\right|^2 \frac{4\alpha^2}{3pq} dJ_{2p}^{(1)}\left(E_q, \hat{\boldsymbol{p}}\right) \qquad (85)$$

$$dJ_{3p}^{(2)}\left(E_p, \hat{\boldsymbol{p}}\right) = n_\alpha n_\beta C^2 \left|L(p)\right|^2 \frac{24q\alpha^2}{3p^3} dJ_{1s}^{(1)}\left(E_q, \hat{\boldsymbol{p}}\right) \qquad (86)$$

Thus, the intensity of the second-order process, $dJ_\alpha^{(2)}(E_p, \hat{p})$, is determined by the core-level ionization intensity, $dJ_\alpha^{(1)}(E_p, \hat{p})$, with $E_p$ replaced by $E_q = E_p - E_\alpha$. The additional factors, $n_\alpha n_\beta C^2$ and $|L(p)|^2$, have been determined in Eq. (79). The intensities of emission of secondary electrons from $L_{2,3}$ and $M_{2,3}$ core levels in the first-order processes and the corresponding intensities of the second-order processes are shown in Fig. 14 at $n_\alpha n_\beta C^2 = 1$. Note that the intensities of these two processes are comparable, and in the case of ionization of $M_{2,3}$ core level, the intensity of the second-order process exceeds the intensity of the first-order process in some regions of the spectrum.

With the intensities obtained in Eqs. (84)–(86) we evaluate the probability of hole annihilation due to the second-order process. The probability of the ionization of the atom by an incident electron is determined by the corresponding integral intensity of the first-order process. An estimate of this probability has been made already. Then the probability of the radiationless annihilation of the electron–hole pair created in the atom upon interaction with the incident electron is determined by the ratio between the integral intensities of the first- and second-order processes, $J_\alpha^{(2)}/J_\alpha^{(1)}$. This probability is determined as $J_\alpha^{(2)}/J_\alpha^{(1)} \propto n_\alpha n_\beta C^2$, up to a constant that depends only slightly on the type of the wave function of the core electron. With the expressions obtained, we have determined the value of the constant of proportionality: 0.6, 0.4, and 0.5 for

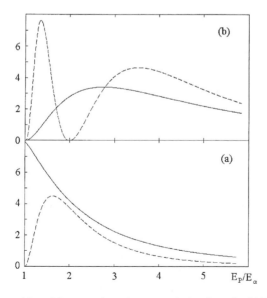

FIG. 14. The intensities of the secondary electron emission from the (a) $L_{2,3}$ and (b) $M_{2,3}$ core levels. Solid line for the first-order process, dashed line for the second-order process.

$|\alpha\rangle = |1s\rangle$, $|2p\rangle$, and $|3p\rangle$, respectively. Thus for $n_\alpha n_\beta C^2 = 1$, the probability of the core-hole annihilation involving the valence electron and the secondary electron created upon excitation of the core level is equal to about 0.5. The magnitude of the parameter

$$n_\alpha n_\beta C^2 = n_\alpha n_\beta \frac{16}{(a_0\alpha)^2} \frac{\beta^3}{\alpha^3} \tag{87}$$

is determined by characteristics of the valence and core states, namely, by the numbers of electrons in these states, $2n_\beta$ and $2n_\alpha$, and by the inverse radii of localization of atomic wave functions, $\beta$ and $\alpha$.

The characteristic feature of the 3d- and 4d-metal electronic structure is a strong localization of the wave functions of valence d-electrons on atoms. The end of the series of transition metals is characterized also by a large number of valence electrons on the atom. That is why the intensities of the second-order processes may be expected to be significant in the 3d- and 4d-metal secondary electron spectra. Considering the EFSs of the secondary electron spectrum—which are located above the low-energy CVV Auger lines (MVV in 3d and NVV in 4d metals), where the inverse radii of localization of the core-electron wave function are rather small—the values of the parameter $n_\alpha n_\beta C^2$ are expected to be sufficiently large; then the intensity of the second-order process should be comparable to the intensity of the first-order processes in the atomic spectrum of secondary electron emission.

As the core-level binding energy increases, the intensity of the second-order process decreases substantially in the atomic secondary electron spectra. However, in the secondary electron spectrum from a solid, taking into account the second-order processes may become necessary also in the high-energy region. The intensity of the interference terms is determined by the intensity of the atomic transitions as well as by the corresponding amplitudes of secondary electron scattering and by the Debye–Waller factors. As a result, in the secondary electron spectra above the high-energy CVV Auger lines, fine structure formed by scattering in the intermediate state of the second-order process may be observed, as discussed in detail in the next section.

## 5 DESCRIPTION OF THE CHARACTERISTIC FEATURES OF THE SEFS SPECTRA OF THE SECONDARY ELECTRONS

The theoretical description of the SEFS of the secondary electron spectrum (Sec. 3) shows that this structure is formed by two oscillating terms determined by two wave numbers. The existence of these two terms is a characteristic feature of SEFS as compared to the EXAFS and other EXAFS-like phenomena. That is, allowance for this feature of SEFS spectra should explain the characteristic experimental features of SEFS that make it distinct from other EXAFS-like phenomena. Moreover just these features may be used for the estimation of the proper behavior of the theoretical description made and analytical expressions obtained.

The extended fine structure of the secondary electron spectrum, though generally similar to other EXAFS-like phenomena, has a number of essential peculiarities. All these peculiarities are a consequence of the characteristic features of the processes that form the fine structure of the secondary electron spectrum.

Thus, the dependence of the SEFS spectrum on the angular orientation of the sample with a monocrystal surface in angle-resolved experiments is a consequence of the influence on electron scattering of the local atomic structure oriented in a specific manner. The angular orientation of the atomic surface plane of the sample can be taken into account if in Eq. (38), instead of $\chi(p, R_j)$, an interference term of the type of Eq. (20) is used. The same approach can be used in the calculation of the FS of the spectrum of secondary electrons from monocrystal surfaces and of their angular dependence. In the present work we have restricted ourselves to the simplest case, namely, the secondary electron emission from a polycrystal. As a result, there is no angular dependence of the SEFS spectrum in Eq. (38).

The strong dependence of the SEFS spectrum on the valence band structure is explained by the essential contribution to the SEFS spectrum of emission with the participation of the valence electron. Taking into account these processes (such as the second-order process in the present work) gives rise to an interference term whose intensity depends on the valence band structure [see, for example, Eq. (87)]. As a consequence of the change of the relative weight of the interference terms in Eq. (38) that are due to the electron transitions involving valence electrons, the result of the superposition of different channels forming the SEFS spectrum can be substantially changed.

In the present section we consider in detail those features of the SEFS spectrum that demonstrate most obviously the importance of allowing for the two oscillations resulting from the scattering in the final and intermediate states, the latter appearing only if the autoionization second-order process is taken into account.

## 5.1 Temperature Dependence of the SEFS Spectrum

The strong temperature dependence of the extended fine structure of the secondary electron spectrum above the 3d-metal MVV Auger line gave impetus to the study of such structures. Variations in the sample temperature within small ranges result in significant changes of the SEFS spectrum, namely, changes of the relative intensity of peaks in the fine structure. It has been shown that they are not connected with the changes in the atomic structure of the surface. Thus, the temperature dependence of the SEFS spectrum is determined especially by the effect of the spectroscopic Debye–Waller factors.

In EXAFS experiments as well as in other EXAFS-like methods, variations in the sample temperature are well described by the Debye–Waller factor and lead to the exponential attenuation of fine structure when the sample temperature increases. The temperature dependence of the SEFS spectrum is also described by the Debye–Waller factor—more exactly, by two Debye–Waller factors corresponding to the interference terms of the final and intermediate states [Eq. (38)]. Since these interference terms are determined by different wave numbers, $p$ and $q$, the change of the sample temperature results in a change of the relative intensity of the oscillating terms, which reveals itself in the unusual dependence behavior of SEFS.

Making use of the SEFS formula obtained [Eq. (38)] and the estimates of the atomic electron transition intensities [Eqs. (70), (78)], we calculate the temperature behavior of the Fe, Ni, and Cu MVV SEFS spectra and compare the obtained results with the experimental data.

For convenience we adduce once more the expression [Eq. (38)] that determines the SEFS spectrum:

$$dJ(E_p, \hat{p}) = dJ_{at}\big[1 + \text{Re}\big(AW(p,T)\chi(p,R_j) \\ + 2B\cos\phi\, e^{i\phi}W(q,T)\chi(q,R_j)\big)\big] \tag{88}$$

The interference terms $\chi(p, R_j)$ and $\chi(q, R_j)$ [Eq. (37)], which describe the oscillating part of the secondary electron spectrum, depend not only on interatomic distances but also on the different wave numbers, $p$ and $q$. The relation between the wave numbers is governed by the energy conservation law $E_q = E_p - E_\alpha$ or $q^2 = p^2 - Z^*\alpha^2$, where $\alpha$ is the inverse radius of localization of the core-electron wave function, and $Z^*$ is an effective charge. Considering the MVV SEFS spectra, for simplicity we shall take $Z^* = 1$, i.e., $E_\alpha = \alpha^2/2$ in the atomic system of units. To describe the temperature behavior of the SEFS spectrum we simplify the calculated expression [Eq. (88)] as much as possible. In the first place, we will restrict ourselves to taking into account only the contributions from the nearest neighbors. In the second place, when considering the weight functions $A$ and $B$, we assume that the ionization of the $M_{2,3}$ core level is determined by the dipole approximation. Moreover, to simplify the calculated expression we consider that the partial phase shifts of electron scattering by the ionized atom are equal to zero, since they differ little from each other and vary slowly compared to other phase functions in Eq. (88). Then, in Eq. (38), $a^{(1)} = -a^{(2)} = b$ and the weight functions are written as

$$A = \frac{dJ_{at}^{(1)}(E_p, \hat{p}) - dJ_{at}^{(2)}(E_p, \hat{p})}{dJ_{at}^{(1)}(E_p, \hat{p}) + dJ_{at}^{(2)}(E_p, \hat{p})},$$
$$B = \frac{dJ_{at}^{(2)}(E_p, \hat{p})}{dJ_{at}^{(1)}(E_p, \hat{p}) + dJ_{at}^{(2)}(E_p, \hat{p})} \tag{89}$$

The dependence of the interference terms on the sample temperature is determined by the Debye–Waller factors, which in turn depending on the mean squared change of the interatomic distance, $\langle \Delta R^2 \rangle$. In the range of room temperatures we consider, it may be supposed that $\langle \Delta R^2 \rangle \propto T$, i.e., the mean squared change of the interatomic distance is proportional to the sample temperature. So the change of the sample temperature from $T_0$ to $T$ leads up to change of the Debye–Waller factor values by factors $[W(p^2, T_0)]^{T/T_0}$ and $[W(q^2, T_0)]^{T/T_0}$ for the $p$- and $q$-type oscillations, respectively. The presence of two oscillating terms, i.e., two Debye–Waller factors in the SEFS formula [Eq. (88)], results in the change of the intensity of the $p$-oscillations as related to the intensity of the $q$- oscillation by the factor $[W(p^2, T_0)/W(q^2, T_0)]^{T/T_0} = [W(p^2 - q^2, T_0)]^{T/T_0} = [W(\alpha^2, T_0)]^{T/T_0}$.

The mean squared change of the interatomic distance is determined by the mean squared deviations of atoms from the equilibrium positions and by the correlation of vibrations of neighboring atoms. In the range of room temperatures the coefficient of proportionality, $\langle \Delta r^2 \rangle / T$, between the value of the mean squared deviation of atoms from the equilibrium position, $\langle \Delta r^2 \rangle$, and the sample temperature $T$ has the following values in units of $10^{-5}$ Å²/K: 5.61 (Fe), 6.22 (Ni), and 7.16 (Cu) for the bulk of the solids [79]. The SEFS spectrum under consideration is formed in superthin surface layers of the sample; therefore we shall consider that $\langle \Delta R^2 \rangle = 2\langle \Delta r^2 \rangle$, i.e., the mean squared change of the interatomic distance in superthin surface layers is twice its value in the bulk. The last approximation does not seem unreasonable if we remember that the interatomic interaction on the surface differs essentially from that in the bulk. It results in a stronger temperature dependence of the amplitude and of the nature (the role of anharmonicities) of oscillations of surface atoms [80–88]. Then the change of the sample temperature from the room value $T_0 = 300$ K to $T = 800$ K results in a change of the $p$-oscillation intensity relative to the $q$-oscillation intensity by the factor $[W(\alpha^2, T_0)]^{T/T_0} = 0.41, 0.30$, and $0.36$ for Fe, Ni, and Cu, respectively.

Within the frame of the approximations made for the weight functions $A$ and $B$ [Eq. (89)], using the estimates of the intensities of atomic electron transitions, $dJ_{at}^{(1)}(E_p, \hat{p})$ [Eq. (62)] and $dJ_{at}^{(2)}(E_p, \hat{p})$ [Eq. (85)] as well as the secondary electron scattering parameters calculated by FEFF7 programs [63–68], we have calculated the oscillating parts (the interference term in Eq. (88)) of the Fe, Ni, and Cu MVV SEFS spectra for two temperatures, $T_0 = 300$ K and $T = 800$ K. In the calculation, the corresponding crystallographic bulk values of the nearest interatomic distances $R_1$ (2.48 Å for Fe, 2.49 Å for Ni, and 2.56 Å for Cu) have been used. The rest of the parameters used in the calculation that determine the intensity of atomic electron transitions, $dJ_{at}^{(1)}(E_p, \hat{p})$ [Eq. (62)] and $dJ_{at}^{(2)}(E_p, \hat{p})$ [Eq. (85)], i.e., the weight functions $A$ and $B$, are given in Table I along with the $W(\alpha^2, T_0)$ values. Estimations of the parameters $n_\alpha n_\beta C^2$ and $\varphi$ have been made on the basis of Eqs. (87) and (79), respectively.

TABLE I.    The Parameters Used for the Calculations of the Temperature
Dependence of the MVV SEFS ($T_0 = 300$ K)

| Element | $2n_\alpha$ | $\alpha^2 a_0^2$ | $2n_\beta$ | $\beta^2 a_0^2$ | $n_\alpha n_\beta C^2$ | $\varphi$ | $W(\alpha^2, T_0)$ |
|---------|-----|-----|-----|-----|------|--------|------|
| Fe | 6 | 4 | 8 | 0.5 | 2.06 | $\pi/4$ | 0.72 |
| Ni | 6 | 5 | 10 | 0.5 | 1.59 | $\pi/4$ | 0.64 |
| Cu | 6 | 5.5 | 11 | 0.5 | 1.36 | $\pi/4$ | 0.68 |

In the calculation, the difference $p^2 - q^2$ determining the relation between the wave numbers of the final- and intermediate-state electrons has served as a fitting parameter. This quantity was taken as $p^2 - q^2 = (1 - 0.05)\alpha^2$ for Fe, $p^2 - q^2 = (1 - 0.07)\alpha^2$ for Ni, and $p^2 - q^2 = (1 - 0.07)\alpha^2$ for Cu; here $\alpha^2 = 2E_{M_{2,3}}$.

A comparison between the calculated results and the corresponding oscillating structures is given in Figure 15 for Fe; Figure 16 for Ni; and Figure 17 for Cu as a plot against the measured electron kinetic energy expressed in units of the core-level binding energy, $E_p/E_\alpha$, which corresponds, in the approximation made ($E_\alpha = \alpha^2/2$ in the atomic system of units) to $p^2/\alpha^2$. In the comparison, we have assumed that the threshold for the appearance of the second-order process, $p^2/\alpha^2 = 1$, corresponds to the position of the $M_{2,3}$VV Auger line of the experimental spectrum.

Under change of the sample temperature from 300 to 800 K, the experimental spectra change importantly in the range about $p^2/\alpha^2 = 2$. Thus with increasing sample temperature, the peak splitting disappears in this range in the oscillating parts of the Ni (Fig. 16) and Cu (Fig. 17) MVV SEFS spectra, while it appears in the oscillating parts of the Fe (Fig. 15) spectra. Such a temperature dependence of the oscillating parts is not characteristic of any other EXAFS-like phenomena, and the result on the $T$-dependence of the Fe MVV SEFS spectra exhibits an anomalous behavior, since with increasing the sample temperature the spectrum structure becomes richer.

A comparison of the calculated and experimental SEFS spectra (Figs. 15, 16, 17) shows quite good agreement between them; in particular, the calculated SEFSs reproduce properly the temperature behavior of the experimental results. The region of the secondary electron spectrum about $p^2/\alpha^2 = 2$, which corresponds to one of the largest temperature changes in SEFS is the range where the intensity of the atomic second-order process, $dJ_{at}^{(2)}(E_p, \hat{\boldsymbol{p}})$ [Eq. (88)], has a sharp minimum. By virtue of this fact, the weight functions $A$ and $B$ that determine the amplitudes of $p$ and $q$ oscillations have an essentially nonmonotonic behavior. Thus the weight function $B$ has a sharp minimum, and in the case that the parameter $n_\alpha n_\beta C^2$ is more than unity, the weight function $A$ changes its sign

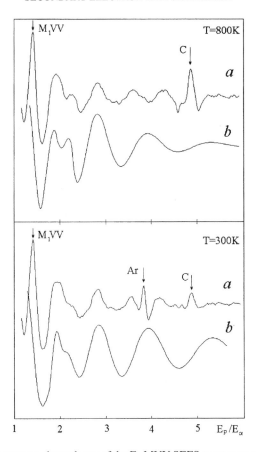

FIG. 15. The temperature dependence of the Fe MVV SEFS spectrum: $a$, the experimental results; $b$, the calculated results.

twice. The form of the weight functions $A$ and $B$ calculated at the parameter value $n_\alpha n_\beta C^2 = 2$ is presented in Figure 18. The essentially nonmonotonic behavior in the neighborhood of $p^2/\alpha^2 = 2$ and the change of the relative intensity of $p$ and $q$ oscillations (due to the effect of the Debye–Waller factors) explain the most gross changes in fine structure with the change of temperature. This manifests itself in the calculated SEFS also, which adequately reflects the experimental results. The qualitative difference in behavior between the Ni and Cu MVV SEFS spectra and the Fe one is a consequence of the fact that in the Ni and Cu spectral regions where the splitting of the peak disappears with increasing temperature, the $p$ and $q$ oscillations add in phase, while in the Fe spectrum they add in antiphase, which leads to the splitting of the peak with increasing temperature. Ultimately

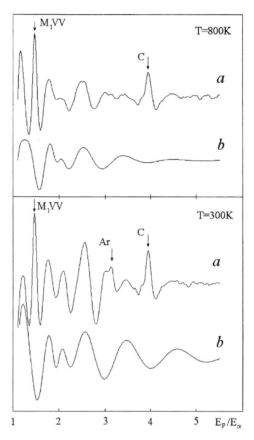

FIG. 16. The temperature dependence of the Ni MVV SEFS spectrum: *a*, the experimental results; *b*, the calculated results.

this difference in behavior between the Ni and Cu and the Fe MVV SEFS is a consequence of the different $M_{2,3}$ core-level binding energies of these elements, since it is just this quantity that determines the difference in energy between the wave number values $p$ and $q$ ($E_p - E_q = E_\alpha$) and consequently the difference in phase between $p$ and $q$ oscillations.

The calculated MVV SEFS spectra represent adequately the temperature dependence of the corresponding experimental SEFS spectra. However, the correspondence between the calculated and experimental fine structure is, at best, semiquantitative. The discrepancy between them may be caused mainly by the roughness of the approximations used in the calculation. In the calculation of the temperature dependence of the SEFS spectrum, we have greatly simplified

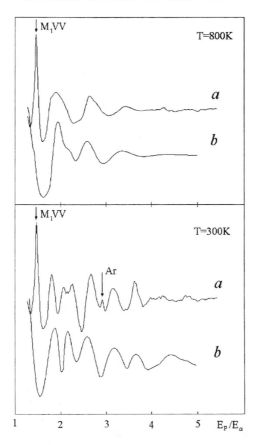

FIG. 17. The temperature dependence of the Cu MVV SEFS spectrum: $a$, the experimental results; $b$, the calculated results.

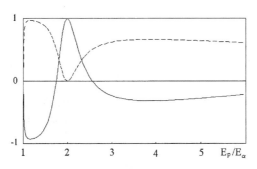

FIG. 18. The weight functions: solid line, $A$; dashed line, $B$. Here $n_\alpha n_\beta C^2 = 2$ is used.

the expressions that describe the oscillations resulting from the scattering of the intermediate- and final-state electrons. Thus in spite of a qualitatively correct description of the temperature dependence of the MVV SEFS spectra, the approximations used are unlikely to be appropriate for the quantitative analysis of the local atomic structure of the samples under study. Nevertheless, the result obtained demonstrates clearly the presence of oscillations of two types in SEFS spectra, i.e., the contributions of both the first- and second-order processes.

One more qualitative verification of the necessity of allowance for the second-order processes in the formation of fine structure above the CVV Auger lines is the oscillating structures above the high-energy Auger lines.

## 5.2 SEFS above the High-Energy CVV Auger Lines

Considering the atomic processes secondary electron emission, we have concluded that an additional threshold process can exist in the electron emission spectrum as a result of a second-order process. In this case, if there is no resonance level in the atomic spectrum of the secondary electron emission, the second-order processes are not practically detectable, since the only characteristic feature of the second-order nonresonance processes, the threshold of their appearance, coincides energetically with the CVV Auger line. However, the second-order processes may turn out to be important in principle in describing the secondary electron spectrum.

In the previous section, we have considered the effect of the scattering of both intermediate- and final-state electrons on the temperature behavior of fine structure in the low-energy region of the secondary electron spectrum. The temperature attenuation of the intermediate- and final-state oscillations becomes all the more evident in the fine structure of the high-energy Auger transition spectra. Unlike the MVV and NVV SEFS spectra of 3d and 4d metals, in the case of the fine structure located above the high-energy Auger lines, the periods of oscillation of intermediate and final states differ considerably, since the wave numbers of $p$ (final) and $q$ (intermediate) electrons differ essentially. The ratio of $T_q$, the period of the intermediate-state oscillation (the $q$-oscillation), to $T_p$, the period of the final-state oscillation (the $p$-oscillation) may be evaluated by the following expression:

$$\frac{T_q}{T_p} = \frac{q}{p} = \sqrt{1 - \frac{E_\alpha}{E_p}} \tag{90}$$

This ratio depends on the value of the binding energy of the core level involved in the second-order process. Thus for the LVV (3d metals) and MVV (4d metals) SEFS spectra, the ratio of the periods of $q$ and $p$ oscillations is about 0.3 and 0.5, respectively, for $E_p = E_\alpha + 100$ eV.

The idea of the correspondence of the low-intensity structure above the Ag MVV Auger series to EXAFS-like phenomena was advanced for the first time in

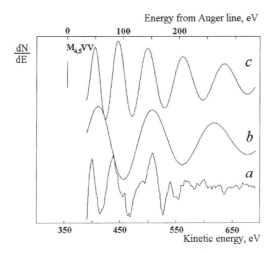

FIG. 19. The Ag MVV SEFS spectrum: the experimental result (curve *a*) in comparison with the oscillations of *p* type (curve *b*) and *q* type (curve *c*).

[16]. The reason for drawing such a conclusion was the temperature dependence of these structures and their characteristic period. In fact, the comparison between the Ag MVV SEFS spectrum and the corresponding calculated oscillations of *q* and *p* types (Fig. 19) shows that the calculated oscillations of *q* type agree well (as to the oscillation period) with the experimental structure. Note that similar structures of low intensity are also observed in the corresponding spectra of other 4d metals [51].

Later, the Fe and Ni LVV SEFS spectra were obtained. A comparison between the experimental oscillating structures of Fe and Ni LVV SEFS spectra and the corresponding calculated oscillations of *p* and *q* types is given in Figure 20. Due to the marked difference between the oscillation periods, it is evident that the oscillating structure of spectra under study is determined by the *q*-oscillation, i.e., by the intermediate-electron coherent scattering in the second-order process.

At first glance, the obtained result seems contradictory to the estimations made in Section 4. There the estimates of the intensities of the first- and second-order processes of the atomic secondary electron emission show that, as the binding energy of the core electron increases, the relative intensity of the second-order process decreases. Here the ratio between the first- and second-order process intensities $(dJ^{(2)}/dJ^{(1)} \approx n_\alpha n_\beta C^2)$ may be estimated as $n_\alpha n_\beta C^2 = n_\alpha n_\beta (16/\alpha^2 a_0^2) \beta^3 / \alpha^3$, a parameter determined by the inverse localization radii of the wave functions, $\alpha$ and $\beta$, and the numbers of electrons, $2n_\alpha$ and $2n_\beta$, in the core level and the valence state. In the study of the processes that determine the SE spectrum above the MVV (Ag) and LVV (Fe and Ni) Auger lines in the atomic

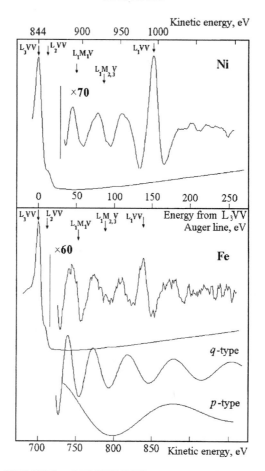

FIG. 20. The Fe LVV SEFS and Ni LVV SEFS spectra: experimental results in comparison with oscillations of $p$ and $q$ type.

spectrum of the secondary electron emission, the second-order processes can be neglected (estimates of the parameter $n_\alpha n_\beta C^2$ are given in Table II). However the intensities of interference terms of the final and intermediate states that result in the fine structure of the secondary electron emission from a substance are determined not only by the atomic electron transition intensities but by other parameters, namely, the amplitudes of the secondary electron scattering by atoms and the values of the corresponding Debye–Waller factors. Then with increasing the core-level binding energy, the intensities of atomic electron transitions decrease the relative weight of the $q$-oscillation (the oscillation caused by scattering in the intermediate state of the second-order process), while the scattering amplitudes

TABLE II.    The Parameters Used for the Calculations of the Fe and Ni LVV
SEFS and Ag MVV SEFS ($T_0 = 300$ K)

| Element | $2n_\alpha$ | $\alpha^2 a_0^2$ | $2n_\beta$ | $\beta^2 a_0^2$ | $n_\alpha n_\beta C^2$ | $\varphi$ | $W(\alpha^2, T_0)$ |
|---------|-------------|------------------|------------|-----------------|------------------------|-----------|--------------------|
| Fe      | 6           | 53               | 8          | 0.5             | 0.0034                 | $\pi/4$   | 0.0147             |
| Ni      | 6           | 64               | 10         | 0.5             | 0.0027                 | $\pi/4$   | 0.0036             |
| Ag      | 10          | 26               | 11         | 0.5             | 0.0041                 | $\pi/4$   | 0.0152             |

and the Debye–Waller factors, on the contrary, enhance the relative intensity of this oscillation. As a result the ratio between the intensities of $q$ and $p$ oscillations can be estimated by the following expression:

$$n_\alpha n_\beta C^2 \cos\phi \frac{W(q^2, T)|f_j(-\widehat{\boldsymbol{q}_j, \boldsymbol{q}_j})|/q R_j^2}{W(p^2, T)|f_j(-\widehat{\boldsymbol{p}_j, \boldsymbol{p}_j})|/p R_j^2}$$
$$= n_\alpha n_\beta C^2 \cos\phi\, W^{-1}(p^2 - q^2, T) \frac{p}{q} \frac{|f_j(-\widehat{\boldsymbol{q}_j, \boldsymbol{q}_j})|}{|f_j(-\widehat{\boldsymbol{p}_j, \boldsymbol{p}_j})|} \qquad (91)$$

Then with the use of the electron backscattering amplitudes calculated by FEFF7 and the estimates of the Debye–Waller factor made in the same approximations as in Section 5.1 (see Table II), the ratio between the intensities of $q$ and $p$ oscillations has the following values: 2.7 for Fe LVV SEFS, 12.3 for Ni LVV SEFS, and 2.1 for Ag MVV SEFS. So if there is fine structure in the secondary electron spectrum above the high-energy Auger line, it is determined by the oscillation resulting from the scattering in the intermediate state in the second-order process, in spite of the low atomic intensity of this process.

All the comments we have made concern the secondary electron spectra obtained from the samples at room temperature. A change of the sample temperature leads to changes of both the absolute intensity of the $q$-oscillation and the intensity of this oscillation relative to the intensity of the $p$-oscillation. Increase of the sample temperature causes the relative intensity of the $q$-oscillation to increase, at the same time its absolute intensity decreases. This may result in reduction of the fine structure to the level of the experimental spectrum noise because of the extremely low intensity of the structures under study. So the increase of the sample temperature up to 500 K doesn't allow us to extract the SEFS from the experimental spectrum noise in the Fe and Ni secondary electron spectrum above the LVV Auger line even with $10^6$ experimental statistics. The increase of the core-level binding energy also leads to an increase of the relative intensity of the $q$-oscillation simultaneously with a decrease of its absolute intensity. Thus, at room temperature we have failed to extract fine structure above the Cu LVV Auger line with $10^6$ statistics.

The experimental results obtained need further discussion. The intensity of the fine structure under study is extremely low: it is about two orders of magnitude lower than that of the main Auger line. Thus the ratio of the structure intensity to the intensity of the LVV Auger line for Fe and Ni LVV SEFS spectra is 1/60 and 1/70, respectively; for Ag MVV SEFS it is 1/30. Therefore, the SEFS obtained with $10^6$ statistics is only a few times greater than the noise amplitude of the experimental spectrum. Besides, low-intensity Auger lines resulting from the radiationless annihilation of a core hole (L for Fe and Ni; M for Ag) may exist in the spectrum region studied. The location of the $L_1VV$ Auger line in the Fe and Ni spectra, as well as the energetic positions of the rest of the Auger lines, are presented in Figure. 20. The conclusion that the fine structure under study is an EXAFS-like phenomenon and not a sequence of low-intensity Auger series is supported by good agreement between all three presented SEFS spectra if they are superimposed in reference to the position of the main Auger line (LVV for Fe and Ni, and MVV for Ag) on the energy scale (Figs. 19 and 20). This agreement is to be expected in the study of the EXAFS-like mechanism of formation of the spectrum fine structure, since the period of the main (the most intense) oscillation is determined by the wave number of the intermediate-state electron, whose zero corresponds to the position of the main Auger line, and by the nearest interatomic distance, which differs little in Fe, Ni, and Ag. At the same time, coincidence of the Auger line positions of different elements appears improbable. Besides, the disappearance of the fine structure into the experimental noise with increasing sample temperature is explained by the EXAFS-like nature of the oscillating structure and cannot be explained within the framework of Auger transitions. Furthermore, if we assume the fine structure to be an Auger series, it cannot be explained why it exists in Fe and Ni and not in Cu. From these qualitative experimental data it may be deduced that SEFSs above the high-energy Auger lines result from the EXAFS-like scattering in the intermediate state of the second-order process.

## 6 ANALYSIS OF THE SURFACE LOCAL ATOMIC STRUCTURE BY THE SEFS METHOD

Like other EXAFS-like spectral features, the SEFS spectrum can be used for the experimental analysis of the local atomic structure. The implementation of extended spectral fine structure as a method of structural analysis came into use in EXAFS spectroscopy nearly 40 years ago. Nowadays EXAFS spectroscopy is a well-developed and intensively used technique for structural analysis. By contrast, the EELFS and SEFS methods are less developed and hence less employed, though the structural information the SEFS contains is often unique, especially for thin surface layers of condensed matter.

As mentioned in the Introduction, there are two approaches to obtaining information on the local atomic structure from SEFS spectra in both EXAFS and EXAFS-like methods. The first approach is based on the calculation of the oscillating structure of the spectrum for a local atomic configuration and subsequent comparison between the obtained result and the experimental data. The second approach consist in solving the inverse problem to obtain the atomic pair correlation function (PCF) from the experimental oscillating part of a spectrum.

Both these approaches were first developed as applied to EXAFS spectra. By now the use of these techniques in EXAFS spectroscopy makes it possible to obtain parameters of the local atomic structure with a high degree of reliability. Both methods have their advantages and disadvantages. The joint use of these complementary techniques allows one to diminish the disadvantages of the individual methods. The use of both methods, however, calls for a detailed theoretical description of SEFS formation.

In the present work, considering the possibilities of the SEFS methods for analysis of the local atomic structure of surface layers of matter, we dwell on the procedures of solving the inverse problem only. The choice of this approach derives mainly from the fact that solution of the inverse problem is in essence a phase analysis of the oscillating structure, i.e., the method of solving the inverse problem is more sensitive to phase parameters of the SEFS under study than to its amplitude characteristics, and the main characteristics of a local atomic structure are interatomic distances, which determine the period and the phase of the oscillation. As for the methods of the direct SEFS calculation, they are more sensitive to correct description of the amplitude part and less sensitive to the phase part of oscillations. In the previous section, in the calculation of SEFS, this made it possible to obtain not only qualitative agreement between calculation and experiment (the $T$-dependence and the character of the fine structure above the high-energy Auger lines), but quite good agreement between the SEFS themselves, in spite of very rough approximations made for the phase parameters of oscillations.

## 6.1 Formalism of the Inverse Problem in the SEFS Method

In Section 3 the formation of the SEFS spectrum was described in the approximation of single scattering of secondary electrons by the nearest atoms to the ionized atom. In the framework of this approximation, the local atomic surroundings of the ionized atom are entirely described by the atomic pair correlation function $g(r)$, which determines the probability of detecting an atom of a specific chemical species at a distance $r$ from the ionized (central) atom (also of a specific chemical species). In the present work we restrict our consideration to one-component systems; thus, the PCF $g(r)$ has no indices denoting the chemical species.

Taking into account the secondary electron scattering in the final and intermediate states [see Eq. (38)], the atomic pair correlation function is related to the oscillating part of the secondary electron spectrum by the following integral equation:

$$4\pi\rho_0 \int_0^\infty dr\, r^2 \operatorname{Re}\big[AW\big(p^2 - q^2, T\big)\chi(p, r)$$

$$+ 2B\cos\phi\, e^{i\phi}\chi(q, r)\big]g(r) = I - I_{\mathrm{bg}} \tag{92}$$

where $\rho_0$ is the atomic density of the material. Note that the atomic pair correlation function in Eq. (92) involves the thermal disorder of interatomic distances. The interference terms $\chi(p, r)$ and $\chi(q, r)$, which determine the oscillations resulting from the backscattering of secondary electrons in the final and intermediate states, are given by Eq. (37) and depend on the wave numbers $p$ and $q$. Recall that

$$\chi(p, r) = f(p, \pi)\exp(-2\gamma_p r)\frac{\exp(i2pr)}{ipr^2}$$

$$\chi(q, r) = f(q, \pi)\exp(-2\gamma_q r)\frac{\exp(i2qr)}{ipr^2} \tag{93}$$

The relation between the wave numbers of the final-state electron $(p)$ and the intermediate-state electron $(q)$ is given by the energy conservation law: $E_q = E_p - E_\alpha$, where $E_\alpha$ is the binding energy of the core electron. As before, we consider the relation between the energies and wave numbers of the secondary electrons to be determined by the free-electron dispersion law, i.e., $p = [(2m/\hbar^2)E_p]^{1/2}$ and $q = [(2m/\hbar^2)E_q]^{1/2}$.

The oscillating part of the experimental spectrum $(I - I_{\mathrm{bg}})$ is the difference between the experimental spectrum itself $(I)$ and its nonoscillating component $(I_{\mathrm{bg}})$, i.e., the background. In the EXAFS method it is the practice to present the oscillating part on the scale of wave numbers. In the SEFS technique, the oscillating part is more conveniently presented on the scale of the wave number $q$ of the intermediate-state electron, the position of the main Auger line being taken as the zero of wave numbers, since it corresponds to the threshold of the second-order process if one can neglect the binding energy of valence electrons. Unlike the EXAFS method, when extracting the oscillating part of the secondary electron spectrum from experiment, attempts to normalize the oscillating part to the intensity of the atomic process fail, since the nonoscillating part of the secondary electron spectrum is formed by both the atomic secondary electron emission and the background of true secondary electrons, and there is no way of separating contributions from these processes.

The amplitude of the oscillating terms is determined by the weight functions $A$ and $B$, which depend on the intensities of the atomic first- and second-order electron transitions and on the secondary electron scattering by the ionized atom, namely, on the partial phase shifts. In addition, the relative intensity of the $p$-type oscillation depends on the sample temperature through the factor $W(p^2 - q^2, T)$, and the $q$-type oscillation depends on $\phi$, the phase of the matrix element describing the atomic electron transition in the second-order process.

## 6.2 Methods of Solving the Inverse Problem

The method of analysis of the local atomic structure from experimental oscillating parts is based on the solution of the integral equation (92), i.e., on the determination of the atomic PCF from the experimental oscillating part of the spectrum. This requires that the integral operator in Eq. (92) be known. The main difference between Eq. (92) and the integral equation of EXAFS spectroscopy is the presence of two oscillating terms determined by different wave numbers in the integral operator [Eq. (93)]. In the case that one of the two oscillations in Eq. (93) can be neglected, the relation between the atomic PCF and the oscillating part of the spectrum is given, up to amplitude factors, by a simple Fourier transformation. In the framework of these approximations the solution of the inverse problem becomes simple, since the inverse operator is known. As a result, at the present time the Fourier transformation is most commonly used for solving the inverse problems in EXAFS and EXAFS-like methods.

As applied to SEFS, the Fourier procedure for solving the inverse problem can be used in principle for analysis of fine structure above high-energy Auger transitions, since the oscillation is formed in these spectra only as a result of the intermediate-state electron scattering in the second-order process. However, because of the extremely low intensity and extent of these SEFSs, they can hardly be considered as a reliable source of experimental information on the surface local atomic structure. Unlike the SEFS above the high-energy Auger transitions, the intensity and extent of the SEFS above the low-energy Auger transitions prove to be quite satisfactory, often better than in the corresponding EELFS spectra. At the same time, the presence of two oscillations forming these SEFSs produces using the Fourier transformation method. But the smaller is the difference between the wave numbers $p$ and $q$, the more justified is the use of the Fourier transformation. The difference in value between the wave numbers of the final- and intermediate-state electrons is determined by the core electron binding energy: $p^2 - q^2 = (2m/\hbar^2)E_\alpha$. Thus, the smaller is the core electron binding energy (i.e., the smaller is the energy of the main Auger transition), the more correct and successful is the use of the Fourier transformation for the SEFS treatment. Moreover, the difference between the wave numbers $p$ and $q$ decreases with increasing energy of the detected (final) electron; i.e., the farther is the SEFS region under consideration, the smaller is the difference between the values of the wave numbers. As a result, in spite of the presence of oscillations of two types, the application of the Fourier transformation to treat SEFS spectra makes it possible to obtain qualitative agreement between the SEFS Fourier transforms and the corresponding Fourier transforms of EXAFS and EELFS spectra. The tendency toward improvement of the Fourier transforms of the SEFS spectra with decreasing core electron binding energy is shown in Figures 21c, 22c, and 23c, namely, the Fourier transforms of the oscillating structures in MVV SEFS spectra qualitatively improve on going from Cu to Fe, i.e., as the binding energy of the $M_{2,3}$ core level decreases.

But, in spite of the simplicity and availability of the Fourier transformation and qualitative agreement between the results obtained by this method and the known experimental data, the use of the Fourier procedure for the determination of the local atomic structure parameters from SEFS spectra turns out to be unsatisfactory even on the semiquantitative level. Taking into account oscillations of two types in the kernel of the integral equation of the SEFS method [Eqs. (92), (93)] calls for direct solution of the inverse problem.

In mathematical terms the integral equation (92) is a Fredholm integral equation of the first kind, and the search for a solution of this equation falls into the class

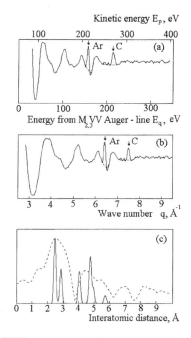

FIG. 21. The Fe MVV SEFS spectrum on (a) an energy scale, (b) a wave-number scale. Panel (c) is a comparison of the Fourier transformation of the spectrum (dashed line) with the Fe crystallographic atomic PCF (solid line).

of ill-posed inverse problems. The Fredholm equations of the first kind have long been under study in mathematics, and powerful methods of their solution have been developed. In particular, the Tikhonov regularization method is successfully employed in solving the inverse problems in EXAFS and EXAFS-like methods.

6.2.1 The Tikhonov Regularization Method. We shall dwell briefly on the mathematical aspect of solving the first-kind Fredholm equation by the Tikhonov regularization method, referring basically to the works that deal with the use of this method in the EXAFS spectroscopy [41–44]. The integral equation (92) can be presented in the operator form

$$\hat{A}g = u \tag{94a}$$

where $\hat{A}$ is the integral operator in Eq. (92), $g$ is the sought solution (in our case, the atomic pair correlation function), and $u$ is the known right side (in our case, the oscillating part of the spectrum). The solution of Eq. (94a) is minimized by

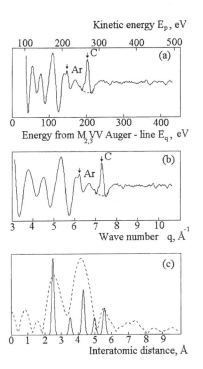

FIG. 22. The Ni MVV SEFS spectrum on (a) an energy scale, (b) a wave-number scale. Panel (c) is a comparison of the Fourier transformation of the spectrum (dashed line) with the Ni crystallographic atomic PCF (solid line).

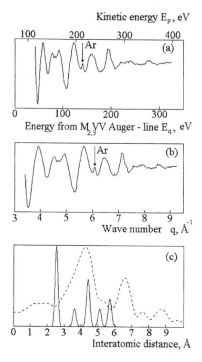

FIG. 23. The Cu MVV SEFS spectrum on (a) an energy scale, (b) a wave-number scale. Panel (c) is a comparison of the Fourier transformation of the spectrum (dashed line) with the Cu crystallographic atomic PCF (solid line).

the functional

$$\left\| \hat{A}g - u \right\|^{2} \tag{94b}$$

where $\| \cdots \|^{2}$ denotes the square of the norm in $L_2$ space. The problem of minimizing the functional (94b) is an ill-posed inverse problem and is the unstable: small changes in the right side (the experimental data) can result in large changes of the solution sought (the atomic pair correlation function). Such a problem can be solved by the Tikhonov regularization method. The regularization of the functional (94b)] reduces to its replacement by the functional

$$\left\| \hat{A}g - u \right\|^{2} + \alpha \left\| g - \tilde{g} \right\|^{2} + \beta \left\| \frac{d}{dr}(g - \tilde{g}) \right\|^{2} \tag{95}$$

where $\tilde{g}$ is a trial function or zero; as a rule $\tilde{g} = 0$ is used in the EXAFS method. In the functional (95) $\alpha$ and $\beta$ are small nonnegative regularization parameters. In this case the parameter $\alpha$ is suited to the requirement of continuity of the solution sought; the parameter $\beta$, to the requirement of continuity of its first derivative. The introduction of these additional requirements is physically reasonable and can be considered as the use of *a priori* information on the solution sought.

After the regularization is performed, i.e., once the functional (94b) is replaced by the regularized functional (95), the problem becomes stable. Direct solution of the inverse problem, namely, the minimization of the functional (95), is carried out numerically. The numerical formalization of the problem is performanced by the method of collocations: the sought solution in $r$-space is considered to be equal to 0 in the range from zero to a certain value $\alpha$, and equal to 1 in the range from a certain $\beta$ to infinity. These additional conditions imposed on the solution sought are a consequence of the use of *a priori* physical information on the atomic pair correlation function. The parameters $\alpha$ and $\beta$ are also chosen from physical considerations. As a result the minimization of the functional (95) reduces to solving the set of linear algebraic equations

$$\left(A_{q,r}^T A_{q,r} + B_{r,r}\right)g_r = A_{q,r}^T u_q \tag{96}$$

where the matrix elements of the matrix $A_{q,r}$ are determined by the integral operator in Eq. (92), and $A_{q,r}^T$ is the transpose of the matrix $A_{q,r}$. The matrix $B_{r,r}$, being in fact the regularizing matrix, has the form of a bi-diagonal matrix:

$$B_{r,r} = \begin{pmatrix} \alpha' + 2\beta' & -\beta' & 0 & \cdots & 0 \\ -\beta' & \alpha' + 2\beta' & -\beta' & \cdots & 0 \\ 0 & -\beta' & \alpha' + 2\beta & \cdots & 0 \\ \vdots & \vdots & \vdots & & \vdots \\ 0 & 0 & 0 & \cdots & \alpha' + 2\beta' \end{pmatrix} \tag{97}$$

where $\alpha' = \alpha \Delta r / \Delta q$, $\beta' = \beta \Delta r / \Delta q$; here $\Delta r$ and $\Delta q$ are step values in the coordinate space ($r$-space) and the reciprocal space (wave-vector space), respectively. The values of the regularization parameters are conveniently prescribed as percentages of the mean value of the diagonal matrix elements of $A^T A$. As a result of all the transformations made, the total matrix operator in Eq. (95) is symmetrical, positive definite, and regular. Thus, the solution of Eq. (94a) is as follows:

$$g_r = \left(A_{q,r}^T A_{q,r} + B_{r,r}\right)^{-1} A_{q,r}^T u_q \tag{98}$$

where standard methods of linear algebra can be used for finding the inverse matrix.

As applied to solving the inverse problem in the EXAFS method, the outlined Tikhonov method was enhanced by an iterative procedure that takes into account that the solution sought is not negative. In fact, the desired atomic pair correlation function is in essence a probability density, i.e., a positive definite function, while the solution of Eq. (98) contains both positive and negative (as a rule, not too large) values. In that case, the negative values are mainly due to incomplete correspondence between the kernel of the integral operator and the physical process that forms the oscillating part of the spectrum. The solution sought is nonnegative, and this is taken into account in the framework of the iterative procedure:

$$g_r^{(n+\alpha)} = P\left(A_{q,r}^T A_{q,r} + B_{r,r}\right)^{-\alpha}\left(A_{q,r}^T u_q + B_{r,r} g_r^{(n)}\right) \qquad (99)$$

where $P$ is the projection operator determined as follows:

$$Pg_r = \begin{cases} g_r & \text{if } g_r \geq 0 \\ 0 & \text{if } g_r < 0 \end{cases} \qquad (100)$$

Experience in using the iterative procedure for solving the inverse problem by the Tikhonov regularization method in EXAFS spectroscopy shows that the obtained solution can be improved by performing a few iterations.

Using the described method for solving the inverse problems in EXAFS spectroscopy, the most significant results were achieved in the study of two-component amorphous solids. Advantages of the suggested procedure over Fourier transformation for determining the local atomic structure from experimental oscillating parts include the possibility of more correctly prescribing the kernel of the integral operator, in particular to take account of the finiteness of the secondary electron mean free path (the photoelectron mean free path in EXAFS). Besides, in the context of this approach, the problem of simultaneous solution of the complete system of integral equations to determine the pair correlation functions is formalized for multicomponent systems in the manner that was successful for the case of two-component systems. Nevertheless, the described procedure for solving the inverse problem has a number of significant disadvantages as compared to the Fourier transformation. First, this is a sophisticated mathematical and algorithmic procedure. In addition, while using this method, one must prescribe as exactly as possible both the form of the integral operator and its parameters. Thus, test solution of the inverse problem by the Tikhonov regularization method becomes necessary in order to adjust both the mathematical procedure and the physical parameters in the kernel of the integral equation.

The Tikhonov regularization method as applied to solving the inverse problem in the SEFS method has the advantage that it allows one to take into account oscillations of two types (oscillations determined by different wave numbers) in the kernel of the integral operator.

## 6.3 Determination of Parameters of the Local Atomic Structure of Fe, Ni, and Cu Thin Surface Layers from the MVV SEFS Experimental Data

After appropriate modification of the program package for solving the inverse one-component EXAFS problem by the Tikhonov method, the atomic pair correlation functions were calculated from Fe, Ni, and Cu experimental MVV SEFS spectra. The changes made are connected with the difference of the kernel of the integral equation (91) from that of the EXAFS equation, and with the procedure of calculating the matrix $A_{q,r}$.

As mentioned above, to solve the inverse problem one should specify all the parameters in the kernel of the integral equation. A number of such parameters appear in the kernel of the integral equation (94a) that are absent in other EXAFS-like methods. First of all, these include the temperature parameter $W(p^2 - q^2, T)$ and the weight functions $A$ and $B$. We shall use the estimates of the temperature parameter $W(p^2 - q^2, T)$ made in Section 5 when solving the inverse problem. The simplest approximation to take into account oscillations of two types in the kernel of the integral equation is the approximation where the weight functions $A$ and $B$ are merely constants and their ratio is a fitting parameter. In the framework of this approach the inverse problem was solved for the Cu MVV SEFS spectrum [37]. The result obtained was in a quite good agreement with the Cu crystallographic atomic PCF: the main PCF peaks proved to be well resolved, and their positions correspond approximately to the crystallographic ones. So the obtained result is essentially better than the result of Fourier transformation. This confirms the necessity to allow for oscillations of two types, but the approximation (constant weight functions $A$ and $B$) is crude. Its crudity is demonstrated, for example, by the fact that in Tikhonov's solution (the zeroth iteration) the amplitude of negative outliers is about 30% of the amplitude of the positive part of the solution. This shows that more accurate account should be taken of the amplitude part of the oscillations. In the study of the Fe, Ni, and Cu MVV SEFS spectra, the estimates (70), (86) were used to specify the weight functions $A$ and $B$. Since the experimental oscillating part in Eq. (92) is unnormalized, the weight functions $A$ and $B$ determined by Eq. (36) are also unnormalized to the intensity of the atomic emission in the integral equation (92), so

$$A = \frac{2}{3} \exp\left(i2\delta_2^0\right) d J_{3p}^{(1)}\left(E_p, \hat{\boldsymbol{p}}\right) - \frac{3}{5} \exp\left(i2\delta_3^0\right) d J_{3p}^{(2)}\left(E_p, \hat{\boldsymbol{p}}\right),$$

$$B = \frac{2}{3} \exp\left(i2\delta_2^0\right) d J_{at}^{(2)}\left(E_p, \hat{\boldsymbol{p}}\right)$$

(101)

The electron transitions are assumed to follow the dipole selection rules, and the transitions with a decrease in the orbital quantum number are ignored in Eq. (101). Hereinafter, to simplify the calculation we consider that $2\delta_3^0 = 2\delta_2^0$. This approximation is not too rough, since the difference between $2\delta_3^0$ and $2\delta_2^0$ is insignificant. Estimates of the intensities of atomic electron transitions in processes of first order $[d J_{3p}^{(1)}(E_p, \hat{\boldsymbol{p}})$, Eq. (70)] and second order $[d J_{3p}^{(2)}(E_p, \hat{\boldsymbol{p}})$, Eq. (86)] have been made in the preceding section, and their relative intensity is given by the parameter $n_\alpha n_\beta C^2 = n_\alpha n_\beta [16/(a_0\alpha)^2]\beta^3/\alpha^3$, depending on the number of valence and core electrons and the inverse radii of localization of their wave functions. When

TABLE III.    The Integral Equation Kernel Parameters That Were Used in
Solving the Inverse Problem

| Element | $E_p - E_q, (E_{M_{2,3}}VV)$ (eV) | $n_\alpha n_\beta C^2$ | $\phi$ (rad) | $W(p^2 - q^2, T)$ |
|---------|-------------------------------------|------------------------|--------------|-------------------|
| Fe | 54, (56) | 1.9 | 0.9 | 0.72 |
| Ni | 62, (68) | 1.1 | 0.1 | 0.64 |
| Cu | 72, (74) | 1.4 | 0.8 | 0.74 |

solving the inverse problem with the use of the estimates made in Section 4, we shall consider $n_\alpha n_\beta C^2$ to be a fitting parameter. In addition, the kernel of Eq. (92) contains $\phi$, the phase of the matrix element of the second-order process. As before, $\phi$ is considered as a parameter rather than a function. Since the inverse problem is more sensitive to the phase parameters, in this case $\phi$ will be considered to be a fitting parameter. The difference $E_p - E_q$ between the energies of the intermediate- and final-state electrons ($E_q$ and $E_p$) is also a fitting parameter that does not essentially deviate from the relation governed by the energy conservation law: $E_q = E_p - E_\alpha$.

In addition to the parameters of the kernel of the integral equation that are unique to the SEFS technique, Eq. (92) involves functions typical of the EXAFS method, namely, the amplitudes of the secondary electron backscattering and the secondary electron mean free path. In calculating the atomic PCF from MVV SEFS spectra, we have used the parameters calculated with the FEFF7 program for the bcc Fe and fcc Ni and Cu crystallographic structures.

Within the approximations described (the values of the fitting parameters are listed in Table III), the inverse problem of finding the atomic PCF from Fe, Ni, and Cu MVV SEFS spectra has been solved using a 5% regularization (see Figs. 24, 25, 26). The results obtained are given in Figures 24, 25, and 26 in comparison with the corresponding crystallographic PCFs. Unlike the solution discussed before with the weight functions being parameters, in this case the magnitudes of negative outliers in Tikhonov's solution (the zeroth iteration) did not exceed 10% of the positive values of the obtained solution, which is a quite satisfactory result in solving the inverse problems in the EXAFS method and indicates that the amplitude functions used are reasonable. A comparison between the calculated and crystallographic PCFs shows that, contrary to the results of the Fourier transformation of the $M_{2,3}VV$ SEFS spectra (see Figs. 21c, 22c, 23c), all main peaks of the calculated atomic PCF are resolved and their positions correspond to those of the crystallographic PCF. However, the roughness of the approximations made in the kernel of the integral operator, evidenced, in particular, by the presence of substantial false peaks, gives no way of studying the parameters of

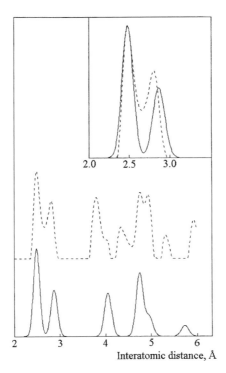

FIG. 24. Comparison of the inverse problem solution for the Fe MVV SEFS (dashed line) with the Fe crystallographic atomic PCF (solid line).

the far range of the calculated PCF, even on the semiquantitative level. The rougheset approximations made in the kernel of integral operator [Eq. (92)], which cause a large error in the far range of the PCF calculated from the experimental data, are those concerning the phase parameters of interference terms ($\phi =$ const and $2\delta_3^0 = 2\delta_2^0$). To improve the results obtained, the amplitudes and intensities of atomic electron transitions in processes of both first and second order should be calculated more accurately than in Section 4. Thus, we restrict the discussion to parameters of the nearest atomic surroundings of the surface layers of samples under study.

## 6.4 Results and Discussion

A comparison between the first maxima of the calculated PCFs and the corresponding crystallographic data is presented in Figures 24, 25, and 26. The main peculiarity of the calculated results is a marked asymmetry of the peaks. For Cu, an slight decrease of the first interatomic distance is also observed as compared to

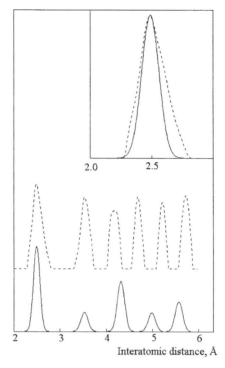

FIG. 25. Comparison of the inverse problem solution for the Ni MVV SEFS (dashed line) with the Ni crystallographic atomic PCF (solid line).

the corresponding crystallographic value. Numerical parameters that characterize the local atomic surroundings and the analyzed layer depth are given in Table IV in comparison with the corresponding crystallographic parameters. Here the interatomic distance values are estimated as the peak maxima of the atomic PCFs, i.e., the most probable locations of the atoms. The asymmetry of a PCF peak is characterized by the ratio between the left-side (LS) and right-side (RS) areas of the peak (relative to the maximum position), and its values for the first peak are also given in Table IV. The obtained asymmetry is likely to be a characteristic feature of the atomic PCF of surface layers. The nature of this phenomenon lies in the influence of surface effects on the atomic structure of matter. Since the obtained experimental results are in fact a PCF averaged over the analyzed layer depth, the surface relaxation and surface anharmonicities of oscillations may cause a deviation of the PCF peak form from the Gaussian function. Moreover, in EELFS experiments with Fe, Ni, and Cu, it has been shown [3–5] that as the analyzed layer depth increases to 20 Å, so that the surface effects on the average PCF can

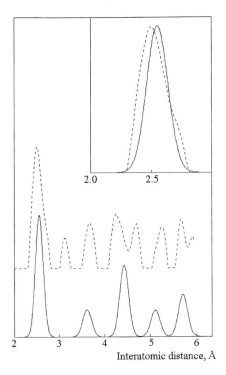

FIG. 26. Comparison of the inverse problem solution for the Cu MVV SEFS (dashed line) with the Cu crystallographic atomic PCF (solid line).

be neglected, the asymmetry of the first PCF peak disappears. Besides, no asymmetry of peaks is observed in EXAFS studies of the appropriate crystal samples. Because of significant asymmetry of the experimental PCFs of superthin surface layers, the analysis of the atomic mean squared deviation from the results of these investigations makes no sense; therefore we do not consider this parameter.

Along with a pronounced asymmetry of the peak in the experimental Cu PCF, a small decrease of the interatomic distance as compared to the crystallographic data is observed. This is a surface effect mentioned by the majority of investigators. However, the obtained values of the decrease in interatomic distance, if any, are much less than the values obtained by other authors with the SEFS method [12–19]. At the same time, our results are in good agreement with the results of LEED studies of the relaxation of Fe, Ni, and Cu pure sample surfaces [88, 89] (see Table V). A $0.05 \pm 0.02$ Å decrease of the nearest interatomic distance (from the crystallographic value 2.56 Å) for the Cu (111) monoplane in the analyzed surface layer of 5–7 Å, averaged over the depth, is in good agreement

TABLE IV.    Local Atomic Environment Parameters for Fe, Ni, and Cu in Comparison with
Crystallographic Ones

| Environment | Surface | Thickness (analyzed) (Å) | Interatomic distance (Å) | $\Delta R$ (Å) | Peak asymmetry LS/RS |
|---|---|---|---|---|---|
| Fe (SEFS) first neighbors | bcc polycrystal | ~5–6 | 2.48 ± 0.02 | 0.00 ± 0.02 | 0.80 ± 0.05 |
| Fe (crystallographic) first neighbors | — | — | 2.48 | — | 1 |
| Fe (SEFS) second neighbors | bcc polycrystal | ~5–6 | 2.82 ± 0.02 | −0.04 ± 0.02 | 1.00 ± 0.05 |
| Fe (crystallographic) second neighbors | — | — | 2.86 | — | 1 |
| Ni (SEFS) | fcc polycrystal | ~5–6 | 2.49 ± 0.02 | 0.00 ± 0.02 | 0.80 ± 0.05 |
| Ni (crystallographic) | — | — | 2.49 | — | 1 |
| Cu (SEFS) | fcc (111) | ~5–6 | 2.51 ± 0.02 | 0.05 ± 0.02 | 0.75 ± 0.05 |
| Cu (crystallographic) | — | — | 2.56 | — | 1 |

TABLE V.    The LEED Data for Fe, Ni, and Cu Surface Reconstructions

| Crystal | Decrease in interatomic distance (%) | | |
|---|---|---|---|
|  | (001) | (110) | (111) |
| bcc Fe | −1.4 | ≈0 | — |
| fcc Ni | +2.5 | −5 | −1 |
| fcc Cu | — | −10 | −4.1 |

with the LEED data (a 4.1% relaxation of this surface). Also, with allowance for the polycrystallinity of the studied sample surface, the absence of a decrease in interatomic distance (within the accuracy of the results obtained) for Fe and Ni agrees well with the LEED investigations. In our opinion, the too great decrease in nearest interatomic distance obtained by the Fourier analysis of SEFS spectra is a consequence of the Fourier procedure being invalid for obtaining the quantitative parameters of the local atomic structure from SEFS spectra.

Since, as mentioned above, at present the normalization of SEFS spectra to the atomic emission intensity is impossible, we could not estimate the coordination numbers from these experimental data.

A rather complicated procedure has been used in calculating the atomic PCFs from the experimental SEFS spectra; besides, a number of parameters have been fitting parameters in the calculation. But the method used does to solve the inverse problem uniquely, which makes it possible to take into account two types of oscillations that form the SEFS spectrum. As a result, we have succeeded in obtaining PCFs much better than the corresponding results of Fourier transformation. In particular, two nearby first peaks are resolved in the Fe PCF obtained from experimental results, which is a good test of both the formalization itself and the method for solving the inverse problem. The fitting parameters used prove to be quite close in magnitude to the corresponding theoretical estimates (see Sec. 4); moreover, the considered inverse problem turns out to be rather stable with respect to the assigned values of fitting parameters: small changes in them have little or no effect on the first peak and affect only slightly the far region of the calculated PCFs, while parameter values significantly different from the ones used result in a physically meaningless PCF.

On the basis of the results obtained, one can say that the SEFS results are very useful in analyzing the atomic structure of superthin surface layers of matter. But whenever studies of the local atomic structure can be performed by other methods, for example by EELFS, the complexity of the SEFS technique becomes an important disadvantage. However, in the experimental study of atomic PCF's of surface layers of multicomponent atomic systems within the formalism of the inverse problem solution, a complete set of integral equations is necessary to provide mathematical correctness. This set of equations can be solved by the methods of direct solution only. In this case the use of the SEFS method may be a necessary condition for obtaining a reliable result. Besides, the calculations made can be used as a test when studying multicomponent systems.

# 7 CONCLUSION

Because the presented results have been covered adequately in each section of this work, here we restrict our consideration to the most important peculiarities of the SEFS method. The presented results can be arbitrarily divided into two groups: in the first the SEFS is considered as a spectroscopic feature in secondary electron spectra of condensed matter; the second concerns the possible use of the SEFS method to determine parameters of the local atomic structure of superthin surface layers.

About SEFS as a spectroscopic feature, it can be said that this structure is a manifestation of effects of the strong electron–hole correlation (with a hole on the atom core level) upon excitation of the atom by an electron impact. As opposed to the classical autoionization effects (Fano effects), where the presence of a resonance level is assumed, in the present case a strongly correlated state is observed between the core hole and an electron over a wide region of the continuous spectrum. Apart from the fact that the presence of this effect in the secondary electron spectrum has its own importance, the existence of SEFSs should be taken into account during the quantitative Auger analysis of materials containing 3d and 4d metals.

While considering SEFS spectroscopy as a method of analysis of the local atomic structure of superthin surface layers, we should dwell on the most essential points. The main disadvantage of the SEFS method as a structural method is its sophisticated nature; namely, it is a two-step process for obtaining the oscillating spectrum. As a result, when analyzing the sample local atomic structure from the experimental data, it is essential to take into account complicated processes of electron transitions, so that the standard EXAFS methods turn out to be inapplicable. In our opinion, just this problem is the main cause of the SEFS technique being still practically unused as a structural method, in spite of the fact that the SEFS itself and its relation to atomic structure were discovered more than 30 years ago. At the same time, there are a number of unique possibilities for SEFS as a structural analysis method. First, the method appears to be the only one that allows the study of the local atomic surface structure in a superthin layer of 5 Å. Second, the standard equipment of Auger spectroscopy will suffice for obtaining the experimental SEFS spectra.

Turning back to the problem of treatment of the SEFS experimental data to obtain information on the local atomic structure of the studied material, we dwell on the presented results in more detail. The relation between the atomic pair correlation function describing the local atomic structure and the oscillating part of the secondary electron spectrum is determined by integral equations of the EXAFS type. The fundamental difference between the integral equations of the EXAFS and SEFS methods is that the kernel of the SEFS integral operator is determined by two oscillating terms that depend on different wave numbers. The presence of two different oscillating terms results in the inapplicability of the Fourier method to the processing of the SEFS spectra. However, integral equations of this type and methods of their solution have long been studied by mathematicians, and by now rather powerful methods and algorithms for solving these equations have been developed. Although the direct methods of solving integral equations are somewhat more complicated than the Fourier procedure, they are being used more and more intensively in spectral structural analysis. Thus, the inverse problem for obtaining atomic pair correlation functions from SEFS experimental data can be solved quite correctly—provided, of course, that the prescribed parameters of the integral operator kernel are correct. The estimates of these parameters given in the present work should be considered as a first approximation, which is not sufficiently exact, although it has allowed a number of quantitative parameters of the local atomic structure to be obtained. In our opinion further development of the method for extracting information on the local atomic structure from SEFS experiments should be focused on obtaining more exact estimates of the integral equation kernel parameters, which will substantially improve the quality of the results obtained.

As for the applicability of SEFS to analysis of the atomic structure of a matter, this method seems to be the most suitable one for analysis of the local atomic structure of superthin surface layers of two-component systems containing a transition element as one of the components. Using the SEFS method in combination with EELFS for such systems allows one to make up a complete set of simultaneous integral equations for obtaining all the atomic pair correlation functions. The information obtained by these methods is of importance in studying disordered (having no long-range order) thin films and surfaces, for which the traditional diffraction methods turn out to be unsuitable. Analysis of the pair correlation functions obtained by the SEFS method makes it possible to study such surface effects as relaxation and reconstruction of the surface, the surface anharmonicity of atomic vibrations, and segregation effects. Thus, along with the traditional diffraction and spectroscopic methods for studying the atomic structure of thin films and surfaces of condensed matter, the SEFS spectroscopy should allow one to obtain additional and often unique information on the atomic structure and the properties of the solids.

## Acknowledgment

This work was supported in part by the Russian Foundation for Fundamental Research under grant 00-03-33049a.

## References

1. P. A. Lee, P. M. Citrin, P. Eisenberger, and B. M. Kincaid, *Rev. Mod. Phys.* 53, 769 (1981).
2. B. K. Teo and D. C. Joy, "EXAFS Spectroscopy and Related Techniques," Plenum, New York, 1981.
3. E. A. Stern, *J. Phys. (Paris)* 12, C8, 3 (1986).
4. M. DeCrescenzi, *CRC Crit. Rev. Solid State Mater. Sci.* 15, 279 (1989).
5. M. DeCrescenzi, *Surf. Sci. Rep.* 21, 89 (1995).
6. M. A. Chesters and J. Pritchard, *Surf. Sci.* 28, 460 (1971).
7. L. H. Jenkins and M. F. Chung, *Surf. Sci.* 26, 151 (1971).
8. L. McDonnell, B. D. Powell, and D. P. Woodruff, *Surf. Sci.* 40, 669 (1973).
9. G. E. Becker and H. D. Hagstrum, *J. Vac. Sci. Technol.* 11, 284 (1974).
10. D. P. Woodruff, *Surf. Sci.* 189/190, 64 (1987).
11. N. K. Singh, R. G. Jones, and D. P. Woodruff, *Surf. Sci. Lett.* 232, L228 (1990).
12. M. DeCrescenzi, E. Chainet, and J. Derreien, *Solid State Commun.* 57, 487 (1986).

13. E. Chainet, J. Derreien, R. C. Cinti, T. T. A. Nguyen, and M. DeCrescenzi, *J. Phys. (Paris)* 47, 209 (1986).
14. G. Chiarello, V. Formoso, L. S. Caputi, and E. Colavita, *Phys. Rev. B* 35, 5311 (1987).
15. J. Derreien, E. Chainet, M. DeCrescenzi, and C. Noguera, *Surf. Sci.* 189/190, 590 (1987).
16. T. Tyliszczak, A. P. Hitchcock, and M. DeCrescenzi, *Phys. Rev. B* 38, 5768 (1988).
17. M. DeCrescenzi, A. P. Hitchcock, and T. Tyliszczak, *Phys. Rev. B* 39, 9839 (1989).
18. J. Y. Veuillen, A. Bensaoula, M. DeCrescenzi, and J. Derreien, *Phys. Rev. B* 39, 10398 (1989).
19. L. S. Caputi, A. Amoddeo, R. Tucci, and E. Colavita, *Surf. Sci.* 251/252, 262 (1991).
20. D. E. Guy, Y. V. Ruts, S. P. Sentemov, V. I. Grebennikov, and O. B. Sokolov, *Surf. Sci.* 298, 134 (1993).
21. I. Davoli, R. Bernardini, C. Battistoni, P. Castucci, R. Gunnella, and M. DeCrescenzi, *Surf. Sci.* 309, 144 (1994).
22. D. E. Guy, Y. V. Ruts, D. V. Surnin, V. I. Grebennikov, and O. B. Sokolov, *Physica B* 208–209, 87 (1995).
23. L. Lozzi, M. Passacantando, P. Picozzi, S. Santucci, and M. DeCrescenzi, *J. Electron Spectrosc. Rel. Phenom.* 72, 223 (1995).
24. L. Lozzi, M. Passacantando, P. Picozzi, S. Santucci, and M. DeCrescenzi, *Surf. Rev. Lett.* 2, 255 (1995).
25. E. G. McRae, *Surf. Sci.* 44, 321 (1974).
26. P. Aebi, M. Erbudak, F. Vanini, and D. D. Vvedensky, *Surf. Sci. Lett.* 264, L181 (1992).
27. R. V. Vedrinskii, A. I. Taranukhina, A. A. Novakovich, and L. A. Bugaev, *J. Phys. Condensed Matter* 7, L181 (1995).
28. P. Aebi, M. Erbudak, F. Vanini, D. D. Vvedensky, and G. Kostorz, *Phys. Rev. B* 41, 11760 (1990).
29. V. I. Grebennikov and O. B. Sokolov, *Fiz. Met. Metalloved.* 78, 113 (1994).
30. V. I. Grebennikov and O. B. Sokolov, *J. Phys. Condensed Matter* 7, 5713 (1995).
31. V. I. Grebennikov, O. B. Sokolov, D. E. Guy, and Yu. V. Ruts, *Fiz. Tverd. Tela (Leningrad)* 40, 1589 (1998).
32. M. Tomellini and M. Fanfoni, *Solid State Comm.* 90, 391 (1994).
33. V. I. Grebennikov and O. B. Sokolov, *Phys. Scripta* 41, 51 (1992).
34. R. G. Agostino, A. Amoddeo, L. S. Caputi, and E. Colavita, *Phys. Scripta* 41, 149 (1992).
35. D. E. Guy, V. I. Grebennikov, Y. V. Ruts, S. P. Sentemov, and O. B. Sokolov, *Jpn. J. Appl. Phys.* 32, 26 (1993).
36. V. I. Grebennikov and O. B. Sokolov, *J. Phys. Condensed Matter* 7, 5713 (1995).
37. D. E. Guy, Yu. V. Ruts, A. N. Deev, V. I. Grebennikov, and O. B. Sokolov, *Surf. Rev. Lett.* 4, 947 (1997).
38. F. W. Lytle, D. E. Sayers, and E. A. Stern, *Phys. Rev. B* 11, 4825 (1975).
39. A. N. Tikhonov and V. Ya. Arsenin, "Methods of Solution of Ill-posed Problem" (J. Filtz, Ed.), Wiley, New York, 1977.
40. A. B. Bakushinsky and A. V. Goncharsky, "Ill-Posed Problems: Theory and Applications," Kluwer, Amsterdam, 1994.
41. V. V. Vasin and A. L. Ageev, "Ill-Posed Problems with A Priori Information," VSP, Utrecht, 1995.

42. Yu. A. Babanov, V. V. Vasin, A. L. Ageev, and N. V. Ershov, 105, 747 (1981).

43. N. V. Ershov, A. L. Ageev, V. V. Vasin, and Yu. A. Babanov, *Phys. Status Solidi B* 111, 103 (1981).

44. N. V. Ershov, Yu. A. Babanov, and V. R. Galakhov, *Phys. Status Solidi B* 117, 749 (1983).

45. D. E. Guy, Y. V. Ruts, D. V. Surnin, V. I. Grebennikov, and O. B. Sokolov, *Surf. Rev. Lett.* 4, 223 (1997).

46. D. E. Guy, Yu. V. Ruts, A. N. Deev, D. V. Surnin, V. I. Grebennikov, and O. B. Sokolov, *J. Electron Spectrosc. Relat. Phenom.* 88–91, 551 (1998).

47. D. E. Guy, D. V. Surnin, A. N. Deev, Yu. V. Ruts, and V. I. Grebennikov, *J. Electron Spectrosc. Relat. Phenom.* (2001) to be published.

48. J. K. Riviere, "Practical Surface Analysis by Auger and X-ray Photoelectron Spectroscopy" (D. Briggs and M. P. Seah, Eds.), Wiley, New York, 1983.

49. D. P. Woodruff and T. A. Delchar, "Modern Techniques of Surface Science," Cambridge University Press, Cambridge, 1986.

50. M. Prutton, "Introduction to Surface Physics," Oxford University Press, New York, 1994.

51. C. L. Hedberg (Ed.), "Handbook of Auger Electron Spectroscopy," Physical Electronics Inc., Minnesota, 1995.

52. V. A. Shamin and Yu. V. Ruts, *Fiz. Met. Metalloved.* 81, 847 (1996).

53. A. Kadikova, V. Shamin, A. Deev, and Yu. Ruts, *J. Phys. (Paris) IV* 7, C2, 579 (1997).

54. D. V. Surnin, Yu. V. Ruts, and D. E. Denisov, *J. Phys. (Paris) IV* 7, C2, 577 (1997).

55. V. A. Shamin, A. H. Kadikova, and Yu. V. Ruts, *Fiz. Met. Metalloved.* 83, 47 (1997).

56. V. A. Shamin, A. H. Kadikova, A. N. Deev, and Yu. V. Ruts, *J. Electron Spectrosc. Relat. Phenom.* 88–91, 57 (1998).

57. V. A. Shamin, A. H. Kadikova, A. N. Deev, and Yu. V. Ruts, *Fiz. Tverd. Tela (Leningrad)* 40, 1156 (1998).

58. E. A. Stern, *Phys. Rev. B* 10, 3027 (1974).

59. P. A. Lee and J. B Pendry, *Phys. Rev. B* 11, 2795 (1975).

60. B. K. Teo and P. A. Lee, *J. Am. Chem. Soc.* 101, 2815 (1979).

61. B. Sinkovic, J. D. Fridman, and C. S. Fadley, *Journal of Magnetism and Magnetic Materials* 92, 301 (1991).

62. A. G. McKale, B. W. Veal, A. P. Paulikas, S. K. Chan, G. S. Knapp, *J. Am. Chem. Soc.* 110, 3763 (1988).

63. J. J. Rehr, J. Mustre de Leon, S. I. Zabinsky, and R. C. Albers, *J. Am. Chem. Soc.* 113, 113 (1991).

64. S. I. Zabinsky, J. J. Rehr, A. Ankudinov, R. C. Albers, and M. J. Eller, *Phys. Rev. B* 52, 2995 (1995).

65. J. J. Rehr, S. I. Zabinsky, and R. C. Albers, *Phys. Rev. Lett.* 69, 3397 (1992).

66. J. J. Rehr, *Jpn. J. Appl. Phys.* 32, 8 (1993).

67. J. Mustre de Leon, J. J. Rehr, S. I. Zabinsky, and R. C. Albers, *Phys. Rev. B* 44, 4146 (1991).

68. J. J. Rehr and R. C. Albers, *Phys. Rev. B* 41, 8139 (1990).

69. U. Fano and J. W. Cooper, *Rev. Mod. Phys* 40, 441 (1968).

70. S. D. Bader, G. Zajac, and J. Zak, *Phys. Rev. Lett.* 50, 1211 (1983).

71. G. Zajac, S. D. Bader, A. J. Arko, and J. Zak, *Phys. Rev. B* 29, 5491 (1984).
72. T. Fujikawa, S. Takaton, and S. Usami, *Jpn. J. Appl. Phys.* 27, 348 (1988).
73. D. K. Saldin and J. M. Yao, *Phys. Rev. B* 41, 52 (1990).
74. D. K. Saldin and Y. Ueda, *Phys. Rev. B* 46, 5100 (1992).
75. Y. Ueda and D. K. Saldin, *Phys. Rev. B* 46, 13697 (1992).
76. B. Luo and J. Urban, *Surf. Sci.* 239, 235 (1990).
77. M. Diociaiuti, L. Lozzi, M. Passacantando, S. Santucci, P. Picozzi, and M. De-Crescenzi, *J. Electron Spectrosc. Relat. Phenom.* 82, 1 (1996).
78. E. Clementi and C. Roetti, *At. Data Nucl. Data Tables* 14, 11760 (1974).
79. P. H. Dederichs, H. Schober, D. J. Sellmyer, "Phonon States of Elements: Electron States and Fermi Surface of Alloys, in "Metals: Phonon States Electron States and Fermi Surfaces", Vol 13a, p. 458. (K.-H. Hellwege, J. L. Olsen, Eds.), Springer-Verlag, and Landolt-Bornstein, Berlin, 1981.
80. R. Schneider, H. Durr, T. Fauster, and V. Dose, *Phys. Rev. B* 42, 1638 (1990).
81. Y. Cao and E. Conrad, *Phys. Rev. Lett.* 64, 447 (1990).
82. P. Statiris, H. C. Lu, and T. Gustafsson, *Phys. Rev. Lett.* 72, 3574 (1994).
83. D. E. Fowler and J. V. Barth, *Phys. Rev. B* 52, 2117 (1995).
84. H. C. Lu, E. P. Gusev, E. Garfunkel, and T. Gustafsson, *Surf. Sci.* 352–354, 21 (1996).
85. E. P. Gusev, H. C. Lu, E. Garfunkel, and T. Gustafsson, *Surf. Rev. Lett.* 3, 1349 (1996).
86. K. H. Chae, H. C. Lu, and T. Gustafsson, *Phys. Rev. B* 54, 14082 (1996).
87. B. W. Busch and T. Gustafsson, *Surf. Sci.* 407, 7 (1998).
88. F. Jona, *J. Phys. C* 11, 4271 (1978).
89. F. Jona, J. A. Stozier Jr., and W. S. Yang, *Rep. Prog. Phys.* 45, 527 (1982).

# 4. PHOTONIC AND ELECTRONIC SPECTROSCOPIES FOR THE CHARACTERIZATION OF ORGANIC SURFACES AND ORGANIC MOLECULES ADSORBED ON SURFACES

Ana Maria Botelho do Rego, Luis Filipe Vieira Ferreira

Centro de Química-Física Molecular, Complexo Interdisciplinar, Instituto Superior Técnico, 1049-001 Lisboa, Portugal

## 1 INTRODUCTION

Solid surfaces are a field of growing interest from both the applied and fundamental points of view. Their quality of being the "face" of any material, their entrance door, gives them a major importance in *applied areas* as different as composite materials, corrosion, adsorption, biocompatibility, dyeing and lightfastness of dyes on fabrics, heterogeneous catalysis, among many other applications.

Also in *fundamental areas*, surfaces are privileged media to serve as substrates for monolayers (or a few monolayers) as in self-assembled on molecular systems or in Langmuir–Blodgett films (LB), for instance.

Microporous solids play a special role due to their large area/volume ratio providing the opportunity for adsorption of large amounts of guest molecules on very small amounts of substrate, due to the very high specific area of the adsorbent. In particular, powdered solids with controlled pore size solids may have specific uses such as molecular sieves, microreactors of controlled microsize, or catalysts. The surface–probe interaction may simply be of electrostatic nature, of hydrogen bonding, or in terms of an acid or basic behavior (Brönsted or Lewis). Substrates may also interact with the adsorbed probe as electronically active supports and have an important role in redox processes (as in advanced oxidative processes for persistent pollutant destruction) [1–5].

The interest in studying surfaces is intimately related to the development and spreading of suitable techniques. In fact, there are a number of conditions to be fulfilled in order to make a technique surface sensitive and surface specific. Apart from the cases of highly porous materials and/or highly divided media, surface usually represents a negligible amount of matter in solids. Therefore, and given the usual proportionality between the number of species in a medium and the signal obtained in a technique, some care has to be taken to be sure that the obtained information comes exclusively (surface specific) or mainly (surface sensitive) from the surface and not from the bulk.

At this point we will distinguish between surface-specific and surface-sensitive techniques: according to Desimoni et al. [6] the former "are defined as those capable of collecting information relevant to the first few atomic layers of the surface

269

EXPERIMENTAL METHODS IN THE PHYSICAL SCIENCES
Vol. 38
ISBN 0-12-475985-8

of a solid specimen," i.e., a maximum sampling depth of 5–10 nm; the latter probe depths up to 100 nm.

For flat surfaces, the selectivity is achieved using low penetrating excitation probes and/or analyzing low penetrating probes. As examples of low penetrating probes, we can point out heavy ions and electrons. X-ray photoelectron spectroscopy (XPS) and high-resolution electron energy spectroscopy (HREELS) are just two spectroscopies based on the low penetration of electrons. XPS uses highly penetrating probes—X-ray photons—but analyzes electrons, and HREELS uses very low energy electrons both as incident and analyzed probes. These two electronic spectroscopies are surface specific, the specificity being larger for HREELS when the impact mechanism is favored—a sampling depth of a few Ångström. In XPS, the sampling depth is of the order of 3 to 10 nm.

The goal of this chapter is to emphasize the synergetic effect obtained on putting together the larger number of available techniques to the service of characterizing surfaces. Particularly, this chapter will deal with optical spectroscopies in the UV-Vis-IR range, used in diffuse reflectance arrangement, combined with electronic spectroscopies, specifically XPS and HREELS.

## 2 DIFFUSE REFLECTANCE TECHNIQUES FOR SURFACE PHOTOCHEMISTRY STUDIES

### 2.1 Ground-State Diffuse Reflectance Absorption Spectra (UV-Vis-NIR)

Ground-state absorption spectra of solid opaque samples can experimentally be obtained using a procedure similar to the one used to perform absorption spectra of transparent samples. In the latter case one has to use the Beer–Lambert law to determine absorbances as a function of wavelength. In the case of solid samples, it is possible to determine the *reflectance* $(R)$ as a function of the wavelength using an integrating sphere, following an initial calibration of the apparatus as Figure 1 shows. An ideal diffuser has a unitary reflectance (in practice pure barium sulphate or magnesium oxide can be used since for both compounds $R \sim 0.98 \pm 0.02$ in the 200 to 900 nm wavelength range). Finely divided carbon particles $(R \sim 0)$ can be used as a black reference. Alternatively, it is possible to use calibrated standards either for white or black, both of which are commercially available.

Ground-state absorption spectrum for powdered microcrystalline cellulose is also shown in Figure 1. It exhibits $R$ values quite close to unity for visible (Vis) and near infrared (NIR) spectral regions and shows a significative absorption in the ultraviolet (UV). Silica and silicalite also have a reflectance close to unity in the Vis and NIR regions and differ from cellulose by the fact that UV absorption is smaller. In the case of silicas with different porosity, but with the same particle size, significant variations of the reflectance were detected [7].

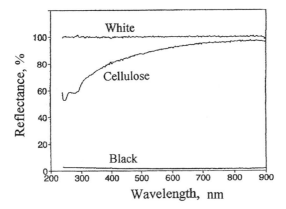

FIG. 1. Reflectance spectra of white and black standards and microcrystalline cellulose.

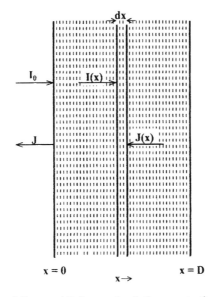

FIG. 2. Incident ($I$) and disperse ($J$) fluxes of radiation penetrating a dispersive medium.

Let us consider now that $I$ and $J$ are the fluxes of incident and disperse light traveling in the forward and reverse directions. The incident flux, $I$, decreases as it penetrates the solid powdered sample because not only is radiation absorbed, but because particles also disperse the incident light. At the same time $I$ increases with $J$ dispersion. The flux of disperse light, $J$, has an analogous variation, although in the opposite direction, as Figure 2 [8] shows.

So we can write

$$dI(x) = -(K + S)I(x)dx + SJ(x)dx \tag{1}$$
$$dJ(x) = +(K + S)J(x)dx - SI(x)dx \tag{2}$$

where $K$ is the absorption coefficient, and $S$ the dispersion coefficient. One can very easily obtain the Beer–Lambert law for transparent and homogeneous media, where no dispersion exists, so $S = 0$. Then,

$$dI(x) = -KI(x)dx \tag{3}$$

and therefore $I = I_0 \exp(-Kx)$, where $K = \varepsilon C$, where $\varepsilon$ is the naperian coefficient for absorption and $C$ is the concentration of the absorbing species.

Kubelka and Munk [8] established that, for an ideal diffuser and optically thick samples (all those where a further increase in thickness does not change the reflectance of the sample), the reflectance is given by

$$R = \frac{J}{I_0} \tag{4}$$

where $R$ is related to $K$ and $S$ by the remission function, $F(R)$

$$F(R) = \frac{(1 - R)^2}{(2R)} = \frac{K}{S} \tag{5}$$

and

$$K(\lambda) = 2\varepsilon(\lambda)C \tag{6}$$

In Eq. (6) the factor 2 takes into account the average increase of distance travelled by the excitation light inside the ideal diffuser. In the Kubelka–Munk theory the radiation is only regarded as scattered when it is backward reflected into a hemisphere whose boundary plane lies perpendicular to the $x$-direction [8c].

The remission function varies linearly with the number of absorbing chromophores in the solid sample, which are considered as being uniformly distributed. $K$ and $S$ are independent of the light penetration into the sample. For optically thick samples we have

$$I(x) = I_0 \exp(-bSx) \tag{7}$$
$$J(x) = RI_0 \exp(-bSx) \tag{8}$$

where $b = [1/(2R) - R/2]$, so that the depth of penetration of the exciting light into the sample ($x_0$) can be given by

$$x_0 = \frac{1}{bS} \tag{9}$$

and $I(x_0) = I_0 e^{-1}$ (the incident radiation with the intensity $I_0$ is reduced in this case to $e^{-1}$ of the initial intensity, i.e., 63.1% of the incident radiation). Another criteria for defining the penetration depth can be $5x_0$, which corresponds to 99.3% of absorption of the incident radiation.

Later (Section 4) we will give some specific examples of penetration depth in different substrates, for photon excitation in powdered solid samples.

In the case of more than one absorbing chromophore and for optically thick samples we can write

$$K(\lambda) = K_B + 2 \sum_i \varepsilon_i(\lambda) C_i \qquad (10)$$

where $K_B$ concerns the background substrate. From this, it follows that a difference spectrum can be used to evaluate the temporal evolution of a photochemical reaction (by comparing spectra before and after irradiation):

$$\Delta K = S\left[ F(R)_{\text{irradiated}} - F(R)_{\text{nonirradiated}} \right] \qquad (11)$$

## 2.2 Time Resolved Laser Induced Luminescence

Time gated and intensified charge-coupled device [9] detectors coupled to pulsed lasers as excitation sources, which are monochromatic and may have short pulses and high fluences, are a very attractive and rigorous way of performing time resolved luminescence studies. Opaque solid samples imply the use of a reflection geometry, as shown in Figure 3. This figure also shows in a schematic form the

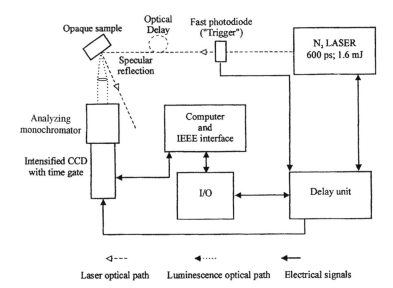

FIG. 3. Schematic diagram of the laser induced luminescence system.

setup used to obtain either fluorescence or phosphorescence spectra. Specular reflection should be avoided in order not to damage the detector, which is the main device of the system.

Since 1997, we have been using in our laboratory an intensified charge-coupled device (ICCD, Oriel model Instaspec V, with a minimum temporal gate of 2.2 ns) in a daily basis for time resolved luminescence studies. The detector has $512 \times 128$ pixels in a maximum spectral range of 200 to 900 nm. With a single laser pulse, a fluorescence or a phosphorescence spectrum can be instantaneously obtained, since the combined use of the delay unit and time gate enables one to separate prompt from delayed emissions.

The fast photodiode detects the zero time, corresponding to the laser pulse. The laser pulse can be optically delayed by the use of an optical fiber to take into account the ICCD and delay unit internal delays. Depending on the signal intensity, the ICCD readings can be intensified or not. The delay unit (picosecond to second range) allows us to perform successive delayed readings according to an initially programmed data acquisition sequence. So both luminescence time resolved spectra and emission decay curves can be obtained with this system. Therefore prompt and delayed luminescence can be obtained without the use of the classical time consuming methods [2].

A good example of room temperature laser induced phosphorescence is presented in Figure 4.

The same time scale was used in both spectra. The decay is slower in the case of inclusion of the ketone into the channel structure of the powdered substrate. Clearly, bimolecular processes as well as nonradiative decay processes were reduced in this case.

## 2.3 Diffuse Reflectance Laser Flash-Photolysis

In the beginning of the 1980s, Frank Wilkinson and co-workers were able to apply the laser flash-photolysis technique to opaque samples [10], thus allowing the study of a considerable number of photochemical processes in heterogeneous systems. Good examples of those studies are, among many others, the photochemical reactions in confined spaces, the study of organic molecules adsorbed or included on substrates with catalytic activity, and dyes adsorbed or covalently bound to natural or synthetic fibres.

Everything points to the laser diffuse reflectance flash-photolysis technique playing in the near future, in regard to transient absorption studies in heterogeneous media, a role at least as important as flash photolysis in the transmission mode has played, until now, for photochemical reactions in homogeneous media.

The technique is based on the study of the temporal evolution of the absorption of excited species, which were created by a laser pulse, through an analyzing light in the diffuse reflectance mode. (See Fig. 5.)

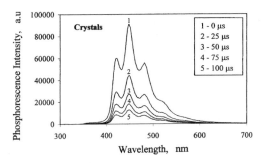

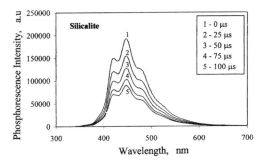

FIG. 4. Room temperature time resolved phosphorescence spectra of benzophenone crystals and benzophenone included into silicalite channels (200 $\mu$mol g$^{-1}$).

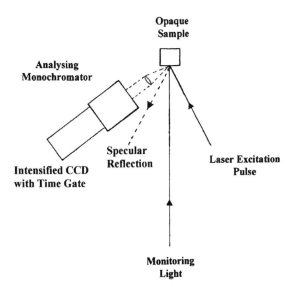

FIG. 5. Laser excitation pulse and monitoring light in the diffuse-reflectance mode.

The setup is identical to the one used for transmission studies, but in this case, a reflection geometry is used, where the monitoring and the detected beams are in the same sample side. As in conventional flash-photolysis, the absorption spectra are difference spectra. This means that the experimentally determined absorption reflects the difference, in terms of absorption, between the excited state (created by laser excitation) and the ground state (detected by the monitoring lamp), for a specific species at a specific wavelength. So the transient absorption (measured as a percentage of absorption, as a hypothesis) depends on both ground-state and excited-state extinction coefficients. In many cases, it is important to consider the emission correction (which is obtained by firing the laser alone) and it is subtractive relative to the transient absorption. (See Fig. 6.)

Apart from being the group responsible for the development of the diffuse-reflectance laser flash-photolysis technique in the temporal range from nanosecond up to seconds [10], Wilkinson et al. were also the first authors to publish transient absorption spectra of opaque materials in the picosecond time domain [11].

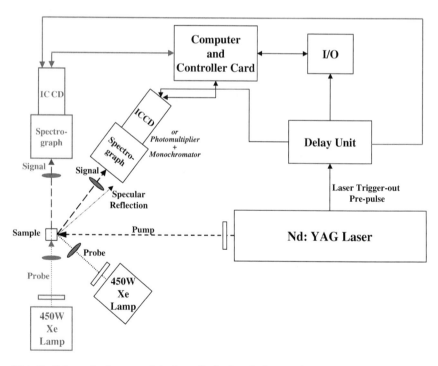

FIG. 6. Schematic diagram of the laser flash-photolysis setup in transmission and diffuse reflectance modes.

In a diffuse-reflectance laser flash-photolysis experiment, the excited chromophores created by the laser excitation may exhibit a nonhomogeneous spatial distribution. Theoretical treatments (Kessler et al. [10b] and Oelkrug et al. [12]) have shown that two extreme types of concentration profiles can be produced. In the first type, the concentration of excited species decreases exponentially as a function of the penetration depth of the excitation radiation whereas, in the second type, the excited species are distributed in a homogeneous manner within a specific width, and then decrease ("plug" profile) as in [10f, 10g].

The latter case can be found for large laser fluences (in terms of moles of photons per square centimeter) and low concentrations of the ground-state absorbers. In this case there is a total conversion of ground-state species into transient excited species. The remission function $F(R)$ can be used for optically thick samples and is a linear function of concentration.

For a small percentage of conversion (high concentration of ground-state absorbers and low laser fluences) the concentration of transients decreases exponentially and a representation of $\Delta R_T$ as a function of time is a good measure of the transient concentration for values of $\Delta R_T < 0.1$.

$$\Delta R_T(t) = \left[ 1 - R_T(t) \right] = \frac{[R_B - R(t)]}{R_B} \tag{12}$$

where $R_B$ is the substrate reflectance before firing the laser, and $R(t)$ is the reflectance at time $t$ after excitation.

For more concentrated samples, the variation of the remission function before and after the laser firing is given by

$$\Delta F(R(t)) = F(R(t)) - F(R_B) \tag{13}$$

and $K(t) = F(R(t))S = K_B + 2\varepsilon_G C_G + 2\varepsilon^* C^*$, $K(0) = F(R_B)S = K_B + 2\varepsilon_G C_o$ and $C_o = C_G + C^*$. $G$ represents ground state, $K_B$ the absorption coefficient, and * means, of course, the excited state. Thus, we can write

$$\Delta F(R(t)) = \frac{2(\varepsilon^* - \varepsilon_G)C^*}{S} \tag{14}$$

and this equation justifies the occurrence of isosbestic points in transient absorption spectra.

Figure 7a presents the four traces needed to obtain a corrected transient absorption decay at a specific wavelength: baseline, top-line, emission, and absorption. By taking into account the entire wavelength range where the excited species absorbs, and also by a suitable choice of the time scale, triplet–triplet time resolved absorption spectra can be obtained, such as the one presented in Figure 7b for benzophenone crystals. These spectra provide both spectroscopic and kinetic information regarding the powdered opaque sample. It is obvious that these spectra will enable one to study the occurrence of a chemical reaction on surfaces, pro-

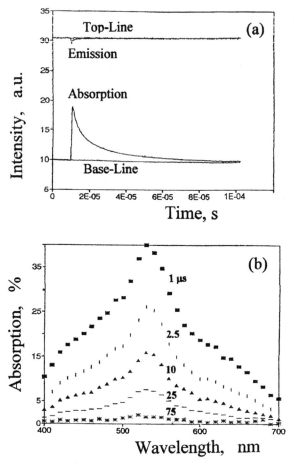

FIG. 7. (a) Data traces for benzophenone microcrystals exciting at 355 nm and (b) time resolved triplet–triplet absorption spectra of microcrystalline benzophenone at room temperature.

vided the time scale is adequately chosen. Figure 7b also shows the triplet–triplet absorption spectra of microcrystalline benzophenone at room temperature, first published by Wilkinson and co-workers in 1984 [10a].

Figure 7b clearly shows that the initial decay is faster than the long time decay, and the excited triplet disappears in the microsecond time scale. The decay is a mixture of first and second order processes, with a maximum absorption at about 540 nm. The phosphorescence emission at room temperature also has an identical kinetic decay, within experimental error.

After the laser pulse, the transient decays. For a unimolecular decay process, the concentration of the excited species decays exponentially, according to

$$C^*(t) = C^*(t = 0) \exp(-k_1 t) \tag{15}$$

And for low fluences of the laser $[1 - R_T(t = 0)] \leq 0.10$ and the transient concentration is proportional to $(1 - R_T)$. So

$$\ln[1 - R_T(t)] = \ln[1 - R_T(t = 0)] - k_1 t \tag{16}$$

For high laser fluences, optically thick layers of homogeneous excited species are produced and one has to use the Kubelka–Munk function to perform the decay analysis; Eq. (16) becomes now

$$\ln\{F[R(t)] - F(R_B)\} = \ln\{F[R(t = 0)] - F(R_B)\} - k_1 t \tag{17}$$

Equation (17) is generally used for data analysis and it predicts a linear relation between $\ln\{F[R(t)] - F(R_B)\}$ with time, from which the rate constant $k_1$ can be determined. For more complex cases one has to use either $[1 - R_T(t)]$ or $F[R(t)]$ depending on the specific concentration profile and on the decay kinetics [12–14].

## 2.4 Other Techniques for Surface Studies

Apart from the techniques described previously for surface photochemistry studies, other approaches can be used.

We would like to stress here the importance of X-ray photoelectron spectroscopy (XPS) for surface characterization, since it analyzes the first 10 or 20 atomic monolayers. It gives information regarding both composition and elemental concentration, as well as the probe–surface interactions. Quite recently [15, 16], this technique allowed the authors to study rhodamine and cyanine dyes physically and/or chemically bound to microcrystalline cellulose.

Also, Fourier transform infrared absorption spectroscopy provides relevant information regarding the specific interactions of different probes within substrates [17], especially in the diffuse-reflectance mode when applied to the study of powdered opaque surfaces that disperse the incident radiation. The extension of this technique to obtain time resolved transient absorption spectra in the IR wavelength range (laser flash-photolysis with IR detection) will certainly play in the near future an important role in terms of clarifying different reaction mechanisms in the surface photochemistry field [17c, 18].

We also want to refer the use of nuclear magnetic resonance, electron paramagnetic resonance, and Raman techniques for solid surface studies [19].

In the next section, the basic principles of XPS will be outlined.

# 3 ELECTRONIC SPECTROSCOPIES

The generic designation of electronic spectroscopy is attributed to every technique that uses electrons as incident (ingoing) probes and/or as analyzed (outgoing) probes. Among the most popular ones, for both the qualitative and quantitative characterization of surfaces, we can find XPS—formerly called electron spectroscopy for chemical analysis—and HREELS.

## 3.1 X-Ray Photoelectron Spectroscopy

### 3.1.1 Basic Principles.
The photoelectric effect is the basis of this technique. When a photon having an energy $E_{\text{photon}}$ impinges on a surface, as schematically displayed in Figure 8, an electron bound to the nucleus with a binding energy $E_b$ is ejected with a kinetic energy $E_k$ related by

$$E_{\text{photon}} = h\nu = E_b + E_k + \phi \tag{18}$$

where $\phi$ is the work function of the spectrometer and needs to be adjusted every time the equipment is vented [20].

Equation (18) can be rewritten in the following ways:

$$E_b = h\nu - E_k - \phi \quad \text{and} \quad E_k = h\nu - E_b - \phi \tag{19}$$

$E_b$ and $E_k$ vary, then, in a symmetrical way. In a photoelectron spectrum, the intensity of ejected electrons as a function of their kinetic energy is registered. With X-ray radiation, $h\nu$ is high enough to eject inner shell electrons. These electrons, contrarily to the valence ones, have their binding energy to the nucleus almost unchanged by the atomic environment. They are, then, a fingerprint of the element where they originate. Since the inner-shell electron binding energy is the variable allowing for the identification of a given element, spectra are usually displayed in the form of photoelectron intensity as a function of the decreasing binding energy rather than the increasing kinetic energy.

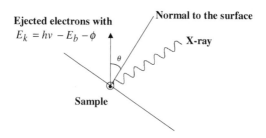

FIG. 8. Schematic representation of an XPS experiment.

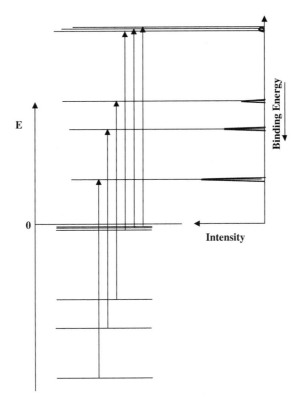

FIG. 9. Schematic representation of an "expected" XPS spectrum.

In a first approximation, we can consider the electron structure as frozen under the photoelectron emission process and identify $E_b$ with the Hartree–Fock energy eigenvalues of the orbitals (Koopman's theorem). A schematic representation of an expected photoelectron spectrum could then be the one in Figure 9.

However, in an accurate calculation of a binding energy, the relaxation energy of the remaining electronic structure to a new hole state has to be included [21]. Besides, many other factors contribute to render the schematic picture displayed in Figure 9 very different in real situations and great care needs to be taken in the interpretation of features appearing in a XPS spectrum. Some of the examples of features other than the main photoelectron peak, which can appear in a XPS spectrum, are shake-up and shake-off peaks, X-ray source satellites (for non-monochromatic X-rays), "cross-talk" peaks, and Auger peaks [22]. Moreover, the ejection of an electron from the inner shell of a given element does not usually give rise to a single peak. Reasons for this are **chemical shift**, orbit–spin cou-

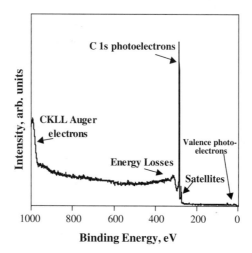

FIG. 10. Survey XPS spectrum of a sample of highly oriented pyrolithic graphite showing several features, other than the photoelectron peak.

pling, and spin coupling (when nonpaired electrons exist in the element). Finally, photoelectrons suffering single or multiple inelastic collisions in the medium lose energy and leave the surface with a lower kinetic energy. This implies that every photoelectron peak has a background at lower kinetic energies (higher binding energies) larger than the background at higher kinetic energies (lower binding energies). In Figure 10, we can see a real survey spectrum of a sample containing a single element from the second period of the periodic table—highly oriented pyrolitic graphite—and exhibiting several of the above mentioned features. The spectrum was obtained with a freshly pealed sample with a Kratos XSAM800 spectrometer operating in a fixed analyzer transmission mode (see Section 3.1.4) under a pressure of the order of $10^{-9}$ mbar and using a pass energy of 10 eV. The magnesium nonmonochromatic radiation was used (main component at $h\nu = 1253.6$ eV).

### 3.1.2 Chemical Shifts.
One of the most useful characteristics of the XPS is the inner-shell binding energy shifts, usually named chemical shifts. A photoelectron ejected from an inner shell feels an attractive action from the nucleus but also a repulsive action from the neighboring electrons, namely the valence electrons. Therefore, a given atom, like carbon, for instance, when bound to a more electronegative atom (O, N, F, Cl, ...) will have its valence electronic density rarefied and so the repulsion felt by its inner electrons will be lower. This will increase the inner electron binding energy. The inverse will occur when it is bound to a less electronegative atom. Also the hybridization state plays a role on the binding energy: the C $1s$ photoelectron binding energy is around 284.7 eV in

TABLE I.    Chemical Shift, in eV, for C 1*s* Phoelectron for the Most
Common Functional Groups in Organic and Organo–Inorganic Polymers

| Functional group | Mean chemical shift (eV) |
|---|---|
| $-C-O-H$ | 1.55 |
| $-C-O-R$ | 1.45 |
| $>C=O$ | 2.90 |
| $-COOH$ | 4.26 |
| $-COOR$ | 3.99 |
| $-OCO-$ | 2.93 |
| $-C-NO_2$ | 0.76 |
| $-C-N<$ | 0.94 |
| $-C=N$ | 1.74 |
| $-CF_3$ | 7.69 |
| $(CF_2)_n$ | 7.48 |
| $-CFH-CH_2-$ | 2.91 |
| $-CSi-$ | -0.67 |
| $-CS-$ | 0.37 |
| $C=C$ | -0.27 |

Reference is C 1*s* in saturated hydrocarbons, $E_b(C1s) = 285$ eV [26].

aromatic compounds and 285 eV in aliphatic chains. Theoretical calculations of chemical shifts appear frequently in literature [23, 24]. Besides general binding energy databases [25], a good database for organic polymers (and a few inorganic ones), was published in the 1990s [26]. In Table I, the chemical shifts for the most common functional groups in organic and inorganic–organic compounds are presented.

### 3.1.3 Charge Shifts.
Most of the samples referred to here—nonconjugated organic films and porous materials like cellulose, silica, zeolites—are electrically insulating. Since the X-ray induces the escape of electrons, the surface of the sample becomes positively charged. A recent paper by Cazaux [27] makes a detailed analysis of the role of various parameters involved in the charging mechanisms of insulating materials, mainly in XPS and Auger spectroscopies. An escaping electron having a kinetic energy, $E_k$, will, then, "feel" an attractive force toward the surface and its movement toward the analyzer is retarded. It is similar to a situation where the electron is ejected with a lower kinetic energy, $E_k - \Delta E$ ($\Delta E > 0$), i.e., an apparent binding energy $E_b + \Delta E$. Therefore, if no charge compensation exists, the charge may increase enormously during the spectrum acquisition. In the limit, no defined peaks are obtained: the apparent binding energy changes continuously. For rapid acquisitions, well-defined peaks are obtained but much

broader than for not charged samples. Hopefully, when nonmonochromatic X-ray sources are used, the sample is very close to the window of the source and the number of electrons emitted by the source filaments, crossing the window and impinging on the sample, is enough to rapidly reach a stationary charge on the surface. The entire spectrum is displaced from its real binding energies but peaks are well defined and their full width at half-maximum (fwhm) has a value very close to the fwhm obtained for a conducting material. In Figure 11, an XPS spectrum of the region Al $2p$ taken on a sample of aluminum covered with a thick film of its native oxide and the same sample freshly chemically etched are shown. This spectrum was acquired in the same conditions as the ones used for the spectrum in Figure 10.

The spectrum drawn in normal line displays two peaks corresponding to the metallic aluminum (lower binding energy) and to the oxidized aluminum ($Al^{3+}$) due to the oxide film that rapidly grows at the surface when it is exposed to the air. These two peaks are a good illustration of the chemical shift cited above. The spectrum drawn with a bold line displays a single peak corresponding to the species $Al^{3+}$ in the oxide film. But, this time, the oxide film is so thick that the metallic aluminum is hidden. Due to its extreme thickness, the film is more insulating than the one existing after the etching and the $Al^{3+}$ peak is charge shifted toward larger binding energies. Simultaneously, the peak is broadened. This example is, therefore, a good example to illustrate the effect of the insulating character of a sample on an XPS spectrum: charge shift and peak broadening. The Al $2p$ peak, as any $np$ peak, has two components – the $2p_{3/2}$ and the $2p_{1/2}$.

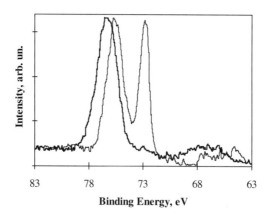

FIG. 11. XPS Al $2p$ region: bold line: Aluminum covered with its native oxide; normal line: the same sample chemically etched to remove the oxide and immediately introduced into the chamber.

However, the energy splitting is very small—~0.4 eV [25]—and they are only separable by curve fitting.

The correction of the binding energy for this charging effect is made using one of the peaks in the spectrum; the most used is the C 1s contamination peak, to which a binding energy equal to 285 eV is assigned [26]. For spectrometers equipped with monochromatic sources, a flood gun for charge neutralization is used [28].

### 3.1.4 Equipment.

A large number of books and book chapters were published containing a description of the X-ray photoelectron spectrometers. The "bible" in this domain is the book edited by Briggs and Seah [29, 30]. The main components in a spectrometer are the X-ray gun, the sample holder, an electrostatic gun, the energy selector, and the electron multiplier. In addition, of course, there is all the electronics needed to apply the voltages in all the lens and energy selector electrodes as well as the software suitable to acquire and treat spectra. It is out of the scope of this chapter to provide a thorough description of the instrumentation details. We just would like to stress the utility of having a multianode X-ray gun, at least a dual-anode one. In fact, in the task of assigning all the features appearing in a spectrum, it helps a lot to have at least two different anodes (the most common combination is magnesium and aluminum). For instance, photoelectron binding energies are independent of the photon energy whereas Auger electrons have constant kinetic energies. Therefore, for spectra of a same sample acquired with a Mg anode ($h\nu = 1253.6$ eV) and with an Al anode ($h\nu = 1486.6$ eV $= (1253.6 + 233)$ eV), in a binding energy scale, photoelectron peaks will remain in the same position. On the contrary, Auger peaks will appear displaced by 233 eV toward larger binding energies in the Al-anode spectrum relative to the Mg-anode one. Also the satellite peaks are at different distances from the main photoelectron peak when different anodes are used, as can be seen in Table II, and the comparison between two spectra allows the complete assignment of satellite features in nonmonochromatic sources.

The last generation of spectrometers is monochromatic, allowing a best resolution in energy (an intrinsic resolution of 0.28 eV is estimated [26]). Also the lateral resolution has increased: in conventional equipment the lateral resolution was of the order of magnitude of 1 mm (in fact, the size of the area "seen" by the monochromator defined as the fwhm of the spatial intensity distribution is about 3 mm × 1 mm) whereas now it is about 10 $\mu$m.

All the spectrometers where electrons are analyzed need to work under conditions of high vacuum. Moreover, to keep surfaces clean (at the atomic level), ultrahigh vacuum conditions are generally used in XPS spectrometers.

The electron energy selection is usually made through two different modes: fixed analyzer transmission and fixed retardation ratio. The first one is the one used when a quantitative analysis is required. It consists of keeping the analyzer voltages constant, the kinetic energy being swept by a variable retardation voltage

TABLE II.    Magnesium and Aluminum X-Ray Spectrum for Magnesium and Aluminum $K\alpha$ Lines

| X-ray | Mg | | | Al | | |
|---|---|---|---|---|---|---|
| | Energy (eV) | Rel. Int. | $\Delta E$ (eV)[a] | Energy (eV) | Rel. Int. | $\Delta E$ (eV)[a] |
| $K\alpha_1$ | 1253.7 | 67 | | 1486.7 | 67 | |
| $K\alpha_2$ | 1253.4 | 33 | 100.0 | – | 1486.3 | 33 | 100.0 | – |
| $K\alpha'$ | 1258.2 | 1.0 | 4.6 | 1492.3 | 1.0 | 5.7 |
| $K\alpha_3$ | 1262.1 | 9.2 | 8.5 | 1496.3 | 7.8 | 9.7 |
| $K\alpha_4$ | 1263.7 | 5.1 | 10.1 | 1498.2 | 3.3 | 11.6 |
| $K\alpha_5$ | 1271.0 | 0.8 | 17.4 | 1506.5 | 0.42 | 19.9 |
| $K\alpha_6$ | 1274.2 | 0.5 | 20.6 | 1510.1 | 0.28 | 23.5 |
| $K\beta$ | 1302.0 | 2.0 | 48.4 | 1557.0 | 2.0 | 70.4 |

[a] $\Delta E$ is computed relative to the $K\alpha_{1,2}$ lines taken respectively at 1253.6 and 1486.6 eV.

between the sample and the analyzer entrance. In this mode the energy resolution, defined as the fwhm of the electron energy distribution leaving the energy analyzer, is kept constant. The second mode is used when the sensitivity at low kinetic energy needs to be increased for identification purposes and is used exclusively in survey spectra. It consists of retarding the photoelectrons between the sample and the analyzer entrance by a constant factor and analyzing their energy by sweeping the voltage difference between the analyzer electrodes.

3.1.5 Electron Escape Mechanisms and Surface Sensitivity.    One important parameter in the XPS technique is the electron **effective atenuation length** (EAL), $\lambda$. It is defined as the distance normal to the surface at which an electron should be ejected from the atom in order for the probability of escaping the surface without losing energy due to inelastic scattering processes to have the value $1/e$ for an angle $\theta = 0$ (see Figure 8). For a good discussion about the concept and some confusion in the literature between this parameter and the inelastic free mean path (IMFP), see, for instance, [31]. If the number of photoelectrons produced at depth $x$ is denoted $n_0(x)$, the number of electrons, $n(x)$, escaping the surface is, therefore, given by

$$n(x) = n_0(x)e^{-x/\lambda} \tag{20}$$

The effective attenuation length is a function of electron kinetic energy and of the medium density. However, when we represent $\lambda$ in number of monolayers, an approximately universal curve is obtained [32] as displayed in Figure 12.

Given the low value of $\lambda$ for the energy range corresponding to photoelectrons generated by Mg or Al sources, this technique analyzes depths of the order of a few atomic monolayers: for an homogeneous material, 63.2% of the signal comes

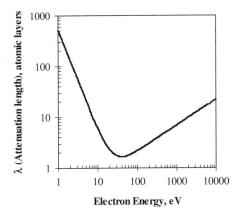

FIG. 12. Schematic representation of electron effective attenuation length as a function of electron energy.

from a depth equal to $\lambda$, 86.5% from $2\lambda$, 95% from $3\lambda$, 98.1% from $4\lambda$, and 99.3% from $5\lambda$. Given these values, it is generally considered that the information depth is $3\lambda$ but some authors take $5\lambda$ as the information depth. EAL are generally measured by the technique of overlayer film [31] which presents many practical problems. However, many semiempirical and empirical formulas exist to compute another related quantity, the IMFP $\lambda_i$. EAL is different from IMFP by about 15–30% depending on the relative importance of the elastic interaction processes [31, 33]. The greatest differences arise for high atomic numbers and low electron energies. One of the most useful relations to obtain IMFP (for kinetic energies, $E_k$, larger than ~500 eV), given its simplicity, is expressed by [34]:

$$\lambda_i = k E_k^m \tag{21}$$

where $k$ and $m$ are empirical parameters. In Table III values of these parameters proposed by Tanuma et al. [35, 36] for some organic polymers and inorganic compounds are presented.

The order of magnitude of IMFP for a given photoelectron depends on the crossed medium. For instance, $\lambda_i$(C 1s) computed using Eq. (21) and parameters in Table III varies between 30 and 34 Å in organic materials and between 18 and 22 Å in inorganic materials. In many applications, the relative IMFPs are sufficient and, contrary to the absolute ones, they are very insensitive to the medium. In Table IV IMFPs relative to C 1s IMFP for several photoelectrons ejected from elements contained in materials here studied are presented.

A NIST database is available in electronic form [37].

### 3.1.6 Quantitative Aspects and ARXPS.
As in any other spectroscopy, the major source of error in the computation of the area of a given peak is the

TABLE III.    Parameters $k$ and $m$ to Compute $\lambda_i$ in Organic Polymers [35]
and Some Inorganic Compounds [36]

| Compound | $k$ | $m$ |
|---|---|---|
| 26-$n$-Paraffin | 0.145 | 0.794 |
| Polyacetylene | 0.107 | 0.790 |
| Poly(butene-1-sulfone) | 0.139 | 0.782 |
| Polyethylene | 0.137 | 0.790 |
| PMMA | 0.149 | 0.786 |
| Polystyrene | 0.138 | 0.790 |
| Poly(2-vinylpyridine) | 0.138 | 0.791 |
| $Al_2O_3$ | 0.112 | 0.760 |
| $SiO_2$ | 0.103 | 0.777 |
| SiC | 0.0936 | 0.770 |
| $Si_3N_4$ | 0.0933 | 0.766 |

TABLE IV.    Inelastic Mean Free Path, $\lambda_i$, Relative to the One for C $1s$, for
Some of the Most Common Photoelectrons Ejected from Organic Polymers
and Some Inorganic Compounds Using the Magnesium $K\alpha_{1,2}$ Line

| Photoelectron | $\lambda_i/\lambda_i(\text{C } 1s)$ |
|---|---|
| O $1s$ | $0.794 \pm 0.005$ |
| N $1s$ | $0.906 \pm 0.005$ |
| F $1s$ | $0.655 \pm 0.005$ |
| Cl $2p$ | $1.068 \pm 0.005$ |
| Si $2p$ | $1.146 \pm 0.005$ |
| S $2p$ | $1.095 \pm 0.005$ |
| Al $2p$ | $1.166 \pm 0.005$ |

background subtraction. In XPS, background is generally curve fitted according to Shirley's method [38]: the background intensity within a peak is assumed proportional to the peak integrated intensity (area between the peak and the background) at higher kinetic energy. The spectrum is constrained to be zero at either end and the background to be subtracted is computed by an iterative method. Another very popular method is due to Tougaard who has developed much work on this subject during the last 20 years [39]. Also Monte Carlo methods can be used to simulate the inelastic background to be subtracted [40, 41]. The problem with these two methods is that they require a very detailed knowledge of all the electron–medium interaction mechanisms and respective cross-sections to electron transport through the medium, a task still in progress [42]. That is why the old Shirley's method cited above keeps its popularity.

Once the background is subtracted, the area of a peak, due to a $w$ photoelectron ejected from an element $B$, is given by [43].

$$I_{Bw}(\theta) = K(\theta) \int_0^{\ell} c_B(x) \exp\left(-\frac{x}{\lambda_{Bw} \cos \theta}\right) dx \qquad (22)$$

where $I_{Bw}(\theta)$ is the intensity of photoelectrons coming from element $B$ and having an effective attenuation length $\lambda_{Bw}$, $c_B(x)$ is the density of the element $B$ as a function of the depth $x$ (concentration profile), $\theta$ is the angle between the normal to the specimen surface and the mean direction of analysis (see Figure 8)[1], and $\ell$ is the sample thickness. $K(\theta)$ contains the $w$ photoelectron ejection cross section from the element $B$, the X-ray intensity,[2] and a response function which depends on the $w$ photoelectron kinetic energy, $E_k$ (usually, $\propto E_k^{-1}$). For a sample homogeneous in depth ($c_B(x) = \text{constant} = c_B$) and having a depth much larger than $\lambda$ ($\ell = \infty$), Eq. (22) becomes

$$I_{wB} = K' Q_{wB} c_B \qquad (23)$$

where $Q_{wB}$ is called the sensitivity factor and $K'$ is a constant independent of the analyzed photoelectron. If two elements, $B$ and $C$, exist in the same sample, we can write

$$\frac{I_{wB}}{I_{yC}} = \frac{Q_{wB}}{Q_{yC}} \frac{c_B}{c_C} \qquad (24)$$

Therefore, only relative values of $Q$ are needed to make quantitative analysis. Usually, sensitivity factors are normalized to fluorine F $1s$ ($Q(\text{F } 1s) = 1$) or to C $1s$ ($Q(\text{C } 1s) = 1$) and are included in libraries furnished with spectrometers.

Equation (22) is particularly useful when a concentration gradient in depth exists. In this case, several spectra at different values of $\theta$ are taken and the analysis is called angle resolved X-ray photoelectron spectroscopy. However, for a maximum efficiency, a flat surface (at an atomic level) is needed to avoid shade effects as shown by Fadley in his early works in the 1970s [44]. An additional problem exists: the extraction of concentration profiles, $c_B(x)$, from Eq. (22) is an inverse problem: the intensity as a function of the analysis angle is the Laplace transform of the composition depth profile of the sample [45] and does not have a unique solution. Several algorithms to solve the inversion problem were developed and tested [46]. They are all very unstable and sensitive to small statistical

---

[1]This angle $\theta$ is the complementary to the take-off angle [29] defined as the angle between the specimen surface and the mean direction of analysis.

[2]Since the attenuation length for X-rays is much larger than for electrons, X-ray intensity is considered constant through the entire analyzed depth.

fluctuations in the photoelectron intensity and to small uncertainties in the analysis angle (smaller than the experimentally available ones). The most usual and practical way of applying Eq. (22) is, therefore, to test concentration profiles, for two or more elements, needing a few parameters (for instance, linear or exponential profiles needing just the atomic densities at the extreme surface and at the bulk and the rate of variation with depth) and to compute them by fitting to the experimental data $I_B(\theta)/I_C(\theta)$. The most simple concentration profile, requiring a single parameter, is enough for example to test if a given sample is made of a material covered by a overlaying film of homogeneous composition and constant thickness. In this case, admitting that element $B$ only exists in the overlaying film and element $C$ only exists in the material underneath, we will have:

$$\frac{I_{wB}}{I_{yC}} = \frac{Q_{wB}}{Q_{yB}} \times \frac{c_B}{c_C} \times \frac{1 - \exp(-\ell/\lambda_{wB})}{\exp(-\ell/\lambda_{yC})} \qquad (25)$$

A single point is enough to obtain $\ell$. However, an angular distribution is needed to confirm the assumption that the thickness film is constant.

## 3.2 High Resolution Electron Energy Loss Spectroscopy

### 3.2.1 Basic Principles.
In a HREELS experiment, a monokinetic electron beam with a primary energy, $E_p$, interacts with the surface region in different ways exciting vibrational modes [47] and electronic states [48]. Intensity of the backscattered electrons is measured versus primary energy for a given direction as shown in Figure 13.

Primary energy can be varied in a continuous way from 0 to a few tens of eV (or even a few hundreds). Energy losses can be measured from 0 to typically 1 eV or, in some modified equipment, from 0 to 15 eV, allowing one to obtain information about excited states from IR to UV, more precisely from vibrations to electronic excitations and ionizations, without changing the probe source or the detection system.

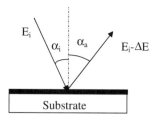

FIG. 13. Schematic representation of an HREELS experiment: $E_i$ is the primary energy, $\alpha_i$ is the angle of incidence, and $\alpha_a$ is the analysis angle, both relative to the surface normal.

**3.2.2 Electron–surface Interaction Mechanisms.** Depending on the mechanism (dipole, impact, or resonance) [47, 49], electron interactions can be produced at different distances. In particular, impact interactions (short range) produced at typical interaction distances of a few Ångström and observed in off-specular conditions become mainly sensitive to molecular groups exposed at the film–vacuum interface [50, 51]. It is also very important to emphasize that HREELS selection rules enable optical forbidden transitions to be observed. In fact, during the interaction, electron exchange between incident and molecular electrons can occur, allowing optically spin forbidden excitations (singlet–triplet transitions, for instance). On the other hand, symmetry modifications of the molecular orbitals induced by the electromagnetic field associated with the incident electrons can also break symmetry selection rules of optical excitations. Furthermore, in HREELS, electrons can exchange the momentum as well as energy with the medium. This fact is important in the case of crystalline materials, since the momentum and energy transfers may lead to the observation of nonvertical electronic excitations, which cannot be detected by optical spectroscopy.

**3.2.3 Equipment.** The pioneering work in HREELS equipment development was made by several groups in parallel in Canada [52], France [53], Germany [54], and in United States [55]. A good review about this item can be found in Chapter 2 of Ibach and Mills' book [56]. In Figure 14, a schematic representation is made of the spectrometer LK 2000-R [57]—an old generation spectrometer—showing the main components: a filament which emits an electron beam broad in energy and two electrostatic 127° cylindrical energy selectors to obtain a beam with a fwhm of about 5 meV. This beam impinges on the sample (which can be rotated).

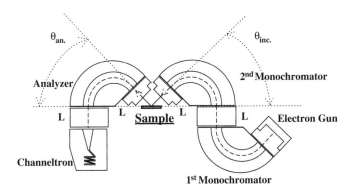

FIG. 14. Schematic representation of LK2000-R spectrometer. Sample can be rotated. The analyzer can also be rotated from $\theta_{an} = 13°$ to $\theta_{an} = 68°$.

Backscattered electrons are analyzed by a third energy selector (in other models, two selectors are also used to analyze backscattered electrons) and multiplied and counted by a channeltron. Between the filament and the first energy selector and between the two energy selectors injection lenses exist. Between the sample and the neighboring energy selectors there are also focusing lenses. All magnetic components must be excluded from the instrument and shielding (assured by cylinders of mu-metal) is required to prevent magnetic deflection from externally based fields (the magnetic earth field, for instance). The new generation models, providing energy resolutions better than 1 meV, have essentially the same components but the injection lenses have different designs [58].

Primary energy can be varied from 0 to 250 eV, in the particular model presented in Figure 14, by varying the voltage applied between the filament midpoint and the sample. For the same model, energy losses can be analyzed from 0 to 15 eV, enabling the acquisition of complete spectra for incident energies lower than 15 eV. However, for the ultimate resolution spectrometer, the range of measurable energy losses is much more limited, ranging from 0 to 1 or 2 eV.

### 3.2.4 Sample Charging.

When a low energy electron impinges on an insulating solid surface, a large probability for penetrating the solid and being trapped exists. Henceforth, the surface charges negatively if the electron primary energy is lower than the ionization energy and may charge negatively or positively for larger primary energies depending on the secondary electron yield. Since the electron energy is determined by the voltage difference, $\Delta V_n$, between the emitting filament and the sample surface, if the surface charges, the primary energy changes continuously during the acquisition of a spectrum. The primary energy, $E_p$, is related to the nominal energy, $E_n$, by

$$E_p = E_n + e\Phi_{ct} + e\Phi_c = e(\Delta V_n + \Phi_{ct} + \Phi_c) \qquad (26)$$

where $\Phi_{ct}$ is the contact potential, $\Phi_c$ is the charge potential, and $e$ is the electron charge. Polymeric films always charge under electron irradiation. For very thick films, a flood gun is frequently used to stabilize the accumulated charge [59], eventually combined with spectral restoration algorithms [60]. For very thin films, with thickness around a few hundreds of Å, this surface charge attains an equilibrium value very rapidly and the only consequence is that the primary energy depends on the final charge. This fact can be used to obtain qualitative information about the relative conductivity of samples. For instance for three different $\alpha$-oligothiophenes films—quaterthiophene (4T), quinquethiophene (5T), and sexithiophene (6T)—deposited on gold samples by a procedure described elsewhere [61] the relation between $E_p$ and $E_n$ becomes increasingly linear and the slope of the variation approaches unity as the chain size increases (see Fig. 15).

However, for the two 6T samples, no difference exists between slopes. The only difference comes from the intercept values. This means that their contact potentials with the spectrometer are different since graphite was replaced by gold.

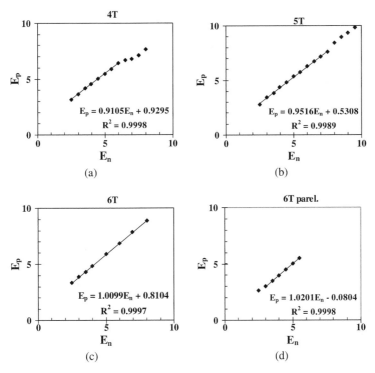

FIG. 15. Extent of HREELS spectra, $E_p$, as a function of nominal energy, $E_n$, for: (a) 4-thiophene; (b) 5-thiophene; (c) 6-thiophene films on gold with chains perpendicular to the substrate; (d) 6-thiophene films on graphite with chains parallel to the substrate. Straight lines fitted by a least square method are also displayed having the following equations: (a) $(0.93 \pm 0.06) + (0.910 \pm 0.015)E_n$; (b) $(0.53 \pm 0.13) + (0.952 \pm 0.024)E_n$; (c) $(0.81 \pm 0.08) + (1.010 \pm 0.016)E_n$; (d) $(0.00 \pm 0.09) + (1.02 \pm 0.02)E_n$. Energies are in eV.

Other sample requirements are low vapor pressure since HREELS spectrometers work under ultra-high-vacuum conditions.

# 4 EXAMPLES OF SYSTEMS STUDIED BY ELECTRONIC AND PHOTONIC SPECTROSCOPIES

## 4.1 Microporous Solid Surfaces and Finely Divided Powders

### 4.1.1 Surface Photochemistry Studies.
There is a growing interest in studies of photoprocesses regarding molecules on surfaces, either physically adsorbed or chemically bound to solid surfaces, grains, powders, or gels [1–5].

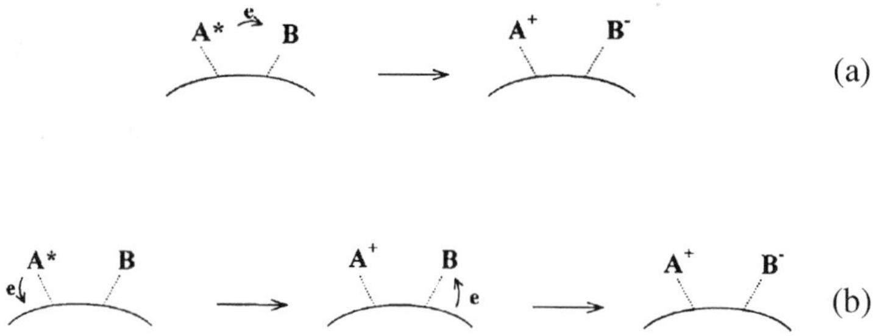

FIG. 16. Electronically inert (a) and electronically active substrate (b).

Common adsorbents are oxides such as silica, alumina, alumino-silicates, and clays, among others. Rarer studies were also presented for cellulose and cellulose derivatives or starch [2a]. Cyclodextrins and calixarenes are good examples of organic host molecules, which can form inclusion complexes with many guest molecules.

In a great variety of solid supports, the number and nature of the surface reactive groups drastically affect the distribution and local organization of the adsorbates. The adsorbent pretreatment and handling, or the solvent used for probe deposition, strongly influence the surface and, as a consequence, the molecule's adsorption. Therefore, their photochemistry and/or photophysics vary according to the surface pretreatment.

Two goals usually exist in these studies: to observe the way interactions with the surface affect the probe's behavior in the excited state, and also how to use photochemistry as a tool to probe the surface of an unknown substrate.

Apart from the interest in the above-mentioned substrates, which we can call "electronically inert substrates" (Fig. 16a), we should also refer to the intense research activity in the field of "electronically active substrates" (Fig. 16b), namely semiconductors.

In these systems, as the solid substrate absorbs the excitation radiation, electrons are promoted to the conduction band and holes are formed in the valence band, which may react with the adsorbate at the solid surface. The excited states of the adsorbed molecule may also be quenched as a consequence of an electron transfer from the probe to the surface. In this text, we shall not consider these cases.

4.1.2 Pioneer Work in This Field. The initial studies on the photochemistry and photophysics of molecules adsorbed onto solid surfaces are relatively recent and were done in the 1960s (apart the work of Boer et al. in the 1930s [62]). Leermakers [63] presented an excellent review of the work produced up to 1970, where he also included his own work concerning ketones adsorbed on silica gel, the *cis–trans* isomerization of stilbene on silica, the spiropyranes photochromism, and the photocleavage of cyclohexadienones.

These studies clearly established that in many cases, after adsorption, energetic changes have occurred, as well as changes in the nature of the electronic excited state. As a consequence, the efficiency of the various photophysical and photochemical processes also changes. The adsorbate–adsorbent interaction may be simultaneously nonspecific and specific, namely, the interactions of the surface-active groups (as an example, the -OH groups on the oxide surfaces). The interaction forces are responsible for adsorption and may have an electrostatic or dispersive nature or hydrogen bonding formation.

Several spectroscopic and nonspectroscopic techniques may be used to study the bonding nature of the adsorbate to the surface [2a, 4]. In the first case we want to emphasize the importance of diffuse reflectance techniques for absorption and emission studies in the ultraviolet (UV), Visible (Vis), and near infrared (NIR) spectral ranges, X-ray photoelectron spectroscopy, and Fourier transform infrared spectroscopy. In the second group, we refer the heat adsorption and the isotherm adsorption techniques, among others.

It is obviously crucial to have a detailed knowledge of the surface structure and its modification under different experimental conditions.

We will refer here to some of the initial studies on silica gel, Vycor porous glass, alumina, and some zeolites. In the 1970s and in the 1980s several groups produced important work in the field, namely de Mayo and Ware et al. for silica surfaces [64]; Oelkrug et al. [65], and Thomas et al. [66] who presented several studies of different probes on alumina surfaces, among others; Turro and Scaiano published very interesting work for silicas, zeolites, and other surfaces [67, 68]. All these studies provided a solid base for this new discipline.

Wilkinson et al. developed, at the end of the 1980s, the diffuse reflectance laser flash-photolysis technique [1, 10], which proved to be crucial for transient absorption and emission studies on surfaces, providing both spectroscopic and kinetical information. This technique for studying solid and opaque media became so important for surface studies as the conventional flash-photolysis was and still is for transparent media, after its discovery by G. Porter in the 1950s.

Pyrene was one of the widely used probes in the initial surface photochemistry studies due to the long lifetime of its monomer, the capacity of excimer formation, and also its spectral sensitivity. The III/I (370 nm/390 nm) vibronic band ratio was successfully used to monitor the microscopic polarity of the adsorbent, either onto silica or alumina [66]. Peak I, the 0–0 band of the $S_0 \rightarrow S_1$ absorption, is symmetry forbidden and grows in polar media. In the case of alumina, this surface exhibits a surface polarity similar to the one presented by polar solvents such as methanol [66a].

On the silica surface, pyrene shows a III/I peak ratio also characteristic of a strongly polar and hydrophilic surface. The derivatized chlorotrimethylsilane surface revealed higher values for the III/I ratio, showing an increase in surface hydrophobicy due to this treatment. Solvent co-adsorption also changes the III/I peak ratio, therefore showing the variation of the hydrophobic and hydrophilic surface characteristics [66b].

de Mayo et al. [64] and other authors [69, 70] used several polycyclic aromatic hydrocarbons (PAHs) such as naphthalene, pyrene, and anthracene as probes to study the surface mobility and also aggregation effects on different surfaces. The percentage of the surface coverage (usually expressed in terms of % of the mono-layer) is crucial with respect to the two above-mentioned aspects. PAHs molecules have delocalized $\pi$ electrons, which strongly interact with the adsorbent surface by forming hydrogen bonds, provided neither steric hindrance nor physically adsorbed water (which may prevent the probe–host direct interaction) acts as a barrier.

This specific host–guest interaction may also occur with nonbonding electron pairs of heteroatoms, as Figure 17 shows.

The fluorescence emission of adsorbed PAHs (either from monomers or excimers) is usually multiexponential, in accordance with the adsorbent heterogene-

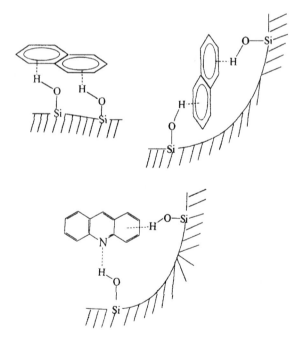

FIG. 17. Schematic representation of naphthalene and acridine adsorption on silica surface.

ity, thus showing different adsorption sites. Usually those multiexponential decays are analyzed with two or three exponential decays, according to

$$I(t) = a_1 e^{-t/\tau_1} + a_2 e^{-t/\tau_2} + a_3 e^{-t/\tau_3} \tag{27}$$

where $\bar{\tau}$ is the average lifetime decay

$$\bar{\tau} = \sum a_i \tau_i^2 / \sum a_i \tau_i \tag{28}$$

and $f_i = a_i \tau_i / \sum a_i \tau_i$ is the fraction of the excited molecules, which have a lifetime of $\tau_i$. Several other kinetic models were presented for data analysis in heterogeneous media [65, 66, 71].

Oelkrug et al. used several silicas and alumina as adsorbents, and different PAHs, diphenylpolyenes, and acridine and derivatives, among others, as adsorbates [65]. They have shown that both spectroscopic and kinetic data were strongly dependent on the surface's pretreatment and also on the adsorption protocol used for sample preparation. In the case of acridine (A) and for moderate pretreatment at low temperatures of the alumina surface (activation temperature $T_a \sim 100\,°C$), the interaction with the surface is essentially composed of hydrogen bonds and protonation of the adsorbate (AH$^+$). For silica activated at $300\,°C$, both ground-state absorption and fluorescence emission are similar to those obtained in solvents such as water or ethanol, thus suggesting that acridine is simply bound to the substrate by hydrogen bonds and is the only emissive species. Al$_2$O$_3$ pretreatment at $600\,°C$ results in A$\sigma^+$ complex formation between the lone electron pair of the nitrogen atom of acridine and the Lewis acidic sites of alumina, whereas AH$^+$ species are unfavored by dehydroxilation.

The acidity of surfaces such as the Vycor porous glass was also studied using appropriate probes: Lin et al. [72] used 9,10-diazofenanthrene as proton acceptor from the Brönsted acidic sites of the surface, which bind to the nitrogen atoms of the probe.

Suzuki and Fujii [73] used acridone as a probe for different pretreatments of silica. The results obtained for acridone adsorbed on moderately pretreated silica (simply warmed at $200\,°C$ under a reduced pressure of $\sim 10^{-6}$ mbar) were compared with the emission in benzene, ethanol, and H$_2$SO$_4$ 18 N. The conclusion was that the main emissive species is the acridone molecule, which forms hydrogen bonds with the silanol surface groups of silica (the carbonyl group of acridone or $N$-methylacridone interacts with the surface hydroxil groups, whereas the amino group hardly interacts). The protonated species also emits on the silica surface, but this component is less important.

It is worthwhile to refer here to the use of rhodamine B and rhodamine 6G and other xanthene dyes as probes for the study of organic crystals, calcium fluoride, and quartz plates [74].

In all the examples quoted until now, molecules with *singlet excited states* were used as probes, with relatively short lifetimes, usually in the nanosecond time range.

The use of *triplet excited states* as probes to study the surface properties of many solids is particulary interesting, due to the fact that they usually exhibit long lifetimes which in most cases come closer to those obtained for rigid matrices. These long lifetimes increase in many cases the efficiency of several photochemical processes. Therefore spectroscopic and kinetic studies can be performed in a wide and interesting variety of situations. As we said before, the development of the diffuse reflectance laser flash photolysis technique [1, 10] by Wilkinson et al. was crucial for the development of these studies on surfaces.

Later we will present some examples of studies of molecules which exhibit high intersystem crossing yields used as probes for surface studies, but before doing that, it is important to describe the methods for sample preparation. We also describe some of the substrates used for surface photochemistry studies.

### 4.1.3 Some Solid Powdered Substrates: Cellulose, Silica Gel, and Silicalite.

*4.1.3.1 Cellulose.*    Cellulose has been used as a solid powdered substrate for the study of photophysical and photochemical studies of several organic probes, mostly dyes. Some of the properties of this substrate, namely the capacity of adsorbing molecules both by entrapment and on the surface of the natural polymer (forming in many cases hydrogen bonds) and also the absence (or extremely reduced) of diffusion of oxygen, make this substrate a particularly attractive one for room temperature luminescence studies [75–83]. We recently published some fluorescence and phosphorescence studies of rhodamine dyes [15, 81, 82], auramine O [81b], 2,3-naphthalimides [84], oxazine [85], acridine orange [86], and cyanine dyes [16, 87, 88] adsorbed on cellulose.

In all these studies, microcrystalline cellulose was used as the solid powdered substrate. The structural formula of cellulose is presented in Figure 18.

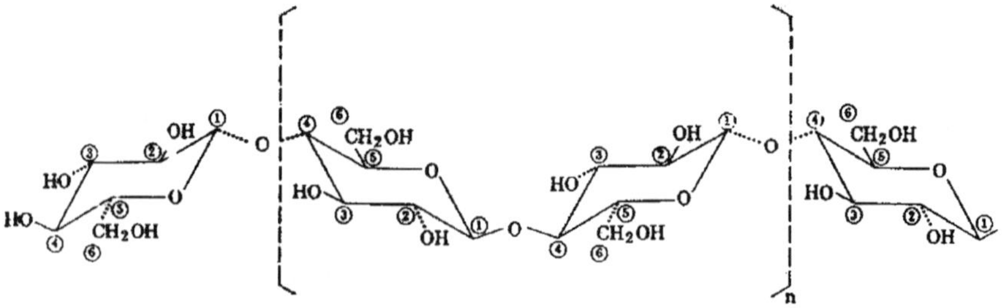

FIG. 18. Cellulose structure.

From a structural point of view, cellulose is a polymer of $D$-glucose in which the individual units are connected by $\beta$-glucoside bonds between the anomeric carbon of one unit and the hydroxyl group in the $C_4$ position of the neighbor unit. Cellulose is probably the most abundant organic compound that exists on earth. It is the chief structural component of vegetal cells. Wood strength is derived from the hydrogen bonds between the hydroxyl groups of neighbor polymer chains. These hydrogen bonds are favored by the linear structure of the natural polymer, which presents conformations that favor those interactions. In the case of starch, although it may also present linear structures, it has no linear conformations; thus the hydrogen bonds are unable to become the main interaction between chains.

X-ray diffraction studies have shown that native cellulose is a two phase system: one is amorphous, with lower order and compactness, and is localized at the elemental fibril surface; the other one is highly ordered and compact (crystallites), where the polymer chains are well organized (crystalline structure) and strongly bound by hydrogen bonds.

Microcrystalline cellulose is simply a pure form of cellulose obtained by an acid treatment of native cellulose. The amorphous regions are preferentially attacked and transformed, and the final residue is highly crystalline.

The swelling of cellulose in the presence of moisture is a well known property of this material. Other protic and nonprotic solvents such as methanol, ethanol, acetonitrile, and acetone also have the capacity to swell microcrystalline cellulose. However, solvents such as benzene, toluene, or dichloromethane do not promote this effect. Therefore it is possible to control adsorption of probes on microcrystalline cellulose, either on the surface or entrapped within the natural polymer chains. After the removal of the solvent used for sample preparation, and for a swelling solvent, a chain–guest–chain interaction is promoted, replacing the previous chain–solvent–chain interaction.

Of particular relevance is the study of dye photodegradation, either on wet or dry cellulose, due to the importance of this effect in the textile industry [89]. Today we know that an increase of the humidity content in the fiber promotes a decrease in the lightfastness of many dyes. This effect exists both for cotton and other polymers where the dye may be adsorbed or covalently bound [89–91]. It also occurs for wool, although to a smaller extent, since in this substrate the mechanism is essentially reductive, while in cellulose it is an oxidative one [89].

*4.1.3.2 Silica Gel.*   Silica gels are porous and granular forms of amorphous silicas, formed by a complex net of microscopic pores which attract and retain water or organic solvents by means of physical adsorption [92–98]. Porous silica has a sponge structure, from which results a very high specific surface area, that varies greatly with pore size (from 20 to 750 $m^2/g$). This surface area is essentially the internal area of the pore walls. The average pore size can be obtained by surface area measurements. Due to the fact that even the smaller pores are larger

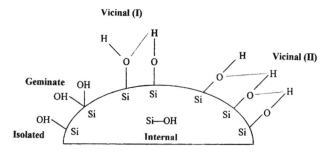

FIG. 19. Schematic representation of silica surface.

than most molecules, although of the same order of magnitude, it is not a surprise that restrictions of mobility do occur for adsorbed molecules [7].

Porous silica surface contains both silanol (Si—OH) and siloxane groups (Si—O—Si). Silanols are considered to be strong sites for adsorption, while siloxanes are hydrophobic sites [92, 96]. Silanols may be isolated, vicinal, or geminate and may be linked by hydrogen bonds to the surface water. Figure 19 shows the silica surface in a schematic way.

In the pioneering study of Snyder and Ward (1966) [92] H-bonded silanols form pairs and these pairs were considered to be the more active sites on a surface (due to an enhanced acidity of the proton not engaged in the H-bond), responsible for probe adsorption at the silica surface. We now know (by temperature-programmed desorption studies) that isolated silanols are the more reactive sites on the silica surface [98]. H-bonded silanols have a desorption energy of 50–60 kcal mol$^{-1}$ and the isolated silanols $\sim$90 kcal mol$^{-1}$ [98].

Moderate heating of silica in vacuum (100–120 °C) ensures a quasi-complete removal of the physically adsorbed water. However, the final water monolayer is removed at activation temperatures of 200 °C. The use of higher temperatures (200–1000 °C) promotes removal of chemisorbed water [96].

The importance of the hydroxyl groups from the point of view of adsorption comes from the fact that the higher the percentage of active silanols per surface area unit, the larger the efficiency of the adsorption process. Small pore size silicas present a higher percentage of active silanols relative to larger pore size silicas [92, 93, 96]. In the former case we detected conformer formation for several dyes, depending both on the pore size and on the dye structure itself [7].

*4.1.3.3 Silicalite.* Silicalites are very specific forms of pentasilic zeolites with small pores and, opposite to alumina–silicate zeolites, they present a strong hydrophobic and organophylic character. Internally they have linear channels of elliptic cross-section (5.7 × 5.1 Å), which intercross with zig-zag channels of an almost circular cross-section (5.4 ± 0.2 Å) in the case of silicalite I or linear ones in the case of silicalite II [99, 100].

Silicalites are characterized by an almost complete absence of aluminum in the structure, while zeolites ZSM-5 and ZSM-11, although with a similar structure, present smaller Si/Al ratios [94, 100]. The hydrophobic characteristic of these materials arises from the absence of $AlO_2^-$ units of the crystalline structure.

Silicalites are used for removal of organic compounds from water or from industrial exhaust smokes. Inclusion of organic molecules into the silicalite channels may impose significant conformational restrictions [83]. The reduced dimension of the channels enables the selective chromatographic use of this material [99b].

### 4.1.4 Sample Preparation: Slurries, Solvent Evaporated Samples and "Equilibrium" Samples.

The adsorption of probes on microcrystalline cellulose or native cellulose, on the surface of different pore size silicas or alumina, and on silicalite surfaces (or other zeolites) has to be made in a differentiated manner, according to each's adsorbent characteristics. Adsorption of probes onto these powdered solids can be performed from a solution containing the probe or from a gas phase.

Generally speaking, one of the simplest and widely used methods is the *solvent evaporation method*. Solvent evaporation methodology means the addition of a solution containing the probe to the previously dried or thermally activated powdered solid substrate, followed by solvent evaporation from the slurry. This can be made in a fume cupboard or by the use of a rotating evaporator. The final removal of solvent can be made under moderate vacuum, $\sim 10^{-3}$ mbar, and the evaluation of the existence of final traces of solvent can be monitored by the use of IR spectra. (See Fig. 20.)

The simplicity of this procedure and the possibility of a simple calculation of the adsorbate concentration are important advantages of this method. Therefore, ground-state diffuse-reflectance absorption spectra of the above-mentioned powdered samples (and also of a blank sample) enable a calculation of the molar extinction coefficients of the probe. These can be compared with the one obtained for transparent samples (homogeneous solutions, films, or solid matrices) by the use of the Beer–Lambert Law.

An alternative procedure for sample preparation is the *equilibrium method:* the solution still containing some probe is removed (by centrifugation, for instance) after prolonged contact with the sample (so that an equilibrium situation can be reached), followed by the residual solvent evaporation. The probe concentration calculation can be made after determination of the amount of probe left in the decanted liquid. The big disadvantage of this method is that large errors can occur in the evaluation of the probe concentration, namely by solvent evaporation. The obvious advantage is that a thermodynamic equilibrium is reached where the probe is shared between the solvent (as solute) and surface (as adsorbate), according to the interactive forces present in each specific case. This is a dynamic equilibrium and the probe molecules keep on going from the solution to the solid surface or

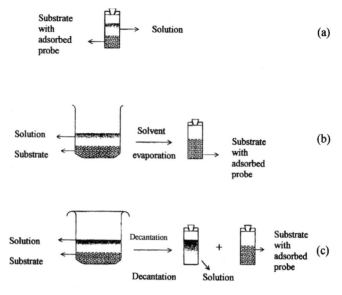

FIG. 20. Sample preparation methods: (a) slurries, (b) solvent evaporation, (c) equilibrium.

from the surface to the solvent. Therefore no "forced" probe aggregation occurs due to the fast solvent evaporation.

In the samples prepared by both methods and after solvent removal, ground-state absorption spectra may exhibit deviations either to the red or to the blue, depending on the surface characteristics, for surface coverages less than the monolayer. Some broadening of the absorption bands can also be detected, or even new bands in the absorption spectra, or significant changes in the extinction coefficients for the different vibronic bands. The formation of "new" species may be related with the "pool" effect which occurs during evaporation (probe molecules may aggregate or form small crystals as the solvent evaporates). Due to the limited number of available adsorption sites at the adsorbent surface, as soon as each site is occupied by a first molecule, the forthcoming molecules have a smaller interaction with the surface. As long as the solvent is being removed, the amount of these weakly interacting molecules increases, and the probability of forming larger ground-state aggregates or small crystals also grows. Some examples of studies on surfaces where absorption spectra exhibit hypsochromic and bathochromic shifts and also data regarding ground state aggregate formation will be given in the next section.

*Slurries* provide mixed information, reflecting both adsorbed molecules and solution molecules. In this sense, this type of sample should always be avoided.

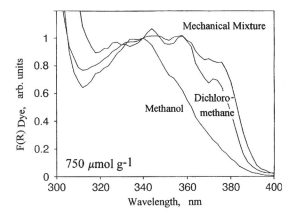

FIG. 21. Remission function for benzophenone adsorbed on microcrystalline cellulose.

In Section 4.1.5 we will describe some of our studies on surfaces using the techniques and substrates described so far.

### 4.1.5 Some Examples of Surface Photochemistry Studies.

*4.1.5.1 Ketones Adsorbed on Microcrystalline Cellulose: Photochemistry and Geminate Radical Pairs Formation Following Laser Excitation.*
Benzophenone.    The use of the diffuse-reflectance ground-state spectra methodology presented in Section 2 has lead us to several interesting results in the study of benzophenone adsorbed on microcrystalline cellulose. Figure 21 shows the ground-state absorption of that aromatic ketone ($n \rightarrow \pi^*$ transition) in samples prepared by both solvent evaporation and the equilibrium method (see [14] and other data not yet published).

The use of polar protic or nonprotic solvents for sample preparation (methanol, ethanol, acetonitrile, acetone) and even a nonpolar solvent as dioxane promotes hypsochromic shifts in the $n \rightarrow \pi^*$ transition of benzophenone, while with solvents such as benzene, isooctane, or even dichloromethane, these bands are shifted to the red and exhibit some vibrational structure. These facts are a consequence of the different swelling capacities of these solvents toward the cellulose polymeric matrix [14, 79b, 80]. The chain–chain interaction is replaced by the chain–solvent–chain interaction, allowing different probe molecules to penetrate within the matrix in different ways, or even the probe simply remains in the external cellulose surface for nonswelling solvents. In this case, increasing concentrations of the probe easily give rise to the formation of ketone microcrystals. Therefore, benzophenone could be used as a probe to evaluate the swelling capacity of several solvents regarding the cellulose matrix.

The consequences of this larger or smaller intimate contact of cellulose with benzophenone are also remarkable from the point of view of the probe photochemical behavior. In fact, the time resolved absorption spectra for benzophenone/cellulose/dichloromethane samples, where the probe is deposited on the surface, present spectroscopic and kinetic characteristics very similar to the microcrystal case. On the contrary, in the case of ethanol or other swelling agents, the ketyl radical of benzophenone is formed with a lifetime longer than the triplet state of benzophenone [14, 87], and the transient absorption peaks at about 550 nm.

Another interesting example of the photochemistry of benzophenone within microcrystalline cellulose is the geminate radical pair formed following laser excitation of samples of coadsorbed benzophenone and 2,4,6-trimethylphenol. The diffuse-reflectance absorption spectrum immediately after laser fire shows both the ketyl radical of benzophenone (BZPH$^\bullet$) and the phenoxyl radical, thus formed (PO$^\bullet$) [13a]. The kinetic study of the geminate recombination of these radicals has shown short ($\sim$10 $\mu$s) and long components ($\sim$100 $\mu$s) and a multiexponential decay pattern which reflects heterogeneous adsorption sites for the radical pair. Cellulose provides very low mobility for both radicals, which form a contact pair. No external magnetic effect was detected in this system.

*β*-Phenylpropiophenone.    Another interesting example of the adsorbent's influence on a specific probe is the case of *β*-phenylpropiophenone included in the channels of silicalite. It is well known that this molecule presents different conformations in a solvent [68] as Figure 22 shows. In one of them (I), the phenyl group may approach the excited carbonyl group; therefore a fast deexcitation of the triplet excited state of this molecule occurs. The triplet lifetime of a benzene solution of *β*-phenylpropiophenone is about one nanosecond. *β*-Phenylpropiophenone inclusion into the silicalite channels increases the lifetime of this molecule five orders of magnitude [83b], as Figure 22 shows.

This phosphorescence emission was obtained at room temperature and with air equilibrated samples. Oxygen diffusion inside the silicalite channels is reduced into the channels which already have *β*-phenylpropiophenone molecules and many triplet excited molecules are not quenched and are therefore emitting. In argon purged samples the triplet lifetime increases about ten times [83c]. *β*-Phenylpropiophenone included within microcrystalline cellulose chains enabled us to detect the ketyl radical of this species and therefore contrast with its solution photochemical inertia [83a].

Several other *diaryl and alkylarylketones* also exhibit room temperature phosphorescence in air equilibrated samples when included in silicalite [83c] or forming inclusion complexes with cyclodextrins [83c, 102], depending on the probe and cavity size. Both substrates provide some degree of protection from oxygen quenching, as well as imposed conformational restrictions that decrease the nonradiative mechanisms of deactivation.

*4.1.5.2 Energy Transfer on Surfaces: Some Examples.*    The first study of a triplet–triplet energy transfer process on surfaces (silicas with different porosity) was reported by Turro and co-workers [67] regarding the benzophe-

β - PP conformations

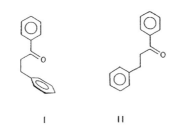

I                    I I

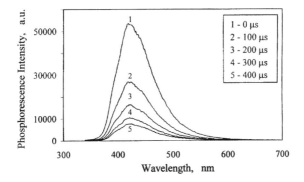

FIG. 22. Time resolved room temperature phosphorescence emission spectra of β-phenylpropiophenone included into silicalite channels.

none/naphthalene system. Following the selective excitation of benzophenone (355 nm), the triplet absorption of naphthalene appeared, peaking at about 400 nm. Increasing amounts of added naphthalene resulted in an increase of the rate constant for the benzophenone decay, presenting evidence for the dynamic nature of the quenching process on the silica surface. The authors also show kinetic data for 255 and 95 Å pore silicas, which seem to evidence larger quenching for the silica with the larger pore. However, by expressing concentrations in terms of moles of probe per surface area unit (the surface area was determined by $N_2$-BET measurements), the rate constants for the quenching process become similar [67].

We performed similar experiments for the systems benzophenone/1-methylnaphthalene [101], benzophenone/oxazine 725 [85], and acetonaphthone/acridine orange [86], as donor/acceptor pairs, all systems co-adsorbed on microcrystalline cellulose. In some of these cases the energy transfer process was also studied in solution (where the quenching process is diffusion controlled) for comparison purposes.

The main conclusion was that on cellulose the quenching process has a static nature. Both direct absorption of the excitation radiation by the acceptor, as well as radiative transfer, have to be quantified for a correct evaluation of the energy transfer efficiency.

A simple model for static quenching was presented, where two types of benzophenone molecules exist: those that have an acceptor as the nearest neighbor and are immediately quenched after laser excitation, and the others that have no such neighbor and are not quenched. This simple model was enough for the interpretation of the results on the cellulose surface. A kinetic analysis showed that the lifetime of the donor remained unchanged within experimental error, only the emission intensity at time zero decreased, again in accordance with the static nature of the quenching process.

In the acridine orange and in the oxazine case, delayed fluorescence was detected, by direct excitation in the first case and sensitization in the second one [85, 86].

Another very important conclusion that emerged from these studies (where microcrystalline cellulose was used as powdered substrate) was that molecular oxygen did not quench the triplet state of molecules entrapped within the polymer chains of this natural polymer. All these studies could be done with air equilibrated samples, and nitrogen or argon purged samples provide similar results for the efficiencies of the quenching process, within experimental error. This property of cellulose makes it a very special substrate to be used for room temperature fluorescence and phosphorescence studies.

### 4.1.5.3 Fluorescence Quantum Yield Determination on Surfaces: The Influence of Aggregation.

Luminescence quantum yield determination of dyes and other organic molecules adsorbed on solid substrate surfaces is an important problem, although with a difficult approach. One has to determine the amount of light absorbed by the sample at the excitation wavelength and also use the appropriate standards. A careful study of the probe aggregation has to be made due to the fact that aggregation may deeply affect emission. Self-absorption also may exist and has to be taken into account. The same applies to concentration quenching effects.

We have studied this problem using first rhodamines 101 and 6G and later sulforhodamine 101, adsorbed onto microcrystalline cellulose. After detailed concentration studies, we concluded that these two compounds could be used as reference compounds for fluorescence quantum yield determination of probes on solid surfaces [81]. Later this method was extended to studies of other surfaces and other molecules [15, 16, 82, 88, 102–104].

The method is based on the comparison of the slopes of the linear part of the curves $I_F$ versus $(1 - R) f_{probe}$ for an unknown sample and for the standard one. These curves may be obtained by plotting the fluorescence emission intensity $(I_F)$ as a function of the absorbed light at the excitation wavelength which is

SOLUTION

$$I_F(\lambda_{em.}) = C \phi_F I_0(\lambda_{exc.})\left(1 - 10^{-\varepsilon cl}\right)$$

$$\phi_F^u = \phi_F^s \frac{I_F^u \ I_0^s(\lambda_{exc.})(\varepsilon cl)^s}{I_0^u \ I_0^u(\lambda_{exc.})(\varepsilon cl)^u}$$

SOLIDS

$$I_F(\lambda_{em.}) = C \phi_F I_0(\lambda_{exc.})(1 - R)f$$

$$\phi_F^u = \phi_F^s \frac{I_F^u \ (1 - R^s)f^s \ I_0^s(\lambda_{exc.})}{I_0^u \ (1 - R^u)f^u \ I_0^u(\lambda_{exc.})}$$

R = Reflectance

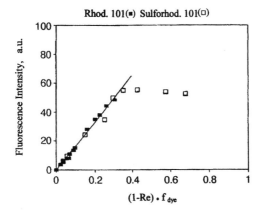

Rhod. 101(■) Sulforhod. 101(□)

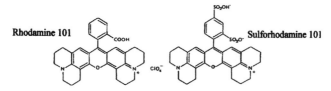

Rhodamine 101          Sulforhodamine 101

FIG. 23. Luminescence quantum yield determination of molecules adsorbed on surfaces.

proportional to $(1 - R)f_{\text{probe}}$ for each sample with a specific dye concentration [81]. The reflectance $R$ has to be determined at the excitation wavelength, with the use of an integrating sphere, as described in Section 2.1.

Figure 23 shows the calibration curves we currently use in our laboratory for luminescence quantum yield determination of probes adsorbed on powdered solids,

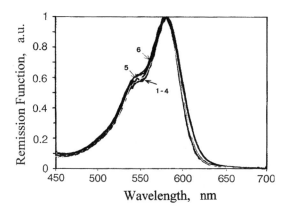

FIG. 24. Ground-state diffuse reflectance spectra of sulforhodamine 101 adsorbed on microcrystalline cellulose. Curve 1—0.01 $\mu$mol/g$^{-1}$; 1—0.05 $\mu$mol/g$^{-1}$; 2—0.10 $\mu$mol/g$^{-1}$; 3—0.25 $\mu$mol/g$^{-1}$ 4—0.50 $\mu$mol/g$^{-1}$; 5—0.75 $\mu$mol/g$^{-1}$; 6—1.0 $\mu$mol/g$^{-1}$.

for optically thick samples. We would like to stress the importance of knowing with accuracy the excitation energy profile for each specific apparatus being used, as we have shown before [104].

This figure also shows the analogies in the main equations used for determining the fluorescence quantum yield in solution and on solids (for optically thick samples) [7, 14, 15, 81–88]. In this figure $u$ stands for unknown sample, $s$ for standard, $G$ for geometrical factor, and $f$ for fraction of the incident light absorbed by the emitting probe under study. All other parameters have the usual meaning.

Several molecules with low fluorescence emission quantum yields in solution, such as several cyanine dye and auramine O, may exhibit a three, four, and sometimes five orders of magnitude increase in the quantum yield of emission at room temperature, simply because the probe was entrapped within microcrystalline cellulose polymer chains. The host imposes severe restrictions to the mobility of the guest molecule, therefore reducing the nonradiative mechanisms of deactivation [81, 83b].

Residual amounts of moisture may have a significant quenching effect in the fluorescence quantum yield of many entrapped molecules [81b, 82].

For most dyes, but not all, aggregation is particularly relevant. A good example is sulforhodamine 101 adsorbed on cellulose, as we show in Figure 24 [81b, 82], where aggregation is only present for the two higher loadings. For auramine O, a very flexible dye, almost no aggregation was detected on cellulose, using ethanol as solvent for sample preparation [81b].

### 4.1.5.4 Photochemistry of Dyes Adsorbed on Microcrystalline Cellulose and Different Pore Size Silicas.

**Cyanines.** Photochemical and photophysical studies of cyanine dyes are an important and up-to-date research domain, due to their use in several relevant applications, such as black and white and color photography, laser dyes, potential sensitizers in cancer photodynamic therapy, and also devices for optical storage of data [88].

Cyanine aggregation in an electronically inert substrate is exemplified in Figure 25, where several diffuse-reflectance spectra of thiacyanines are shown, namely 3,3'-diethylthiacarbocyanine (TCC) and 3,3'-diethyl-9-methylthiacarbocyanine (9-MeTCC) entrapped within the polymer chains of microcrystalline cellulose.

Kubelka–Munk remission function curves are also shown in Figure 26, clearly showing for the second dye that the methyl group in the ninth position of the carbocyanine favors the formation of sandwich aggregates ($H$ aggregation), which absorb at energies higher than the monomer absorption. Head to tail aggregates ($J$ aggregation) are also formed, in the 5 to 15 $\mu$mole of dye per gram of cellulose concentration range, with an absorption peaking at about 610 nm. 3,3'-Diethylthiacarbocyanine no longer exhibits a clear $J$ absorption band, showing in this way that small structural differences may play a major role toward aggregation.

The influence of concentration, and as a consequence, of aggregation in the fluorescence emission intensity is presented in Figures 27 and 28.

Figure 27 shows that, for low loadings of the dye (up to about 0.5 $\mu$mole of dye per gram of cellulose) the fluorescence emission intensity increases with the increase of the light absorbed by the sample, or better, with the fraction of the light which is absorbed by the dye alone (the substrates also absorb at the excitation wavelength). In the 0.5 to 5 $\mu$mole/g$^{-1}$ concentration range a clear decrease in the fluorescence emission intensity is detected, due to the fact that a fraction of the photons goes to the aggregated forms of the dye from which no emission was detected.

Due to the overlap of the monomer fluorescence emission and absorption from the aggregates, a quenching process may occur by a resonance interaction. However, in other cases [85, 86] where we determine the extinction coefficients for ground-state monomers and dimers, by simply taking into account the fraction of the excitation radiation absorbed by the monomers (which are emissive species) and aggregates (from which we could not detect any fluorescence emission), a good superposition between the calculated and experimental fluorescence intensity curve was obtained.

Both TCC and 9-MeTCC in ethanolic solution present a fluorescence emission quantum yield ($\phi_F$) of about 0.05. When these molecules are adsorbed on microcrystalline, $\phi_F$ increases to about 0.95 as Figure 28 shows.

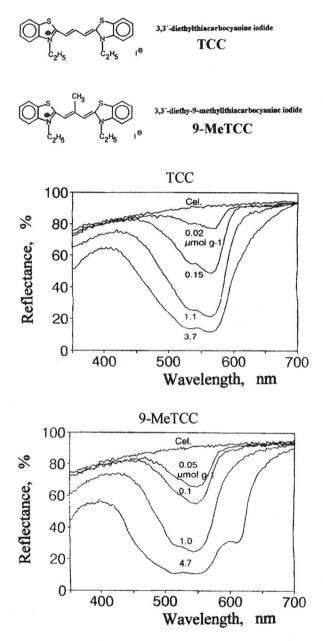

FIG. 25.  Diffuse reflectance spectra for TCC and 9-MeTCC adsorbed on microcrystalline cellulose as a function of the dye concentration.

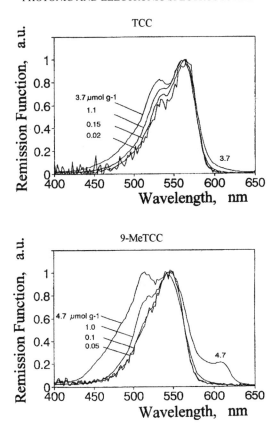

FIG. 26. Remission function for the same samples as in Figure 25.

It is a well documented fact that the main pathway for the $S_1$ state of polyme-thinic dyes at room temperature is the *trans-cis* isomerization (see [16, 88] and references quoted therein). For monomeric cyanines and in the absence of steric hindrance, both fluorescence quantum yields and intersystem crossing quantum yields are usually very low. For TCC and 9-MeTCC entrapped within the poly-mer chains of microcrystalline cellulose $\phi_F$ becomes close to unity, evidence of the decrease of the nonradiative pathways of deactivation.

In sharp contrast with the steady-state behavior described so far, pulsed laser excitation of the same samples produces quite different results: for concentrated samples and for high laser fluences (above about 10 mJ per pulse and per square centimeter) a second emission appears in the case of TCC, as Figures 29 and 30 show. This emission is sharp and appears for lower energies relative to the monomer.

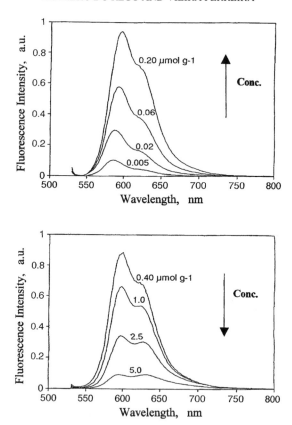

FIG. 27. Fluorescence emission of TCC for low and high dye concentrations.

The cyanine that forms $J$ aggregates does not exhibit this new emission or, at least, it is very much reduced. The new emission occurs in the nanosecond time range following laser excitation and has its origin in the cyanine monomers only.

This new emission was also detected in other cyanines [16, 88], such as 2,2'-cyanine [88b] and oxacyanine dyes [88a]. Energy studies revealed a supralinear dependence on laser energy.

From the above-mentioned studies with several cyanine dyes, it has been possible to establish the origin of this new emission. It occurs as a consequence of a two photon absorption process, the first photon creating a photoisomer which may be excited by the absorption of a second photon if the laser fluence is enough and the energy appropriated for the photoisomer to absorb. This second excited species then emits its own fluorescence [88a].

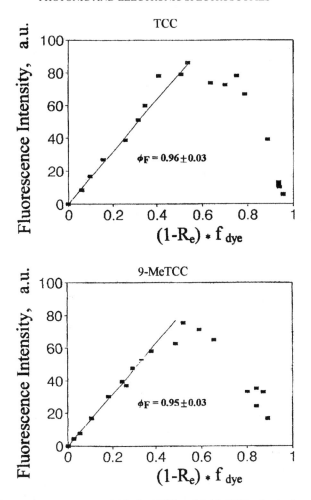

FIG. 28. Fluorescence quantum yields for TCC and 9-MeTCC adsorbed on microcrystalline cellulose.

Picosecond laser excitation ("pump and probe" experiments) allowed us to determine lifetimes of the singlet excited states of these dyes in some cases [88c, 88e].

Rhodamines.   The luminescence quantum yield study for molecules adsorbed on surfaces, which started with rhodamine 6G and 101 [81a], was later extended to zwitterionic rhodamines (sulforhodamine 101 and B) and to other nonrigid rhodamines (rhodamines B and 3B [7, 82]. Important conclusions arising from those studies are that apart from rhodamines 6G and 101, only sulforhodamine

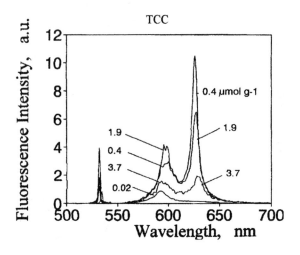

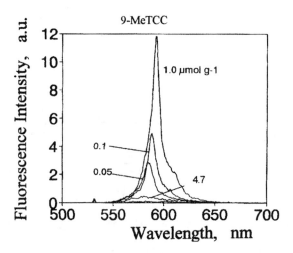

FIG. 29. Laser induced fluorescence for TCC and 9-MeTCC as a function of the sample concentration.

101 has a unitary $\phi_F$, and also that moisture has a strong quenching effect in all cases [7, 82].

Sulforhodamine 101 and rhodamine 6G were further used for fluorescence quantum yield determination on different silica surfaces [7]. Silicas with controlled pore (22, 25, 40, 60, 100, 150 Å), and particle sizes were used. A system-

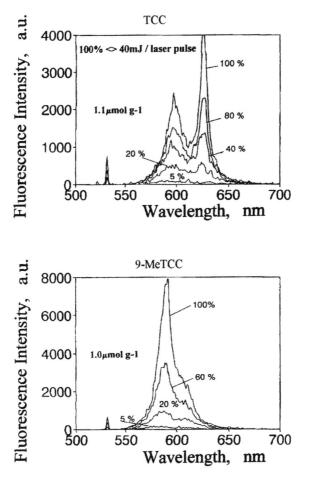

FIG. 30. Laser induced fluorescence for TCC and 9-MeTCC as a function of the laser energy.

atic study of the influence of pore size and silica pretreatment on the emission properties of the two adsorbed probes has shown that sulforhodamine 101 fluorescence emission is severely affected by those factors, as Figure 31 shows, while rhodamine 6G is rather insensitive.

The combined information from photonic and electronic techniques was used in this particular study, which will be described in Section 5. These techniques proved to be complementary in the study of the specific interactions of the two rhodamines with this substrate.

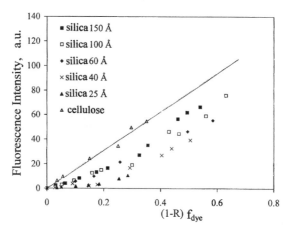

FIG. 31.  Fluorescence intensity (measured as the total area under the corrected emission spectra) as a function $(1 - R) f_{dye}$ (see text) for sulforhodamine 101.

## 4.2  Flat Surfaces. HREELS and XPS Studies on Organic Films

Organic films surfaces, and particularly polymeric ones, can be studied by a great variety of analytical techniques and microscopies.

The association of several techniques to study the same system is always advantageous. For the specific case of surfaces, techniques like XPS, contact angle, diffuse reflectance infrared with Fourier transform, surface area and pore size measurements, scanning electron microscopy appear frequently associated. This is the case for the study of carbon or glass fibers [105–109], activated carbons [110], and polymeric films obtained by plasma polymerization [111]. In Section 5, the specific combination of diffuse reflectance spectroscopies in the region of UV/Vis and XPS will be emphasized.

In this section the application of HREELS to the study of polymeric surfaces will be reviewed with relevance for some of its specific aspects: quantitative application of the vibrational studies and electronic studies.

### 4.2.1  Sample Requirements.
Both HREELS and XPS work under ultrahigh-vacuum conditions [29]. Therefore, organic films need to have low vapor pressures. Among organic films, polymeric ones always fulfill this condition. However, these techniques induce a charge on the surface (see Section 3) if the sample is insulating, which is the case for a large part of organic polymers. With films of molecules of a long oligomer, the hexatriacontane ($C_{36}H_{74}$), it has been possible to record HREELS spectra without charge compensation for up to thicknesses around 1000 Å [112]. Thicknesses lower than this value are obtainable by spin-coating, dipping, or deposition, followed by evaporation, from a polymer dilute solution. Provided that a conducting or a semiconducting substrate is used,

in principle, the film will naturally discharge to the earth the charges trapped on its surface [49]. The same occurs with Langmuir–Blodgett (LB) films [113]. Experiments with perdeuterated polystyrene films on silicon wafers showed that a single dipping is not enough to get a substrate entirely covered by the film independent of the concentration of the solution or the contact time between the substrate and the solution. A second dipping is necessary to assure a complete substrate covering [114]. This result is compatible with a mechanism with a first step consisting of a fast nucleation followed by slow growth of the small domains formed during nucleation—as verified for the deposition of films of poly($o$-methoxyaniline) (POMA) on glass substrates [115]. During the second dipping, another fast nucleation seems to occur on the uncovered substrate, the dissolution of the first domains being slower than the nucleation. With Langmuir–Blodgett films of perdeuterated stearic acid $CD_3(CD_2)_{16}COOH$ on silanized surfaces (hydrophobic) studied by vibrational HREELS, it was found that the film evaporated under ultra-high-vacuum conditions. However, the hydrogen bonding to the substrate, present when the substrate surface is polar, is enough to keep the film on the substrate. As a result of this, many studies were carried on LB films using XPS [116] and HREELS (one of the first works of HREELS on LB films can be found in [113]).

**4.2.2 Polystyrene.** HREELS is usually applied in a qualitative way for the study of adsorbed molecules. The theoretical frame for these studies was established some 20 years ago [47]. However, its application to polymer surfaces is more recent [117, 118]. Polystyrene is a very common polymer and has been extensively studied in optical absorption spectroscopy and is often taken as a standard sample. It was extensively used as a standard sample to evidence the ability of HREELS to study polymeric surfaces, namely its quantitative capabilities.

*4.2.2.1 Vibrational HREELS Studies.* Concerning the vibrational region of the spectrum, the use of selectively deuterated polystyrenes [119] allows a better separation of the main-chain contributions and of pendant-group vibrational modes. End group segregation was also evidenced through the intensity ratio C—D/C—H for the stretching modes in completely deuterated polystyrenes, where just the chain ending was hydrogenated (see Fig. 32).

One of the most striking pieces of evidence of the great sensitivity of this technique to the extreme surface was obtained by the comparison of spectra obtained with two normal polystyrene films made of isotactic and atactic polystyrenes, which is displayed in Figure 33.

Figure 32 shows that, even for a polystyrene with two hydrogen atoms per chain, there is an intense peak in the region around 3000 cm$^{-1}$ corresponding to the C—H stretching modes, aliphatic and aromatic. This is due to multiple losses, which are much more important than in infrared spectra. Blends of completely deuterated and normal polystyrene, having the same molecular weight, in several compositions were used to show how quantitative the technique was and also

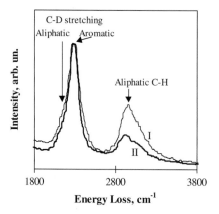

FIG. 32. C—D and C—H stretching vibrations in polystyrene (symmetric and asymmetric modes are not resolved). Comparison of HREELS spectra from two deuterated polystyrenes which differ only in the terminal groups: (I) one end group is $CH_3CH(C_6H_5)$ the other one is C—H (normal line); (II) both end groups are C—H (bold line).

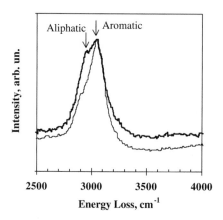

FIG. 33. C—H stretching vibrations in polystyrene (symmetric and asymmetric modes are not resolved). 70% isotactic film (normal line) and atactic film (bold line) show that the first one is much richer in the aromatic component than the second one. Isotactic film exposes preferentially the phenyl groups at the surface whereas the atactic one exposes phenyl groups and chain segments in comparable amounts. Spectra were normalized to the aromatic component.

to try to find relative excitation cross sections for the vibrational modes corresponding to C—D and C—H bonds [120]. Films were prepared by dipping the substrate—films of gold 1000 Å thick evaporated on glass plates—in $CCl_4$ solutions.

Denoting by $x_D$ and $x_H$ the molar fractions of deuterated and hydrogenated monomers, the following relation is expected, similar to the observed for solutions in $CCl_4$:

$$\frac{A_H}{A_D} = \frac{\sigma_H}{\sigma_D} \frac{x_H}{x_D} \qquad (29)$$

$\sigma_D$ and $\sigma_H$ are the excitation cross sections and $A_D$ and $A_H$ are the peak areas for the C−H and C−D stretching vibrations, respectively. Surprisingly, instead of a straight line, a curve with a plateau was obtained, suggesting the existence of deuterated chain segregation at the surface. A thorough study of the surface segregation in blends of perdeuterated and perhydrogenated polystyrenes as a function of relative molecular weights, annealing time, and film thickness was later carried on using surface-enhanced Raman scattering [121]. The segregation of deuterated chains at the surface when chains have comparable length was also verified for a different substrate—silicon wafer—and a different film preparation method—spin-coating.

Later, randomly deuterated polymers were used with the same purpose and a linear relation between $A_D/A_H$ and $x_D/x_H$ was finally obtained as expected [122]. From Eq. (33) the relative excitation cross-sections $\sigma_H/\sigma_D$ were obtained as a function of the primary energy and were used to deduce the ratio $x_D/x_H$ for block copolymers of hydrogenated polystyrene terminated by deuterated blocks at both ends as a function of the electron primary energy [50]. Results were compatible with a segregation of the end groups at the extreme surface (first 5–10 Å) followed by a layer with end group depletion and were among the first experimental evidence of this surface segregation predicted a few years before by de Gennes [123]. Further experimental evidence of this end group segregation at the surface came from neutron reflectometry on the same system [124].

Still with polystyrene (PS) but, this time, integrated in a diblock copolymer with polyethyleneoxide (PEO), it has been possible to quantitatively estimate the relative amounts of PS monomers and PEO monomers at the extreme surface through a method consisting of a previous normalization of spectra to the background [125]. This normalization, previously suggested by Ibach and Mills [47], allows the use of HREELS spectra acquired in different occasions with different films (having, for instance, different roughnesses), one from pure PS film and the other one from PEO film, to be combined, after being multiplied by factors whose sum is unitary, to yield the normalized HREELS copolymer film [125]. Annealing and substrate effects were studied by this method and results were obtained for films spread on two type of substrates: silicon wafers (keeping their native oxide film and with their surface silanized) and spin-coating and casting from diluted solutions (0.1 g/l) in two different solvents (carbon tetrachloride and tetrahydrofuran). These results are presented in Figure 34.

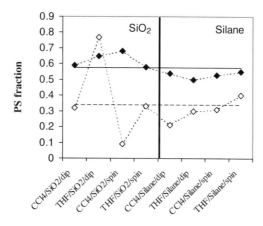

FIG. 34. Surface PS fractions (symbols) for PS-PEO copolymer films prepared by dipping (dip) and by spin coating (spin) from solutions in CCl$_4$ and tetrahydrofuran (THF) on two different substrates: silicon wafer covered by its native oxide (denoted SiO$_2$) and silanized silicon wafer (denoted Silane) for nonannealed (empty squares) and annealed films (full squares). The gray line represents copolymer stoichiometric composition. Stoichiometric composition corresponds to a PS fraction of 0.09. Horizontal lines represent average values for annealed (——) and non-annealed (- - - -) films.

Results clearly show that polystyrene segregates at the surface. The driving force for this segregation is the lower solid state surface tension of PS (36 mJ/m$^2$) compared to the PEO (44 mJ/m$^2$) value [126]. This was confirmed by Thomas and O'Malley for several PS-PEO diblock [127] and PEO-PS-PEO triblock [128] copolymers films cast from different solvents using XPS. However, as already pointed out, techniques have different surface sensitivities: with HREELS we measure the composition of the very first atomic layers (about 5 Å) whereas with XPS the extreme surface composition is always obtained by indirect modes from angle resolved measurements and results are therefore less reliable. The danger of confusing contaminant C—H vibrations with those intrinsic to the system is avoided by using the whole spectrum from both the PEO and PS to curve fit the copolymer spectrum, since the vibrational spectra from both polymers are completely different from saturated hydrocarbon, the usual contaminant. Figure 34 also shows that the amount of segregation at the surface, before annealing, is highly sensitive to substrate, preparation method, and solvent. However, after annealing, only slight differences are noticed.

*4.2.2.2 Electronic HREELS Study.* HREELS is mostly used as a vibrational spectroscopy and the extremely high resolution requirements are a consequence of that use. However, low energy electrons can be very useful to study the energy position and relative excitation probabilities of electronic states optically forbidden for spin or symmetry reasons.

Polystyrene is also, in this respect, a good material to test the capabilities of this technique. Its molecular structure is made of an aliphatic chain and phenyl pendant groups. It has been extensively studied in optical absorption spectroscopy and their singlet excited electronic states are well known: they are essentially the same detected in a monosubstituted benzene [129]. In one of the first applications of low energy backscattered electron spectroscopy to polymers [118], polystyrene electronic losses corresponding to electronic excited states, singlets and triplets were studied with a low resolution spectrometer (fwhm $\sim$ 80 meV). Higher quality spectra and in a wider range of primary energies and geometries were later published confirming that all the excitations detected for energy losses lower than 7 eV correspond to the electronic excited states in the side groups [130].

Polystyrene films were prepared by doubly dipping the substrate—a silicon wafer previously cleaned by pure solvents and etched by hydrogen fluoride—into solutions of 1 $gL^{-1}$ of pure polystyrene in carbon tetrachloride (spectroscopic grade) and allowing the solvent to evaporate. Thickness of the films thus obtained was estimated to be of the order of 100 Å by elastic recoil diffusion analysis (ERDA) measurements [122].

Figure 35 shows HREELS spectra recorded at an incidence angle of 60° and an analysis directions of 50° (both directions are relative to the normal to the film surface) in a loss region between 3 and 11 eV. Spectra correspond to primary energies ranging from 4.7 (bottom) to 11.2 eV (top) by steps of 0.5 eV, which are here set off for clarity of presentation.

As in any complete spectrum (loss energy, $\Delta E$, from 0 to $E_p$) two kinds of features are observed:

- those remaining at a constant energy loss in the different spectra, located at fixed positions from the elastic peak, corresponding to the excitation of electronic individual states and indicated in Figure 35 by vertical lines;
- those corresponding to fixed kinetic energies, located at fixed positions from the end of the spectrum (vacuum level) which are usually associated to density of states maxima in the conduction band [112].

For the assignment of the electronic energy losses, experimental and theoretical data for the electronic excitation of simple molecules having electronic structure comparable to polystyrene were used: benzene and toluene for the side groups and polyethylene or one of its oligomers, such as hexatriacontane, for the chain. For incident energies lower than 7 eV, no electronic excitations are induced by electrons in polyethylene-like compounds. Henceforth, in the range of incident energy from 3 to 7 eV, all the comparisons were done with benzene and toluene. Singlet excitations are well known from UV absorption spectra of benzene and toluene liquid solutions [129, 131]. Triplet and singlet states of benzene were also known by electron impact experiments in the gas phase [132–135], adsorbed on metals [136], or included in xenon matrices or multilayers [137]. In Table V, the

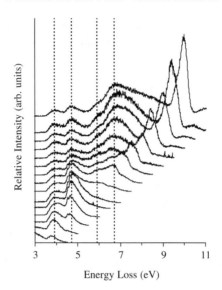

FIG. 35. HREELS spectra for polystyrene films prepared by dipping a silicon wafer in a polymer solution in CCl4 1.0 gL$^{-1}$. Spectra were recorded at an incidence angle of 60° and an analysis direction of 50° (both directions are relative to the normal to the film surface) in a loss region between 3 and 11 eV. Spectra correspond to primary energies ranging from 4.7 (bottom) to 11.2 (top) in steps of 0.5 eV, which are here set off for clarity of presentation.

TABLE V.    Assignment of the Principal Features Present in HREELS
Spectra of Polystyrene Films in Figure 35

| Energy loss (eV) | Assignment |
| --- | --- |
| 3.9 | $T_1(^3B_{1u})$ |
| 4.7 | $S_1(^1B_{2u}) + T_2(^3E_{1u})$ |
| 5.9 | $S_2(^1B_{1u}) + T_3(^3B_{2u}) + T_4(^3E_{2g})$ |
| 6.7 | $S_3(^1E_{1u})$ |

assignment of electronic excitations, clearly visible in spectra of Figure 35, is presented.

Comparison between HREELS spectra and optical absorption spectra of polystyrene was revealed to be very useful in improving the assignment of the peak centered at 4.7 eV. In fact, two hypotheses were plausible: to assign that peak to the $S_1 \leftarrow S_0$ or to the $T_2 \leftarrow S_0$ transitions. The comparison allowed us to establish that it should be a mixture of both with a major triplet character. The comparison was also helpful for the assignment of the peak at 5.9 eV. All the electronic transitions are meant as occurring from the highest occupied molecular orbital to different localized molecular electronic states coexisting in the gap of the insulator.

Features corresponding to fixed kinetic energies, located at fixed positions from the end of the spectrum, are assigned to the accumulation of electrons relaxed in high density of state levels in the conduction band, above the vacuum level following the multiple relaxation processes suffered by incident electrons.

Also the analysis of relative intensities (as a function of primary energy and geometric conditions) is important to assess the type of mechanism acting on the excitation. In polystyrene and for electronic excitations, that analysis showed that the behavior of relative intensities as a function of primary energy is typical of resonant interactions. This is compatible with the fact that triplet excitation (optically forbidden for spin reasons) intensity is favored relative to singlet excitation.

**4.2.3 Oligothiophenes.** Thiophene oligomers are compounds of the form

where $n$ is the number of thiophene rings, and the compound name is usually abbreviated to $n$T.

Apart from being good models for the polythiophene, oligothiophenes occupy, by themselves, a special place among materials suitable for opto-electronics, namely in the fabrication of light emitting devices and field-effect transistors [138]. In fact, these molecules constitute interesting materials for electronic devices as they can be easily used as blocks in the formation of good quality, well-organized, thin organic films obtained by evaporation. For this application, an accurate knowledge of their electronic structures, excitation, and relaxation mechanisms is needed. Many theoretical and experimental studies on isolated molecules (or in dilute solution) [139] and also on LB films [140] were published. However, a very limited knowledge exists on their solid-state electronic structure [141]. In fact, intermolecular interactions may play a crucial role in the optical and electronic properties of oligothiophenes in particular and of conjugated systems in general. For instance, until recently, the picture for the electronic structure of sexithiophene (6T) in solid phase was mainly based upon optical data [61, 141] and could be summarized as follows:

(i) Oligothiophene films present a large $u$ excitonic band.

(ii) Transitions to the bottom of the band at 2.275 eV are optically forbidden by symmetry reasons.

(iii) The top of this band corresponds to the intense peak at 3.5 eV in the absorption spectrum obtained with UV radiation polarized in a direction parallel to the main molecular axis.

(iv) The 0–0 transition determines the position of the exciton band bottom. This was observed at 18335 cm$^{-1}$ (2.288 eV) in the single crystal absorption spectrum recorded at low temperature [61, 141].

More recently, the polarized absorption and fluorescence spectra obtained at 4.2 K from an oriented 6T single crystal allowed the assignment of the lowest singlet electronic transition (located at 18360 cm$^{-1}$) to the $a_u \leftarrow 1^1A_g$ transition [142] and a qualitative scheme of the $1^1B_u$ molecular level Davydov splitting could be drawn as represented in Figure 36 [142a].

Theoretical calculations using a quantum chemical model, which considers the total molecular wavefunctions for each transition, shows good agreement with the experimental findings for the energy and polarization of the optically allowed crystal levels [142b].

In Figure 37 a series of complete HREELS spectra (from $\Delta E = 0$ to $\Delta E = E_p$) is shown for a quinquethiophene film surface for primary energies ranging from $E_p = 2.5$ eV to $E_p = 7$ eV.

For energy losses near zero, a very intense elastic peak appears, presenting a very sharp angle distribution contained in a lobe of fwhm around 12°. Such a high intensity and directionality associated with the elastic peak are characteristic of very well organized and atomically flat surfaces [143].

For low energy losses, below 1 eV, vibrational structures induced by the incident electrons in interaction with the surface of the film appear. For higher energy losses above 1 eV, a structured region, corresponding to electronic losses with an edge above 2 eV, emerges. Different thresholds assigned to exciton and electronic gaps were extrapolated from this wide band [144–146] and are presented in Table VI.

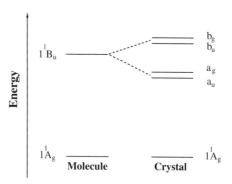

FIG. 36. Energy level diagram of the $1^1B_u$ exciton band structure of $\alpha$-sexithiophene (adapted from [142a]). Copyright 1998, American Institute of Physics.

All the thresholds present in HREELS spectra are also present in UV-Vis optical absorption spectra except the second threshold, named CT in the table; it is therefore assigned to electronic gaps, via charge transfer levels [145]. Similar thresholds were detected in spectra of similar samples using a single and higher primary energy (15 eV) [147]. However, different assignments were proposed which were not compatible with observations in optical absorption and fluorescence spectra.

The observation of strong secondary emissions led to the evaluation of ionization potentials of $4.7 \pm 0.5$ eV for 5T and 6T films. A large difference between optical absorption and HREELS spectra was noticed [145]. The large difference between the interaction of electrons with the surface and of photons with the surface was related to crystalline effects on spectra: they strongly affect the optical absorption spectra shape whereas they are completely absent in HREELS spectra. This is certainly due to the fact that, for equal energies, the wavelength associated with electrons is much smaller than for photons and, therefore, electrons do not probe large delocalization effects.

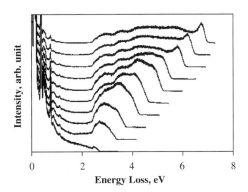

FIG. 37. Complete spectra (from $\Delta E = 0$ to $\Delta E = E_p$) for surfaces of quinquethiophene films for primary energies ranging from $E_p = 2.5$ eV (bottom) to $E_p = 7$ eV (top) by steps of 0.5 eV. Spectra were recorded at an incidence angle of 60° and an analysis directions of 30° (both directions are relative to the normal to the film surface) and were set off for clarity of presentation.

TABLE VI.    Main Features Positions (in eV and a Standard Deviation of $\pm 0.02$ eV) Assignment in HREELS Spectra of 4T, 5T, and 6T Films Surfaces

| Compound | $A_u$ | $B_u$ | CT | $2B_u$ |
|----------|-------|-------|------|--------|
| 6T       | 2.27  | 2.55  | 2.97 | 3.65   |
| 5T       | 2.38  | 2.65  | 3.24 | 4.05   |
| 4T       | 2.64  | 2.85  | 3.60 | 4.55   |

In addition, HREELS vibrational spectra have shown that molecular orientation is not the same when graphite or gold substrates are used for the deposition of $\alpha$-6T evaporated films [148]. HREELS studies did corroborate reflection-absorption infrared spectroscopy (RAIRS) analysis [149], confirming that graphite substrates induce films where molecules lay flat on the surface, whereas gold substrates generate a molecular orientation close to perpendicularity. Complete HREELS spectra recorded with incident electron energy between 2 and 7 eV demonstrated the existence of two relaxation channels in the electron surface interaction—electronic excitation and ionization. In fact, secondary electron emission originating from ionizations is always more efficient for escaping directions perpendicular to the plane of the molecule. This result is associated with the $\pi$-orbital origin of this emission. Oppositely, the efficiency of the electronic excitation is higher for electrons emerging in directions parallel to the plane of the molecule.

## 5 COMBINED STUDIES INVOLVING PHOTONIC AND ELECTRONIC SPECTROSCOPIES

The combination of several techniques to study a specific system, as already pointed, may provide a better insight and understanding of its constitution. In particular, photonic spectroscopies associated to X-ray photoelectron spectroscopy are being used more and more to study the adsorption of dyes on the most varied substrates [150].

In this section, special emphasis will be given to the study of the adsorption of dyes belonging to the cyanine and to the rhodamine families onto microporous substrates (cellulose, silicas, and cyclodextrins).

All the dyes presented here have a common feature: the presence of nitrogen atoms in the conjugated system. On the other hand, the substrates do not contain nitrogen. The organic substrates (cellulose and cyclodextrins) contain carbon, also present in the dye, and oxygen. For the X-ray photoelectron spectroscopy studies, the XPS N $1s$ region is therefore expected to be particularly useful. On one hand, concerning the quantitative evaluation of the amount of dye adsorbed on the substrates by XPS, the N/O (more suitable than the N/C one) and N/Si (for the adsorption on silicas) ratios should be considered. On the other hand, for qualitative studies, the N $1s$ region is again advantageous. For a completely planar molecule, an equivalent electronic distribution around the two nitrogen atoms exists, therefore a unique XPS N $1s$ photoelectron binding energy should be detected. In contrast, in a slightly distorted conformer the two nitrogen atoms are not equivalent since there is a break in the conjugation and they should have different binding energies. As a consequence, a broadening of the N $1s$ peak should occur. The XPS N $1s$ peak is, therefore, expected to be a good probe for the planarity of this kind of dye molecules. Moreover, it can also be used to probe the interaction between the dye molecules and the substrate. In fact, if a nitrogen atom is involved in strong hydrogen bonding, as an electron donor, the XPS N $1s$ binding energy increases [151]. In the cases where the molecule also contained sulphur, the XPS S $2p$ region was also revealed to be interesting as we will show later.

## 5.1 Rhodamine Dye Covalently Bound to Microcrystalline Cellulose

The most important application of reactive dyes is the dyeing and printing of cellulose fibers and cellulose fiber-based materials. These dyes form covalent bonds with the substrate that is to be colored in the dyeing process [15]. The dye molecule contains specific functional groups that can undergo addition or substitution reactions with the –OH, –SH, or –NH$_2$ groups that are present in the textile fibers. According to the number of reactive systems they contain, reactive dyes can be classified as mono-, double-, and multiple-anchor dyes. Rhodamine B isothiocyanate is a reactive monoanchor dye which can easily be chemically bound to cellulose in the presence of a base [15].

The information on the nature of the interactions of rhodamine B with the natural polymer chains, namely to establish clear differences between physical and chemical adsorbed rhodamine B molecules within cellulose, is therefore needed.

Rhodamine B isothiocyanate was adsorbed onto microcrystalline cellulose by two different methods: deposition from ethanolic and aqueous solutions followed by solvent evaporation (Type I) and also from aqueous solutions in equilibrium with the powdered solid and following a dyeing protocol (Type II). Figure 38 displays the scheme of the dyeing procedure used to bind rhodamine molecules to microcrystalline cellulose [15].

After carefully washing the above mentioned samples, their ground state absorption spectra presented clear differences as Figure 39 shows.

Also the fluorescence quantum yields ($\phi_F$) determined were about $0.40 \pm 0.03$ and $0.28 \pm 0.03$ for ethanol and water respectively (solvents which efficiently swell cellulose), when these two solvents are used for Type I sample preparation. For dyed samples, $\phi_F$ is only $0.10 \pm 0.05$. These values for $\phi_F$ can be compared with $0.70 \pm 0.03$ obtained for rhodamine B entrapped into the polymer chains of microcrystalline cellulose as Figure 40 shows.

X-ray photoelectron spectroscopic studies of the same set of samples were essentially centered in the nitrogen as explained above.

Figure 41 shows the XPS spectra for the N $1s$ region of adsorbed (Type I) and chemically bound (Type II) samples of rhodamine isothiocyanate as well as rhodamine B. Clear differences between them are observed: curve a is from a type II sample and is centered around 399.5 eV; curves b and c show increasing shifts toward larger binding energies. They correspond respectively to a type I sample (physically adsorbed dye from aqueous solutions), subsequently washed with water, and to rhodamine B type I sample from an ethanolic solution.

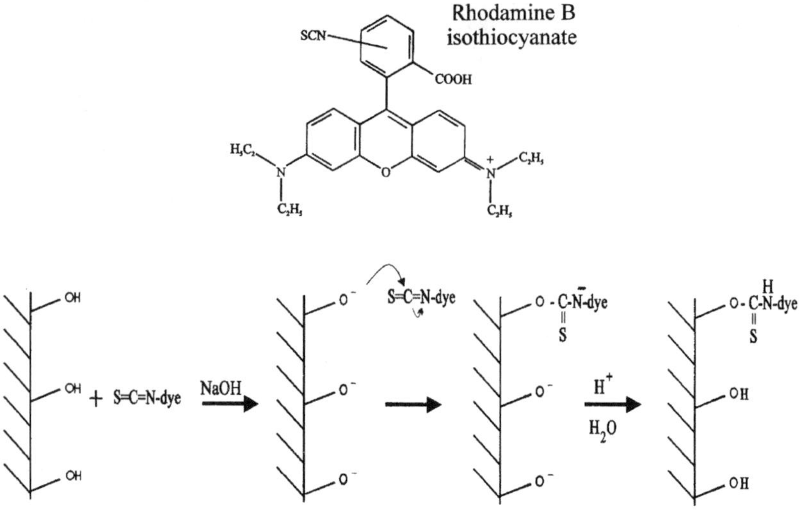

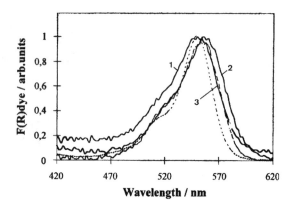

FIG. 38. Schematic representation of the dyeing procedure of cellulose with rhodamine B isothiocyanate.

FIG. 39. Remission function values of rhodamine B isothiocyanate onto microcrystalline cellulose, normalized to the maximum of the absorption of the dye. Curve (1) is a Type II sample with 0.035 $\mu$mol/g$^{-1}$. Curve (2) is a Type I sample 0.030 $\mu$mol/g$^{-1}$. Curve (3) is a 0.035 $\mu$mol/g$^{-1}$ rhodamine B physically adsorbed sample. The dashed curve is a diluted aqueous solution of the dye containing only monomers.

These results show that the positive charge density on the nitrogen atom is smaller for dyed samples when compared with the adsorbed ones. This is compatible with nitrogen atoms that do not participate in the conjugated system. On the other hand, this fact is indicative of the nonplanarity of at least a fraction of

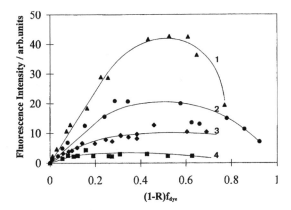

FIG. 40. Variation of the fluorescence intensity of rhodamine B and rhodamine B isothiocyanate adsorbed onto microcrystalline cellulose measured as the total area under the corrected emission spectrum, $I_F$, as a function of $(1 - R)f_{dye}$. Curve (1)—Type I samples of rhodamine B prepared from ethanolic solutions. Curve (2)—Type I samples of the reactive dye prepared from ethanol. Curve (3)—Type I samples of the reactive dye prepared from water. Curve (4)—Type II or dyed samples. Samples from curves 3 and 4 were repeatedly washed after the initial solvent evaporation.

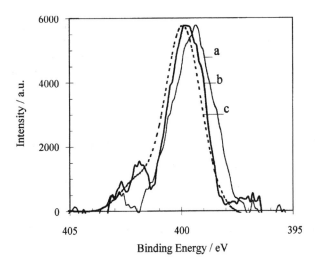

FIG. 41. XPS spectra for the N $1s$ region of samples with loadings of 3.5 $\mu$mol of rhodamine B isothiocyanate per gram of cellulose. (a) Type II sample repeatedly washed with water. (b) Type I sample from aqueous solution after washing with water. (c) Rhodamine B physically adsorbed from ethanolic solutions and subsequent removal of solvent by evaporation.

the dye molecules. This is due to the fact that the chemical reaction of the isoth-iocyanate group occurs with cellulose deep sites and since that molecular end group stays constrained by the chemical bond, geometrical hindrance to the pla-narity of the whole molecule exists, putting the positive charge in the xanthene moiety of the molecule instead of the nitrogen atoms. The sample's behavior in which the dye is physically adsorbed from an aqueous solution and subsequently washed is also very interesting: it presents a shift toward higher binding ener-gies (BE).

These data are consistent with the UV-Vis diffuse reflectance absorption and fluorescence data: in aqueous solution and in the absence of dyeing conditions, dye molecules adsorb on the most favorable sites, being free to acquire a pla-nar configuration. Nitrogen atoms (or, at least, one of them) become, in this way, an active part of the conjugated system. With time evolution, some of the dye molecules react with the substrate (otherwise, they would be removed by wash-ing). However, in the absence of the dyeing procedure, the reaction occurs slowly after the equilibrium adsorption is established and has very little or no effect on the final conformation of the dye molecule. In Figure 41 (curve c) the spectrum of a sample where pure rhodamine B is physically adsorbed from an ethanolic so-lution is also included for comparison purposes. It displays a slight shift towards higher BE. This is assigned to the fact that, for this high loading of the dye, some aggregation on the external surface exists and, as a consequence, the amount of planar conformers increases relatively to curve b case.

These findings indicate that rhodamine B has different conformers in dyed sam-ples as compared to samples where it is adsorbed. In the former case the chemical bond, anchoring the dye to microcrystalline cellulose, leads to nonplanar con-formers with smaller $\phi_F$ and $\tau_F$ values [15]. In the latter case planar conformers predominate, with the consequent increase of both lifetime and fluorescent quan-tum yield.

## 5.2 2,2′-Cyanines Adsorbed onto Microcrystalline Cellulose

The photochemistry and photophysical studies of cyanine dyes are an important research field owing to the economical importance of these substances in color and black and white photography, in dye lasers, and as potential sensitizers for photodynamic therapy.

Polymethine cyanine photochemistry in fluid solution is dominated by *trans-cis* isomerization as the main $S_1$ state decay route.

By studying simple cyanines adsorbed onto microcrystalline cellulose it was possible to get some insight into the nature of the interactions of these cyanines with the polymer chains, namely in regard to the importance of hydrogen bonding of the probe to the substrate, which is of major importance in the immobilization process, and in minimizing the nonradiative transitions of the excited states.

2,2'-Cyanines are dyes with a general formula as follows:

2,2´-Cyanine

2,2´-Carbocyanine

1,1'-Diethyl-2,2'-cyanine iodide and 1,1'-diethyl-2,2'-carbocyanine iodide were adsorbed onto microcrystalline cellulose by two different methods: by deposition from ethanolic solutions followed by solvent evaporation (Type I) and also from ethanolic solutions in equilibrium with the powdered solid (Type II). Both methods provided the same fluorescence quantum yield of the adsorbed dyes in the 0.01 to 5.0 $\mu$moles of dye per gram of cellulose concentration range.

Ethanol swells cellulose and some dye molecules stay entrapped into the natural polymer chains and in close contact with the substrate. The use of dichloromethane, a solvent that does not swell microcrystalline cellulose, provides samples that exhibit a smaller fluorescence quantum yield. This is consistent with a larger degree of mobility (and also formation of nonplanar and less emissive conformers) of the cyanines adsorbed on the surface of the solid substrate, while entrapment provides more rigid, planar, and emissive fluorophors.

For 2,2'-cyanine the fluorescence quantum yields ($\phi_F$) determined were about 0.08 whenever dichloromethane (solvent which does not swell cellulose) was used for sample preparation, while, with ethanol, $\phi_F$ was approximately 0.30. Similar values were also obtained for 1,1'-diethyl-2,2'-carbocyanine as Figure 42 shows. These values are about three orders of magnitude higher than in solution, showing the importance of the rigid dry matrix in reducing the nonradiative pathways of deactivation of the $(\pi, \pi^*)$ first excited singlet state of this cyanine.

The adsorption isotherms of 2,2'-cyanine on cellulose from alcoholic and dichloromethane solutions using a Langmuir adsorption model [16] showed that cellulose surface area accessible to dye adsorption is about twice larger when ethanol is used for sample preparation instead of dichloromethane, confirming the larger swelling power of cellulose by ethanol than by dichloromethane [14].

Figure 43 shows the XPS spectra for the N 1s region of several samples of 2,2'-cyanine adsorbed onto microcrystalline cellulose from an ethanolic solution after total solvent evaporation. We can clearly see that the N 1s peak has a main

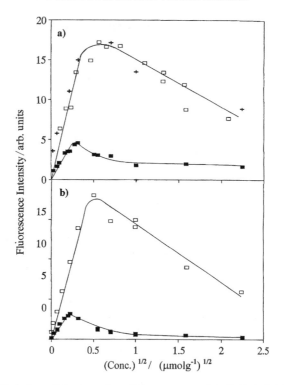

FIG. 42. (a) Variation of the intensity of fluorescence of 1,1′-diethyl-2,2′-cyanine adsorbed onto microcrystalline cellulose (steady state) measured as the total area under the corrected emission spectrum, $I_F$, as a function of the square root of the concentration of dye adsorbed onto microcrystalline cellulose. The solvents used for sample preparation were: □—ethanol, ■—dichloromethane. Full squares and open squares are used for type I and type II samples, respectively. (b) Same data as in (a) but now for 1,1′-diethyl-2,2′-carbocyanine.

component centered at a binding energy of 399.9 ± 0.1 eV for all concentrations under study.

This component is assigned to the nitrogen in a molecule with a planar conformation free from interactions apart from those with other dye molecules, as occurs in H aggregates. Further support for this assignment was provided by the XPS analysis of thick films (∼10 monolayers) of 2,2′-cyanine deposited onto silicon wafers as well as mechanical mixtures of 2,2′-cyanine and cellulose containing dye loadings equivalent to 1.0 and 5.0 $\mu$mol/g. The peak at higher binding energies (∼405 eV), usually associated with nitrogen bound to highly electronegative atoms [25], is here assigned to strongly hydrogen bonded nitrogen. Its intensity decreases with increasing dye loading. This fact suggests that at low loadings

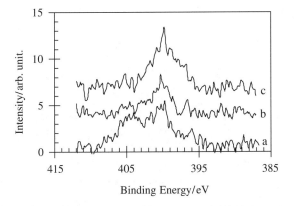

FIG. 43. X-ray photoelectron spectra of the N 1*s* region for three samples of 2,2′-cyanine adsorbed on cellulose from alcoholic solutions and allowing for the complete solvent evaporation. Curve (a) 0.1, curve (b) 1.0, curve (c) 10.0 $\mu$mol of dye per gram of cellulose.

a high percentage of molecules is intimately surrounded by substrate hydroxyl groups (entrapped) and also that with increasing dye loading that percentage decreases.

At lower binding energies, another new component develops as dye concentration increases, indicating that an increasing fraction of molecules has a larger electron density near the nitrogen atoms (on both or at least on one of them). The assignment of this component is rather difficult. However, by comparison with results from optical absorption in the visible region we can say that it exists whenever *J* aggregates are present; one possibility therefore is to assign that component as arising from this kind of aggregate. An alternative explanation could be the existence of nonplanar conformers in which the loss of symmetry of the molecule causes the appearance of different electron densities around the nitrogen atom: one strongly bound to the substrate and the other dangling free above the surface.

In all samples where the cyanine was adsorbed to cellulose from dichloromethane solution, the same effects as for ethanolic samples were observed (Fig. 44b). However, some quantitative differences exist: (i) the low binding energy component appears for lower dye loadings. This fact also supports the assignment of this component to the existence of *J* aggregates, which begins at lower concentrations in the dichloromethane case. (ii) The high binding energy component is less intense and centered at energies slightly lower than in the ethanolic case.

A third set of samples prepared by mechanically mixing cyanine with cellulose was studied. In this case, no intimate contact between the dye and the substrate exists: dye is mainly in the form of tiny crystals. We can see (Fig. 44, curve a)

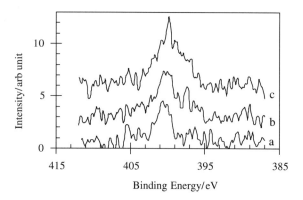

FIG. 44. X-ray photoelectron spectra of the N 1*s* region for three samples of 2,2′-cyanine adsorbed on cellulose in three different conditions: (a) mechanical mixture; (b) from dichloromethane solutions and allowing for complete solvent evaporation; (c) from alcoholic solutions and allowing for complete solvent evaporation. The concentration in all samples is 5.0 $\mu$mol of dye per gram of cellulose.

that the N 1*s* peak becomes narrower, but is centered at a higher binding energy ($\sim$400.4 eV), than the central component of the samples prepared from the dissolved dye. This fact is attributed to a change of the electron density around the nitrogen atoms when the dye molecule is included within a microcrystal.

XPS N 1*s* spectra for 2,2′-carbocyanine also for increasing loadings and for samples prepared by three different methods (adsorption from ethanolic solution, adsorption from dichloromethane, and mechanical mixture) are qualitatively not very different from the ones for 2,2′-cyanine in Figures 43 and 44. However, a few differences were noticed. They can be summarized as follows: (a) with increasing dye loading, the high energy component relative intensity increases; (b) the comparison of samples where the dye was adsorbed from two different solvents and also mechanically mixed with cellulose showed that only "ethanolic" and "dichloromethane" samples exhibited a strong interaction of the nitrogen atom with the substrate which obviously could not exist in the mechanical mixture; (c) 2,2′-carbocyanine nitrogen atoms are not equivalent as they are in 2,2′-cyanine.

Quantitative XPS studies performed through the evaluation of the atomic nitrogen/oxygen (N/O) ratio also allowed us to establish some important differences between 2,2′-carbocyanine and 2,2′-cyanine: (i) for 2,2′-cyanine molecules adsorbed onto cellulose from both solvents, the N/O ratio was invariant, within the experimental error. However, it increased with dye loading for the mechanical mixture. These results strongly suggest that 2,2′-cyanine molecules occupy deep sites in the cellulose, most probably pores, when they are adsorbed from solution.

In the case of the mechanical mixture, the absence of a close dye–substrate contact precludes that type of occupancy. (ii) In the case of 2,2′-carbocyanine there is a different behavior for ethanolic and dichloromethane samples: in the ethanolic case the invariance of the N/O ratio breaks down for the highest dye loading (10 $\mu$mol/g) whereas for dichloromethane samples the invariance disappears or it breaks down for much lower dye loadings. This points to a generally greater difficulty for this molecule in occupying deep sites in the substrate as compared with the smaller 2,2′-cyanine molecule. The difference between the two kinds of samples is consistent with the fact that ethanol is a better swelling solvent than dichloromethane.

X-ray photoelectron spectroscopic studies present evidence for hydrogen bonding of 2,2′-cyanine to cellulose for low loadings and aggregates formation for higher loadings in the ethanol and dichloromethane cases. This hydrogen bonding is assigned to the situation where the dye molecule is entrapped into cellulose chains. On the other hand, for 2,2′-carbocyanine, evidence exists for an increase of hydrogen bonding with dye loading. This result together with ground-state diffuse reflectance absorption and luminescence results is compatible with dye molecules being firmly bound to the substrate by one of the nitrogen atoms, the other one dangling over the substrate's free surface.

## 5.3 Inclusion Complexes of $\beta$-Cyclodextrin and Cyanine Dyes

Cyclodextrins are compounds with a torus-shaped hydrophobic cavity, and with a limited number of D-(+)- glucopyranose units joined by $\alpha$-(1,4) linkages. $\beta$-Cyclodextrin ($\beta$-CD) in particular is composed of seven of these units while $\alpha$-CD and $\gamma$-CD have six and eight, respectively. Figure 45 shows the shape and internal dimensions of these olygosaccharides.

They constitute therefore a substrate chemically similar to cellulose but provide a better characterized environment from the point of view of possible sites for adsorption. It should be a good medium to check for the correctness of the interpretation about the dye/cellulose interaction.

### 5.3.1 2,2′-Cyanine and 2,2′-Carbocyanine. 
XPS studies on the N 1$s$ region show that the behavior of 2,2′-cyanine dyes adsorbed onto cellulose in low concentration (0.1 $\mu$g/g of cellulose) is similar to the behavior of the same dyes included in $\beta$-cyclodextrin. Figure 46 displays the first example.

Figure 46 shows that the N 1$s$ region of the XPS spectrum of 2,2′-cyanine included in $\beta$-cyclodextrin (1 : 50) is very similar to the same spectrum for 2,2′-cyanine adsorbed on cellulose from an alcoholic solution with a very low concentration—0.1 $\mu$g per gram of cellulose. As seen in Section 5.2, one of the peaks, located at 399.9 ± 0.1 eV, is assigned to the nitrogen in a molecule with a planar configuration free from interactions, apart from those with other dye molecules, as occurs in $H$ aggregates. The other one, at higher binding energies

| CD | $d_1/Å$ | $d_2/Å$ | $d_3/Å$ | $h/Å$ |
|---|---|---|---|---|
| α | 5.6 | 4.2 | 8.8 | 7.8 |
| β | 6.8 | 5.6 | 10.8 | 7.8 |
| γ | 8.0 | 6.8 | 12.0 | 7.8 |

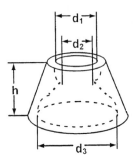

FIG. 45. Schematic representation of cyclodextrins and internal dimensions of the cavities.

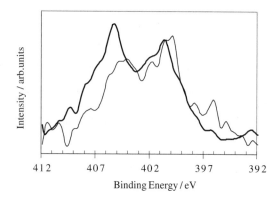

FIG. 46. N $1s$ region X-ray photoelectron spectrum: comparison of 2,2'-cyanine adsorbed on cellulose from an alcoholic solution, with a concentration of 0.1 $\mu$g per gram of cellulose (thin line), with 2,2'-cyanine included in β-cyclodextrin (1 : 50, mol : mol) (bold line).

($\sim$405 eV), is assigned to strongly hydrogen bonded nitrogen and its intensity decreases for lower loadings. When compared to the lowest loaded sample, the 2,2'-cyanine included in β-cyclodextrin (1 : 50) presents also these two components but their relative intensity is different: all the spectrum is displaced toward higher binding energies and the component at higher binding energy has a larger intensity relative to the component at lower BE. These two facts denote a larger interaction between nitrogen and the substrate's hydroxyl groups when the substrate is β-cyclodextrin than when it is cellulose.

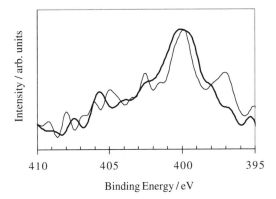

FIG. 47. N 1*s* region X-ray photoelectron spectrum: comparison of 2,2′-carbocyanine adsorbed on cellulose from an alcoholic solution, with a concentration of 0.1 $\mu$g per gram of cellulose (thin line), with 2,2′-carbocyanine included in $\beta$-cyclodextrin (1 : 50, mol : mol) (bold line).

Another striking difference is the existence of a small component at lower binding energy in the case of the adsorption on cellulose which is absent when the substrate is $\beta$-cyclodextrin. This reinforces the assignment of this component to nitrogen belonging to molecules in *J* aggregates. Figure 47 displays the same comparison as Figure 46 but the adsorbed dye is the 2,2′-carbocyanine.

Figure 47 shows that the N 1*s* region of the XPS spectrum of 2,2′-carbocyanine included in $\beta$-cyclodextrin (1 : 50) is very similar to the same spectrum for 2,2′-carbocyanine adsorbed on cellulose from an alcoholic solution with a very low concentration—0.1 $\mu$g per gram of cellulose. Also in this case the greater similarity between the two cases arises for the least concentrated sample adsorbed on cellulose. (We remember that, contrary to the case of 2,2′-cyanine, the high binding energy component increases with dye loading on cellulose: see Section 5.2). This means that the interaction of the substrate's hydroxyl groups with nitrogen in 2,2′-carbocyanine is much weaker than it is with nitrogen in 2,2′-cyanine, as had been already evidenced with the substrate cellulose.

A common point with the 2,2′-cyanine case is the presence of a low binding energy component ($\sim$397 eV) when cellulose is the substrate and its absence when the substrate is the $\beta$-cyclodextrin.

5.3.2 3,3′-Ethylthiacarbocyanine. Also the inclusion of 3,3′-ethylthiacarbocyanine (TCC) in $\beta$-cyclodextrin shows a greater similitude with the cellulose low loading sample, as displayed in Figure 48.

However, in the case of TCC, similar to the case of the carbocyanine, the N 1*s* region does not show evidence for a large interaction between the nitrogen atoms and the substrate (comparable to the one detected in the case of the 2,2′-cyanine).

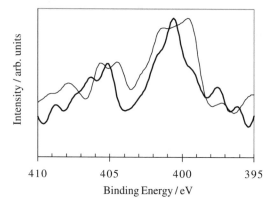

FIG. 48. N 1$s$ region X-ray photoelectron spectrum: comparison of TCC adsorbed on cellulose from an alcoholic solution, with a concentration of 0.1 $\mu$g per gram of cellulose (thin line) and TCC included in $\beta$-cyclodextrin (1 : 50, mol : mol) (bold line).

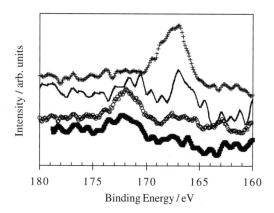

FIG. 49. XPS S 2$p$ region for TCC. From bottom to top: included in $\beta$-CD 1 : 100, mol : mol (■), adsorbed on cellulose (0.1 $\mu$g/g of cellulose) (○), adsorbed on cellulose (1 $\mu$g/g of cellulose) (——), and mechanically mixed with cellulose (+).

In this case, the most interesting XPS region is not the nitrogen region but the XPS sulphur region. Figure 49 displays the XPS S 2$p$ region for four different samples: TCC adsorbed on cellulose with two different concentrations (0.1 and 1 $\mu$g per gram of cellulose), a mechanical mixture of TCC with cellulose, and TCC included in $\beta$-CD with the concentration 1 : 100.

The XPS S 2$p$ region for TCC mechanically mixed with cellulose shows a single component, as expected, since no intimate contact exists between the dye molecule and the substrate. The peak is asymmetric since it is a doublet (S 2$p_{3/2}$

and S $2p_{1/2}$) with an energy splitting of about 1 eV [26]. For TCC adsorbed on cellulose at least two components are exhibited: one at about the same energy as the one detected for the mechanical mixture and another one at higher binding energies usually assigned to sulphur bound to highly electronegative atoms [25]. Decreasing the dye loading, the low BE component almost disappears and only the high BE component is left. Similar to the observation made on the XPS nitrogen region, these results point to a strong interaction between the sulphur atom and the substrate, the sulphur atom acting as a strong electron donor. An obvious candidate is the hydrogen bonding interaction between hydroxyl groups in the substrate and sulphur lone pairs. The same kind of interaction exists also in the inclusion complex $\beta$-CD:TCC. This confirms that this interaction occurs only when molecules are deeply entrapped into the cellulose chains.

In parallel, $\phi_F$ for TCC entrapped within the cellulose polymer chains is about 0.95 (see Figure 28). For TCC included in $\beta$-CD, $\phi_F$ is much smaller, around 0.20. This shows that molecular mobility is larger in the latter case.

## 5.4 Sulforhodamine 101 and Rhodamine 6G Adsorbed on Different Pore Size Silicas

Powdered solids with controlled porosity and particle size are privileged media for the study of reactions in restricted environments or the study of interactions of probes with specific surfaces.

Sulforhodamine 101 and Rhodamine 6G were chosen as probes for the latter study [7]. They were adsorbed onto silicas with different pore sizes ranging from 22 to 150 Å. Ground state diffuse reflectance absorption spectra revealed the formation of different forms of adsorbed sulforhodamine 101 depending on concentration and on the pore size of the silica. For low loadings (0.001 to about 0.025 $\mu$mol of dye per gram of silica) the absorption spectra are broad, hypsochromically shifted in relation to the monomer spectra, and quite different from the ethanolic solution spectra. For high loadings (0.050 to about 0.20 $\mu$mol/g) they are similar to the solution spectra with a small shift of about 7 nm. For rhodamine 6G spectra are much more of the "solution type" in the entire range of concentrations under study (0.001 to about 0.20 $\mu$mol/g). (See Figure 50.)

The "weighed" fluorescence quantum yields ($\sum_i f_i \phi_{F_i}$) determined for sulforhodamine 101 were $0.10 \pm 0.03, 0.35 \pm 0.05$, and $0.50 \pm 0.10$ for low loadings and for 25, 60, and 150 Å silicas, respectively. For high loadings $\sum_i f_i \phi_{F_i} = 0.70 \pm 0.10$, as Figure 31 in Section 4.1 shows. These values for $\sum_i f_i \phi_{F_i}$ can be compared with a value of about 0.70 obtained for rhodamine 6G in all silicas (low loadings) and a unitary $\phi_F$ for high loadings.

XPS experiments were also performed with sulforhodamine 101 and rhodamine 6G on silicas. The dye loadings under study were 0.002, 0.025, 0.10, and 1.0 $\mu$mol/g onto silica gel with 25 Å pores, and 0.002, 0.04, 0.10, and 1.0 $\mu$mol/g

## Rhodamine 6G

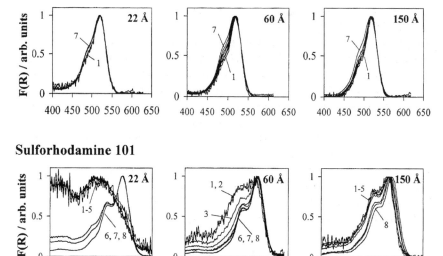

## Sulforhodamine 101

Fig. 50. Remission function values for sulforhodamine 101 and rhodamine 6G adsorbed on silicas with 22, 60, or 150 Å pore diameters for the range of concentrations under study: 1—0.004 $\mu$mol/g to 8—0.1 $\mu$mol/g. Spectra are normalized to the maximum of the absorption of the dye. (All samples are Type I.)

onto silica gel having pores of 150 Å diameter. Within the error associated with XPS quantifications, it was shown that the atomic ratio N/Si was constant (of the order of $10^{-3}$) and independent of dye loading (although a slight variation is detected for sulforhodamine adsorbed on 25 Å pore silica). That ratio was also larger in the silica with the 25 Å pore than in the silica with the 150 Å pore (by a factor around 1.5) as displayed in Figure 51.

These findings are compatible with the following picture for these systems: the dye must be mainly adsorbed near the extreme surface of the pores. Moreover the adsorption must occur essentially in depth (filling the pores) and not with an increase in the number of the occupied pores with increasing loadings. Dye molecules penetrate deeper into the pores in the larger pore size silica, the constraints concerning penetration being more important in the 25 Å pore silica than in the 150 Å pore silica.

To try to understand why the adsorption process occurs following this pattern, it is important to make an estimation of the specific number of pores, $n_p$, which allows computation of the available number of pores per adsorbed molecule.

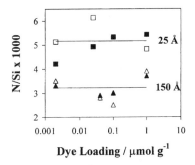

FIG. 51. Atomic ratio N/Si for sulforhodamine 101 (full symbols) and rhodamine 6G (open symbols) adsorbed on 25 or 150 Å pore silicas as a function of dye concentration. Straight lines in the plot represent average values.

The experimental available data for powdered microporous materials are the average particle size $L$, obtained by sieving, the total specific area, $A_{sp}$, and the average pore radius, $r_p$, or pore volume, $v_p$, obtained by gaseous adsorption isotherms or specific porosimetry methods [152]. These three parameters are not independent and are usually related by

$$2\frac{dv_p}{dA_{sp}} = r_k \tag{30}$$

derived with the aid of the conventional Kelvin formula, where $r_k$ is the so-called "Kelvin radius" which coincides with the pore radius, for nonintersecting pores of circular cross-section [152], provided $L \gg r_p$.

For a rough estimation of $n_p$ we will use two very simple models. In the first one, here called model 1 and schematically represented in Figure 52, pores are considered to be a system of nonintersecting cylindrical capillaries and the particles are cubes, with an edge $L$. Given the absence of shared volumes, it will give the **minimum number of pores** for a given pore volume. Moreover, this model is compatible with the fact that highly porous silica present self-similar surfaces with a fractal dimensionality approaching 3 [152, 153].

For a powdered solid containing particles with that shape, and assuming that all the pores are available even in the regions of contact between particles, the specific number of pores, $n_p$, is given by equation

$$n_p = \frac{n_{ps}}{m_p} = \frac{2n_{pp}}{m_p} \tag{31}$$

where $n_{pp}$ is the number of nonended cylindrical pores per particle, $n_{ps}$ is the number of pores emerging at the extreme surface, and $m_p$ is the mass of a particle. On the other hand, the specific total surface, $A_{sp}$, neglecting the contribution of

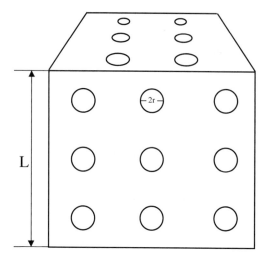

FIG. 52. Schematic representation of cubic particles having diameter $L$ and nonintersecting cylindrical capillaries of radius $r_p$.

the extreme surface, is

$$A_{sp} = \frac{n_{pp} 2\pi r_p L}{m_p} \quad \Rightarrow \quad n_{pp} = \frac{A_{sp} m_p}{2\pi r_p L} \tag{32}$$

From Equations (31) and (32) we have

$$n_p = \frac{A_{sp}}{\pi r_p L} \tag{33}$$

Equation (33) shows that in this system the specific number of pores emerging at the extreme surface is independent of the mass of the particle and is inversely proportional to the particle size as it is intuitively expected. This gives, for the silicas used in this work, the parameters contained in Table VII.

Obviously, if we assume other pore systems, slightly different figures for $n_p$ are obtained. Namely, if the pores cross each other, the number of pores, for a given pore volume, should increase relative to the value from expression (33). Anyway, there is an **upper** unreachable **limit**: the entire external surface occupied by tangent pores of circular cross-section in a close-compact arrangement, here called model 2. In this case,

$$n_{ps} = \frac{6L^2}{2\sqrt{3} r_p^2} \quad \text{and} \quad n_p = \frac{6L^2}{m_p 2\sqrt{3} r_p^2} \tag{34}$$

TABLE VII.    Typical Experimental Parameters for Two of the Silicas Used
in This Work Together with the Specific Number of Pores Computed from
Model 1

| Pore size (Å) | Particle size ($\mu$m) | Specific area (m$^2$/g) | Pore volume (cm$^3$/g) | Specific number of pores ($\mu$mol/g) |
|---|---|---|---|---|
| 25 | 75 | 600 | 0.38 | $3.38 \times 10^{-3}$ |
|  | 150 |  | 0.38 | $1.69 \times 10^{-3}$ |
| 150 | 75 | 300 | 1.13 | $2.82 \times 10^{-4}$ |
|  | 150 |  | 1.13 | $1.41 \times 10^{-4}$ |

In this model, $n_p$ depends on the inverse of the particle mass. The mass of the particle, $m_p$, was estimated by two methods:

(I)  Assume that the particle—a cube of edge $L$—is made of pores and massive amorphous silica with a mass density $d_{Silica} = 2.2$ g/cm$^3$ [154]. In this case, the specific volume is given by $1/d_{Silica} + v_p$. Therefore, the particle mass is given by

$$m_p = \frac{\text{Particle volume}}{\text{Specific volume}} = \frac{L^3}{1/d_{Silica} + v_p}$$

(II)  Use "experimental" mass density, $d_{exp}$, values determined by weighting a given volume of dry powder ($d_{exp} = 0.72$ g/cm$^3$ for the 25 Å silica and $d_{exp} = 0.38$ g/cm$^3$ for the 150 Å silica). In this case, the particle mass is simply

$$m_p = d_{exp}L^3$$

These two values are clearly upper and lower limits: it is unexpected that the walls between pores do have the same density as the massive silica or larger; on the other hand, the experimental value considers the volume between particles as particle volume, thus underestimating the particle mass. The real value should, then, lie between these two extreme values. In order to be sure to compute an upper limit for $n_p$, displayed in Table VIII, the lower mass value was taken into account.

As a conclusion, we can say that the real number of pores should be between the values presented in Tables VII and VIII.

This estimation of the boundaries for the number of pores emerging at the extreme surface shows that all the dye loadings used in this study are larger than the number of pores except, perhaps, for the lowest loading in the case of the silica

TABLE VIII. Estimated Particle Mass and Values for the Specific Number
of Pores for Two of the Silicas Used in this Work, Assuming Model 2

| Pore size (Å) | Particle size ($\mu$m) | Particle mass ($\mu$g) | | Specific number of pores ($\mu$mol/g) |
| | | Assuming $d_{SiO_2} = 2.2$ g/cm$^3$ | Assuming $d_{exp}$ | |
|---|---|---|---|---|
| 25 | 75 | 0.51 | 0.30 | $34.1 \times 10^{-3}$ |
| | 150 | 4.1 | 2.4 | $17.0 \times 10^{-3}$ |
| 150 | 75 | 0.27 | 0.16 | $1.79 \times 10^{-3}$ |
| | 150 | 2.1 | 1.3 | $0.90 \times 10^{-3}$ |

having a 25 Å pore diameter. This means that conclusions about pore occupancy by dye molecules based on XPS results are reliable.

The qualitative analysis of the N $1s$ peak (see Figure 53) does not reveal any measurable change in shape and position, suggesting that the interaction between dye molecules and the substrate leading to the dye molecule deformation, and the consequent $\pi$ delocalization extent decrease, does not involve the nitrogen atom as much as it did in the case of the rhodamine dyes covalently described in Section 5.1.

However, and since optical studies of the same system undoubtedly point to the existence of distorted molecules, the invariance of the XPS N $1s$ peak may be due to the fact that XPS only "sees" surface layers of about 100 Å width. Therefore, we mainly see probe molecules poorly interacting with the silica surface, and the strongest interactions occur deeply inside the pore.

These studies indicate that sulforhodamine 101 forms nonplanar conformers in small pore size silicas as compared to large pore silica samples where the amount of conformers being formed is reduced. Rhodamine 6G samples exhibit very little conformer formation but their $\phi_F$ are still slightly dependent on pore size. Both rhodamines exhibit smaller fluorescence quantum yields when compared to the case of adsorption onto microcrystalline cellulose, this effect being more relevant in the sulforhodamine 101 case.

# 6 CONCLUSIONS

X-ray photoelectron spectroscopy and UV/Vis absorption and luminescence studies have proved to be complementary techniques in the study of dyes in several environments: when adsorbed onto or bound to a natural polymer, microcrys-

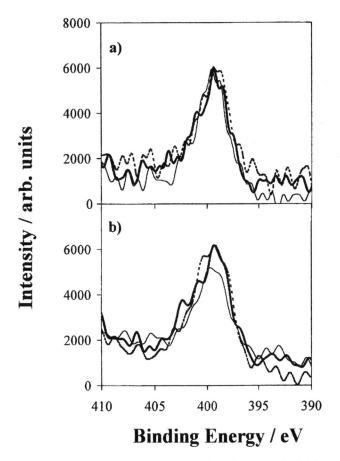

FIG. 53. XPS spectra for the N 1*s* region of samples of the two rhodamines adsorbed on 25 Å silica. (a) Sulforhodamine 101 (——) 0.002 $\mu$mol/g, (—) 0.025 $\mu$mol/g, and (---) 0.1 $\mu$mol/g. (b) Rhodamine 6G (——) 0.002 $\mu$mol/g, (—) 0.020 $\mu$mol/g, and (- - -) 0.1 $\mu$mol/g. (All samples are Type I.)

talline cellulose, when forming inclusion complexes with cyclodextrins, and also when interacting with silicas with different pore sizes.

UV/Vis ground-state diffuse-reflectance absorption and luminescence studies showed in most cases a huge increase of fluorescence emission quantum yield when compared to solution studies for several dyes, when they are entrapped into microcrystalline cellulose, due to the reduction of mobility and formation of planar and emissive conformers and to the amount and type of aggregates which are formed. The emission depends on the solvent used to adsorb the probes. In some cases a photoisomer fluorescence emission was detected following a two photon absorption process, the first one used for the formation of the photoisomer and the second one creating the excited state responsible for this special second emission from the cyanines which coexists with the monomer emission.

A complementary view is given by X-ray photoelectron studies which provide evidence for hydrogen bond formation as well as preferential attachment by one or two nitrogen atoms per cyanine molecule to the hydroxyl groups in the substrates. In the thiacarbocyanine case, hydrogen bond formation was detected for microcrystalline cellulose and $\beta$-cyclodextrin cases which involved the sulphur atom of the cyanine.

The combined use of diffuse reflectance techniques and XPS allowed us to establish of a clear picture of the specific interactions of dyes and substrates in several cases and enabled a much deeper insight than the sum of information provided by each technique alone.

## Acknowledgments

We thank Drs. M. Rei Vilar, A. S. Oliveira, and O. Pellegrino and Mrs. M. J. Lemos for their collaboration in some of the work described here. We also thank Dr. A. S. Oliveira for a critical reading of the manuscript and Dr. O. Pellegrino and Mrs. M. J. Lemos for helping us with some of the drawings.

## References

*1.* F. Wilkinson and G. P. Kelly, in "Handbook of Organic Photochemistry" (J. C. Scaiano, Ed.), Vol. 1, p. 293. CRC Press, Boca Raton, 1989, and references quoted there.

*2a.* R. J. Hurtubise, "Phosphorimetry: Theory, Instrumentation and Applications," VCH Publishers, New York, 1990.

*2b.* R. J. Hurtubise, "Solid Surface Luminescence Analysis. Theory, Instrumentation and Applications," Dekker, New York, 1981.

*2c.* R. J. Hurtubise, *Anal. Chem.* 61, 889A (1989).

*3.* "Surface Photochemistry" (M. Anpo, Ed.), Wiley, Baffins Lane, 1996; "Photochemistry on Solid Surfaces" (M. Anpo and T. Matsuara, Eds.), Elsevier, Amsterdam, 1989.

*4.* C. Bohne, R. W. Redmond, and J. C. Scaiano, in "Photochemistry in Organized and constrained Media" (V. Ramamurthy, Ed.), Chap. 3, p. 79. VCH Publishers, New York, 1991.

*5.* L. J. Johnston, in "Photochemistry in Microheterogeneous Systems" (K. Kalyanasundaram, Ed.), Chap. 8, p. 359. Academic Press, Orlando, 1987; P. V. Kamat, *Chem. Rev.* 93, 267 (1993).

 *6.* E. Desimoni and P. G. Zambonin, in "Surface Characterization of Advanced Polymers" (L. Sabbatini and P. G. Zambonin, Eds.), p. 5. VCH, Weinheim, 1993.

 *7.* L. F. Vieira Ferreira, M. J. Lemos, M. J. Reis, and A. M. Botelho do Rego, *Langmuir* 16, 5673 (2000).

 *8a.* P. Kubelka and F. Munk, *Z. Tech. Phys.* 12, 593 (1931).

 *8b.* W. Wendlandt and H. G. Hecht, in "Reflectance Spectroscopy," Wiley, New York, 1966.

 *8c.* G. Kortum, "Reflectance Spectroscopy. Principles, Metods, Applications," Springer-Verlag, Berlin, 1969.

 *9.* P. M. Epperson, J. V. Sweedler, R. B. Bilhom, G. R. Sims, and M. B. Denton, *Anal. Chem.* 60, 327A (1988); J. V. Sweedler, R. B. Bilhom, P. M. Epperson, G. R. Sims, and M. Bonner Denton, *Anal. Chem.* 60, 283A (1988).

 *10a.* F. Wilkinson and C. J. Willsher, *Chem. Phys. Lett.* 104, 272 (1984).

 *10b.* R. W. Kessler and F. Wilkinson, *J. Chem. Soc. Faraday Trans. I* 77, 309 (1981).

 *10c.* F. Wilkinson, *J. Chem. Soc. Faraday Trans. II* 82, 2073 (1986).

 *10d.* R. W. Kessler, G. Krabichler, S. Schafler, S. Uhl, D. Oelkrug, W. P. Hagan, J. Hyslop, and F. Wilkinson, *Opt. Acta* 30, 1099 (1983).

 *10e.* F. Winkinson and G. P. Kelly, in "Photochemistry on Solid Surfaces" (M. Anpo and T. Matsuara, Eds.), p. 31. Elsevier, Amsterdam, 1989.

 *10f.* F. Wilkinson, *Tetrahedron* 43, 1197 (1987).

 *10g.* F. Wilkinson and D. R. Worrall, *Proc. Indian Acad. Sci. (Chem. Sci.)* 104, 287 (1992).

 *11.* G. P. Kelly, P. A. Leicester, F. Wilkinson, D. R. Worrall, L. F. Vieira Ferreira, R. Chittock, and W. Toner, *Spectrochim. Acta A* 46, 975 (1990).

 *12.* D. Oelkrug, W. Honnen, C. J. Willsher, and F. Wilkinson, *J. Chem. Soc. Faraday Trans. I* 81, 2081 (1987).

 *13a.* P. P. Levin, L. F. Vieira Ferreira, and S. M. B. Costa, *Chem. Phys. Lett.* 173, 277 (1990).

 *13b.* P. P. Levin, L. F. Vieira Ferreira, S. M. B. Costa, and I. V. Katalnikov, *Chem. Phys. Lett.* 193, 461 (1992).

 *13c.* P. P. Levin, L. F. Vieira Ferreira, and S. M. B. Costa, *Langmuir* 9, 1001 (1993).

 *13d.* P. P. Levin, S. M. B. Costa, and L. F. Vieira Ferreira, *J. Phys. Chem.* 100, 15171 (1996).

 *14.* L. F. Vieira Ferreira, J. C. Netto-Ferreira, I. V. Khmelinskii, A. R. Garcia, and S. M. B. Costa, *Langmuir* 11, 231 (1995).

 *15.* L. F. Vieira Ferreira, P. V. Cabral, P. Almeida, A. S. Oliveira, M. J. Reis, and A. M. Botelho do Rego, *Macromolecules* 30, 3936 (1998).

 *16.* A. M. Botelho do Rego, L. Penedo Pereira, M. J. Reis, A. S. Oliveira, and L. F. Vieira Ferreira, *Langmuir* 13, 6787 (1997).

 *17a.* L. M. Ilharco, A. R. Garcia, J. Lopes da Silva, and L. F. Vieira Ferreira, *Langmuir* 13, 4126 (1997).

 *17b.* L. M. Ilharco, A. R. Garcia, J. Lopes da Silva, M. J. Lemos, and L. F. Vieira Ferreira, *Langmuir* 13, 3787 (1997).

 *17c.* L. M. Ilharco, *Química* 69, 34 (1998).

 *18.* N. Câmara de Lucas, J. C. Netto-Ferreira, J. Andros, J. Lusztyk, B. D. D. Wagner, and J. C. Scaiano, *Tetrahedron Lett.* 38, 5147 (1997).

19. "Perspectives in Modern Chemical Spectroscopy" (D. L. Andrew, Ed.), Springer-Verlag, Berlin/Heidelberg, 1990.

20. See, for instance, XSAM 800, "Operators Handbook," Vol. 2. Kratos Analytical, Manchester, 1990.

21. K. Siegbahn, "Electron Spectroscopy for Atoms, Molecules and Condensed Matter," Nobel Lecture, 1981; N. Martensson and A. Nilsson, *J. Electron Spectrosc. Relat. Phenom.* 75, 209 (1995); T. L. Barr, E. Hoppe, T. Dugall, P. Shah, and S. Seal, *J. Electron Spectrosc. Relat. Phenom.* 99, 95 (1999).

22. D. Briggs and J. C. Rivière, in "Pratical Surface Analysis by Auger and X-Ray Photoelectron Spectroscopy" (D. Briggs and M. P. Seah, Eds.), Chap. 3. Wiley, New York, 1983.

23. See, for instance, M. Sastry and P. Ganguly, *J. Phys. Chem.* 102, 697 (1998) and references therein.

24. Z. B. Maksic, D. Kovacek, K. Kovacevik, and Z. Medven, *J. Mol. Structure* 304, 151 (1994); D. Kovacek, K. Kovacevik, D. Korenic, and Z. B. Maksic, *J. Mol. Structure* 304, 163 (1994); D. Kovacek, Z. B. Maksic, I. Petanjek, and I. Basic, *J. Mol. Structure* 304, 261 (1994).

25. See, for instance, C. D. Wagner, "NIST X-Ray Photoelectron Spectroscopy Database," version 1, U.S. Dept. of Commerce, NIST, 1989, available on diskette. Version 2 is now available on the Web site http://srdata.nist.gov/xps/.

26. G. Beamson and D. Briggs, "High Resolution XPS of Organic Polymers. The Scienta ESCA300 Database," Wiley, New York, 1992.

27. J. Cazaux, *J. Electron Spectrosc. Relat. Phenom.* 105, 155 (1999).

28. D. A. Huchital and R. T. Mckeon, *Appl. Phys. Lett.* 20, 158 (1972).

29. D. Briggs and M. P. Seah (Eds.), "Practical Surface Analysis," Wiley, New York, 1983.

30. D. Briggs and M. P. Seah (Eds.), "Practical Surface Analysis," 2nd ed., Vol. 1. Wiley, New York, 1996.

31. C. J. Powell, *J. Electron Spectrosc. Relat. Phenom.* 47, 197 (1988).

32. L. C. Feldman and J. W. Mayer, "Fundamentals of Surface and Thin Film Analysis," p. 129. North-Holland, New York, 1986.

33. C. J. Powell. A. Jablonski, I. S. Tilinin, S. Tanuma, and D. R. Penn, *J. Electron Spectrosc. Relat. Phenom.* 98-99, 1 (1999).

34. C. D. Wagner, L. E. Davis, and W. M. Riggs, *Surf. Interface Anal.* 2, 53 (1980).

35. S. Tanuma, C. J. Powell, and D. R. Penn, *Surf. Interface Anal.* 21, 165 (1994).

36. S. Tanuma, C. J. Powell, and D. R. Penn, *Surf. Interface Anal.* 20, 77 (1993).

37. C. J. Powell, "NIST Inelastic Electron-Mean-Free-Path Database," available in diskette.

38. D. A. Shirley, *Phys. Rev. B* 5, 4709 (1972).

39. See, for instance, P. C. J. Graat, M. A. J. Somers, and A. Böttger, *Surf. Interface Anal.* 23, 44 (1995).

40. See, for instance, E. Puppin and E. Ragaini, *J. Electron Spectrosc. Relat. Phenom.* 73, 53 (1995).

41. R. Shimizu and Z.-J. Ding, *Rep. Prog. Phys.* 55, 487 (1992); J.-Ch. Kuhr and H. J. Fitting, *J. Electron Spectrosc. Relat. Phenom.* 105, 257 (1999).

42. A. Dubus, A. Jablonski, and S. Tougaard, *Prog. Surf. Sci.* 63, 135 (2000).

43. C. S. Fadley, R. J. Baird, W. Siekhaus, T. Novakov, and S. A. L. Bergstrom, *J. Electron Spectrosc. Relat. Phenom.* 4, 93 (1977).

44. C. S. Fadley, "Progress in Surface Science" (S. G. Davison, Ed.), Vol. 16, p. 275. Pergamon, Oxford, 1984, and references therein.

45. J. E. Fulghum, *Surf. Interface Anal.* 20, 161 (1993).

46. P. J. Cumpson, *J. Electron Spectrosc. Relat. Phenom.* 73, 25 (1995).

47. H. Ibach and D. L. Mills, "Electron Energy Loss Spectroscopy and Surface Vibrations," Chap. 3. Academic Press, New York, 1982.

48. M. Rei Vilar, M. Heyman, and M. Schott, *Chem. Phys. Lett.* 94, 552 (1983).

49. J.-J. Pireaux, in "Surface Characterization of Advanced Polymers" (L. Sabattini and P. G. Zambonin, Eds.), Chap. 2. VCH, Weinheim, 1993.

50. M. Rei Vilar, A. M. Botelho do Rego, J. Lopes da Silva, F. Abel, V. Quillet, M. Schott, S. Petitjean, and R. Jérôme, *Macromolecules* 26, 5900 (1993).

51. M. Schreck, M. Abraham, W. Göpel, and H. Schier, *Surf. Sci. Lett.* 237, L405 (1990).

52. D. Roy and J. D. Carette, in "Electron Spectroscopy for Surface Analysis" (H. Ibach, Ed.), Topics in Current Physics, Vol. 4. Springer-Verlag, New York, 1977.

53. Y. Ballu, *Rev. Phys. Appl.* 3, 46 (1968).

54. H. Froitzheim, H. Ibach, and S. Lewhald, *Rev. Sci. Instrum.* 46, 1325 (1975).

55. L. L. Kesmodel, *J. Vac. Sci. Technol. A* 1, 1456 (1983).

56. H. Ibach and D. L. Mills, "Electron Energy Loss Spectroscopy and Surface Vibrations," Chap. 2. Academic Press, New York, 1982.

57. "Owner's Manual for the Model LK2000," LK Technologies, Inc., Bloomington, IN.

58. H. Ibach, "Electron Energy Loss Spectrometers: The Technology of High Performance," Springer Series in Optical Sciences, Vol. 63. Springer, Berlin, 1991.

59. M. Liehr, P. A. Thiry, J.- J. Pireaux, and R. Caudano, *Phys. Rev. B* 33, 5682 (1986).

60. L. L. Kesmodel, S. Wild, and G. Apai, *Surf. Sci.* 429, L475 (1999).

61. F. Deloffre, F. Garnier, P. Srivastava, A. Yassar, and J.-L. Fave, *Synth. Met.* 67, 223 (1994).

62. J. H. de Boer, *Z. Phys. Chem. Abt. B* 14, 163 (1931).

63. C. H. Nikolls and P. A. Leemakers, *Adv. Photochem.* 8, 315 (1971) and references therein; L. D. Weis, B. W. Bowen, and P. A. Leemakers, *J. Amer. Chem. Soc.* 88, 3176 (1966).

64. P. de Mayo, L. V. Natarajan, and W. R. Ware, in "Organic Phototransformations in Nonhomogeneous Media" (M. A. Fox, Ed.), ACS Symp. Series, Vol. 278, p. 1. Am. Chem. Soc., Washington, DC, 1984, and references therein; P. de Mayo, *Pure Appl. Chem.* 54 1623 (1982); P. de Mayo, L. V. Natarajan, and W. R. Ware, *Chem. Phys. Lett.* 107, 187 (1984); P. de Mayo, L. V. Natarajan, and W. R. Ware, *J. Phys. Chem.* 89, 3526 (1985).

65. D. Oelkrug, M. Plauschinat, and R. W. Kessler, *J. Lumin.* 18/19, 434 (1979); R. W. Kessler, S. Uhl, W. Honnen, and D. Oelkrug, *J. Lumin.* 24/25, 551 (1981); W. Honnen, G. Krabichler, S. Uhl, and D. Oelkrug, *J. Phys. Chem.* 87, 4872 (1983); K. Kempfer, S. Uhl, and D. Oelkrug, *J. Mol. Stucture* 114, 225 (1984); D. Oelkrug, W.

Flemming, R. Fullerman, R. Gunter, W. Honnen, G. Krabichler, S. Schafler, and S. Uhl, *Pure Appl. Chem.* 58, 1207 (1986) and references therein.

66a. K. Kalyanasumdaram and J. K. Thomas, *J. Amer. Chem. Soc.* 99, 2039 (1977); G. Beck and J. K. Thomas, *Chem. Phys. Lett.* 94, 553 (1983); J. K. Thomas, *Chem. Rev.* 93, 301 (1993) and references therein; J. K. Thomas, *Chem. Rev.* 80, 283 (1980).

66b. L. Francis, J. Lin, and L. A. Singer, *Chem. Phys. Lett.* 94, 162 (1983).

67. N. J. Turro, M. B. Zimmit, and I. R. Gold, *J. Amer. Chem. Soc.* 107, 5826 (1985); N. J. Turro, I. R. Gold, and M. B. Zimmit, *Chem. Phys. Lett.* 119, 484 (1985); N. J. Turro, C. C. Cheng, and W. Mahler, *J. Amer. Chem. Soc.* 106, 5022 (1984); N. J. Turro, *Pure Appl. Chem.* 58, 1219 (1986).

68. J. C. Scaiano, H. L. Casal, and J. C. Netto-Ferreira, *Amer. Chem. Soc. Sympos. Ser.* 13, 211 (1985); H. L. Casal and J. C. Scaiano, *Canad. J. Chem.* 62, 628 (1984); H. L. Casal and J. C. Scaiano, *Canad. J. Chem.* 63, 1308 (1985); H. L. Casal, J. C. Netto-Ferreira, and J. C. Scaiano, *J. Inclusion Phenom.* 3, 395 (1985); R. Boch, C. Bohne, and J. C. Scaiano, *J. Org. Chem.* 61, 1423 (1996).

69. H. Ishida, H Takahashi, and H. Tsubomura, *J. Amer. Chem. Soc.* 92, 275 (1970); H. Ishida and H. Tsubomura, *J. Photochem.* 2, 285 (1973).

70. R. Dabestani, *Inter-Amer. Photochem. Soc. Newslett.* 20, 24 (1997).

71. W. J. Albery, P. N. Bartlett, C. P. Wilde, and J. R. Darwent, *J. Amer. Chem. Soc.* 107, 1854 (1985); K. F. Scott, *J. Chem. Soc. Faraday Trans. I* 76, 2065 (1980); A. Habti, D. Keravis, P. Levitz, and H. Van Damme, *J. Chem. Soc. Faraday Trans. II* 80, 67 (1984).

72. C. T. Lin and W. C. Hsu, *J. Phys. Chem.* 92, 1889 (1998); C. T. Lin, W. L. Hsu, and M. A. El-Sayed, *J. Phys. Chem.* 91, 4556 (1987).

73. S. Suzuki and T. Fujii, in "Photochemistry on Solid Surfaces" (M. Anpo and T. Matsuara, Eds.), p. 77. Elsevier, Amsterdam, 1989.

74. K. Kemnitz, T. Murao, I. Yamazaki, N. Nakashima, and K. Yoshihara, *Chem. Phys. Lett.* 101, 337 (1983); K. Kemnitz, N. Tamai, I. Yamazaki, N. Nakashima, and K. Yoshihara, *J. Phys. Chem.* 91, 1423 (1987); K Kemnitz, N. Tamai, I. Yamazaki, N. Nakashima, and K. Yoshihara, *J. Phys. Chem.* 91, 5094 (1986).

75. A. D. French, N. R. Bertoniere, O. A. Battista, J. A. Cuculo, and D. G. Gray, in "Ullmann`s Encyclopedia of Industrial Chemistry," 4th ed., Vol. 3, p. 476. VCH, New York, 1993.

76. O. A. Battista, in "Encyclopedia of Polymer Science and Technology," (H. F. Mark, N. G. Gaylord, and N. M. Bikales, Eds.), Vol. 3, p. 285. Wiley, New York, 1965.

77. T. Vo-Dinh, "Room-Temperature Phosphorimetry for Chemical Analysis," Wiley, New York, 1984.

78. E. M. Schulman and C. Walling, *Science* 178, 53 (1972); E. M. Schulman and C. Walling, *J. Phys. Chem.* 77, 902 (1973); E. M. Schulman and R. T. Parker, *J. Phys. Chem.* 81, 1932 (1977).

79a. T. Vo-Dinh, E. L. Yen, and J. D. Winefordner, *Anal. Chem.* 48, 1186 (1976).

79b. D. L. McAlease and R. B. Dunlap, *Anal. Chem.* 56, 2244 (1984).

80. J. Murtagh and J. K. Thomas, *Chem. Phys. Lett.* 148, 445 (1988).

81a. L. F. Vieira Ferreira, M. R. Freixo, A. R. Garcia, and F. Wilkinson, *J. Chem. Soc. Faraday Trans.* 88, 15 (1992).

*81b.* L. F. Vieira Ferreira, A. R. Garcia, M. R. Freixo, and S. M. B. Costa, *J. Chem. Soc. Faraday Trans.* 89, 1937 (1993).

*82.* M. J. Lemos and L. F. Vieira Ferreira, in "Eurolights II—Light on Organized Molecular Systems," Hengelhoef, Belgium, 1995, p. 90; L. F. Vieira Ferreira, M. J. Lemos, M. J. Reis, and A. M. Botehlo do Rego, *Langmuir* 16, 5673 (2000).

*83a.* J. C. Netto-Ferreira, L. F. Vieira Ferreira, and S. M. B. Costa, *Quim. Nova* 19, 230 (1996).

*83b.* L. F. Vieira Ferreira, J. C. Netto-Ferreira, and S. M. B. Costa, *Spectrochim. Acta A* 51, 1385 (1995).

*83c.* L. F. Vieira Ferreira, A. S. Oliveira, and J. C. Netto-Ferreira, in "Proceedings of the Third Conference on Fluorescence Microscopy and Fluorescent Probes," Prague, June 1999, pp. 199–208.

*84.* L. F. Vieira Ferreira, M. J. Lemos, V. Wintgens, and J. C. Netto-Ferreira, *Quim. Nova* 22, 522 (1999); L. F. Vieira Ferreira, M. J. Lemos, V. Wintgens, and J. C. Netto-Ferreira, *Spectrochim. Acta A* 55, 1219 (1999).

*85.* F. Wilkinson, P. A. Leicester, L. F. Vieira Ferreira, and V. M. M. Freire, *Photochem. Photobiol.* 54, 599 (1991).

*86.* L. F. Vieira Ferreira, A. S. Oliveira, I. V. Khmelinskii, and S. M. B. Costa, *J. Lumin.* 60&61, 485 (1994); F. Wilkinson, D. R. Worrall, and L. F. Vieira Ferreira, *Spectrochim. Acta A* 48, 135 (1992).

*87.* L. F. Vieira Ferreira, J. C. Netto-Ferreira, A. S. Oliveira, and S. M. B. Costa, *Química* 60, 50 (1996).

*88a.* A. S. Oliveira, L. F. Vieira Ferreira, F. Wilkinson, and D. R. Worrall, *J. Chem. Soc. Faraday Trans.* 92, 4809 (1996).

*88b.* L. F. Vieira Ferreira, A. S. Oliveira, F. Wilkinson, and D. R. Worrall, *J. Chem. Soc. Faraday Trans.* 92, 1217 (1996).

*88c.* L. F. Vieira Ferreira, A. S. Oliveira, K. H. Henbest, F. Wilkinson, and D. R. Worrall, "RAL CLF Annual Report," p. 143, 1997.

*88d.* A. S. Oliveira, P. Almeida, and L. F. Vieira Ferreira, *Coll. Czech. Chem. Comm.* 64, 459 (1999).

*88e.* L. F. Vieira Ferreira, A. S. Oliveira, P. Matousec, M. Towrie, A. W. Parker, F. Wilkinson, and D. R. Worrall, "RAL CLF Annual Report," p. 81, 1998.

*89.* H. E. A. Kramer, *Chimia* 40, 160 (1986).

*90.* N. S. Allen, *Rev. Prog. Col.* 17, 61 (1987).

*91.* L. M. G. Jansen, I. P. Wilkes, D. C. Greenhill, and F. Wilkinson, *J. Soc. Dyes Col.* 114, 327 (1998).

*92.* L. R. Snyder and J. W. Ward, *J. Phys. Chem.* 70, 3941 (1966).

*93.* N. J. Turro, *Tetrahedron* 43, 1589 (1987).

*94.* W. J. Leigh and L. Johnson, "Handbook of Organic Photochemistry" (J. C. Scaiano, Ed.), Vol. 2, Chap. 22, p. 401. CRC Press, Boca Raton, 1989, and references therein.

*95.* Aldrich Technical Information Bulletin, AL-144.

*96.* J. Nawrocki, *J. Chromatogr.* 779, 29 (1997).

*97.* K. K. Unger, "Porous Silica," Elsevier, Amsterdam, 1979; R. Iler, "The Chemistry of Silica," Wiley, New York, 1979.

*98.* P. Van de Voort, I. Gillis-D`Hamers, and E. F. Vansant, *J. Chem. Soc. Faraday Trans.* 86, 3751 (1992); I. Gillis-D`Hamers, I. Cornelissens, K. C. Vrancken, P. Van de Voort, and E. F. Vansant, *J. Chem. Soc. Faraday Trans.* 88, 723 (1992).

*99a.* E. M. Flanigen, J. M. Bennett, R. W. Grose, R. L. Patton, R. M. Kirchner, and J. V. Smith, *Nature* 271, 512 (1978).

*99b.* G. M. W.Shultz-Sibbel, D. T. Gjerde, C. D. Chriswell, J. S. Fritz, and W. E. Coleman, *Talanta* 29, 447 (1982).

*99c.* G. T. Kokotailo, S. L. Lawton, and D. H. Olson, *Nature* 272, 437 (1978).

*100.* D. M. Bilby, N. B. Milestone, and L. P. Aldridge, *Nature* 280, 664 (1979).

*101.* F. Wilkinson and L. F. Vieira Ferreira, *J. Lumin.* 40&41, 704 (1988).

*102.* J.C. Netto-Ferreira, L. M. Ilharco, A. R. Garcia, and L. F. Vieira Ferreira, *Langmuir* 16, 10392 (2000).

*103.* M. Gabriela Lagorio, L. E. Dicerio, M. I. Litter, and H. San Roman, *J. Chem. Soc. Faraday Trans.* 94, 419 (1998).

*104.* L. F. Vieira Ferreira, S. M. B. Costa, and E. J. Pereira, *J. Photochem. Photobiol. A* 55, 361 (1991); L. F. Vieira Ferreira and S. M. B. Costa, *J. Lumin.* 48&49, 135 (1991).

*105.* M. Desaeger, M. J. Reis, A. M. Botelho do Rego, J. D. Lopes da Silva, and I. Verpoest, *J. Mater. Sci.* 31, 6305 (1996).

*106.* U. Zielke, K. J. Huttinger, and W. P. Hoffman, *Carbon* 34, 983 (1996).

*107.* H. Zhuang and J. P.Wightman, *J. Adhes.* 62, 213 (1997).

*108.* A. E. E. Jokinen, P. J. Mikkola, J. G. Matisons, and J. B. Rosenholm, *J. Colloid Interf. Sci.* 196, 207 (1997).

*109.* A. Provatas, J. G. Matisons, and R. S. Smart, *Langmuir* 14, 1656 (1998).

*110.* S. H. Park, S. Mcclain, Z. R. Tian, S. L. Suib, and C. Karwacki, *Chem. Mater.* 9, 176 (1997).

*111.* G. A. Hishmeh, T. L. Barr, A. Sklyarov, and S. Hardcastle, *J. Vacuum Sci. Technol. A* 14, 1330 (1996).

*112.* M. Rei Vilar, M. Schott, and P. Pfluger, *J. Chem. Phys.* 92, 5722 (1990).

*113.* See, for instance, M. Schreck, M. Abraham, A. Lehmann, H. Schier, and W. Göpel, *Surf. Sci.* 262, 128 (1992).

*114.* A. M. Botelho do Rego, unpublished results.

*115.* M. Raposo, R. S. Pontes, L. H. C. Mattoso, and O. N. Oliveira, Jr., *Macromolecules* 30, 6095 (1997).

*116.* See, for instance, A. Maroufi, L. J. Mao, and A. M. Ritcey, *Macromol. Sympos.* 87, 25 (1994).

*117.* J.-J. Pireaux, P. A. Thiry, R. Caudano, and P. Pfluger, *J. Chem. Phys.* 84, 6452 (1986).

*118.* A. M. Botelho do Rego, M. Rei Vilar, J. Lopes da Silva, M. Heyman, and M. Schott, *Surf. Sci.* 178, 367 (1986).

*119.* M. Rei Vilar, M. Schott, J.-J. Pireaux, C. Grégoire, P. A. Thiry, R. Caudano, A. Lapp, A. M. Botelho do Rego, and J. Lopes da Silva, *Surf. Sci.* 189/190, 927 (1987).

*120.* M. Rei Vilar, M. Schott, J.-J. Pireaux, C. Grégoire, R. Caudano, A. Lapp, A. M. Botelho do Rego, and J. Lopes da Silva, *Surf. Sci.* 211/212, 782 (1989).

*121.* P. P. Hong, F. J. Boerio, and S. D. Smith, *Macromolecules* 27, 596 (1994)

*122.* M. Rei Vilar, A. M. Botelho do Rego, J. Lopes da Silva, F. Abel, V. Quillet, M. Schott, S. Petitjean, and R. Jérôme, *Macromolecules* 27, 5900 (1994).

*123.* P.-G. de Gennes, *C. R. Acad. Sci. B* 307, 1841 (1988).

*124.* W. Zhao, X. Zhao, M. H. Rafailovich, J. Sokolov, R. J. Composto, S. D. Smith, M. Satkovski, T. P. Russell, W. D. Dozier, and T. Mansfield, *Macromolecules* 26, 561 (1993).

*125.* A. M. Botelho do Rego, O. Pellegrino, J. M. G. Martinho, and J. Lopes da Silva, *Langmuir* 16, 2385 (2000); ibid, *Surface Science* (2001), in press.

*126.* "Polymer Handbook" (J. Brandrup and E. H. Immergut, Ed.), Wiley, New York, 1989.

*127.* H. R. Thomas and J. J. O'Malley, *Macromolecules* 12, 323 (1979).

*128.* J. J. O'Malley, H. R. Thomas, and G. M. Lee, *Macromolecules* 12, 996 (1979).

*129.* "UV Atlas of Organic Compounds," Vol. 1. Butterworths, London/Verlag Chemie, Weinheim, 1966.

*130.* A. M. Botelho do Rego, M. Rei Vilar, and J. Lopes da Silva, *J. Electron Spectrosc. Relat. Phenom.* 85, 81 (1997).

*131.* H. H. Jaffe and M. Orchin, "Theory and Applications of Ultraviolet Spectroscopy," Wiley, New York, 1962.

*132.* E. Yamamoto, T. Yoshidome, T. Ogawa, and H. Kawazumi, *J. Electron Spectrosc. Relat. Phenom.* 63, 341 (1993).

*133.* A. Skerbele and E. N. Lassettre, *J. Chem. Phys.* 42, 395 (1965).

*134.* E. N. Lassettre, A. Skerbele, M. A. Dillon, and K. J. Ross, *J. Chem. Phys.* 48, 5066 (1968).

*135.* J. P. Doering, *J. Chem. Phys.* 51 2866 (1969).

*136.* Ph. Avouris and J. E. Demuth, *J. Chem. Phys.* 75, 4783 (1981).

*137.* L. Sanche and M. Michaud, *Chem. Phys. Lett.* 80, 184 (1981).

*138.* A. Dodabalapur, L. Torsi, and H. E. Katz, *Science* 268, 270 (1995); B. Servet, G. Horowitz, S. Ries, O. Lagorse, P. Alnot, A. Yassar, F. Deloffre, P. Srivastava, R. Hajlaoui, P. Lang, and F. Garnier, *Chem. Matter.* 6, 1089 (1994); P. Ostoja, S. Guerri, S. Rossini, M. Servidori, C. Taliani, and R. Zamboni, *Synth. Met.* 54, 447 (1993).

*139.* D. Birnbaum and B. E. Kohler, *J. Chem. Phys.* 96, 2492 (1992); D. Birnbaum, D. Fichou, and B. E. Kohler, *J. Chem. Phys.* 96, 165 (1992); R. S. Becker, J. Seixas de Melo, A. L. Maçanita, and F. Elisei, *J. Phys. Chem.* 100, 18683 (1996), and references therein.

*140.* See, for instance, M. Schmelzer, S. Roth, P. Bäuerle, and R. Li, *Thin Solid Films* 229, 255 (1993).

*141.* M. Muccini, E. Lunedei, C. Taliani, F. Garnier, and H. Baessler, *Synth. Met.* 84, 863 (1997).

*142a.* M. Muccini, E. Lunedei, A. Bree, G. Horowitz, F. Garnier, and C. Taliani, *J. Chem. Phys.* 108, 7327 (1998);

*142b.* M. Muccini, E. Lunedei, C. Taliani, D. Beljonne, J. Cornil, and J.-L. Brédas, *J. Chem. Phys.* 109, 10513 (1998).

*143.* P. Dannetun, M. Rei Vilar, and M. Schott, *Thin Solid Films* 286, 321 (1996).

*144.* O. Pellegrino, M. Rei Vilar, G. Horowitz, F. Kouki, F. Garnier, J. D. Lopes da Silva, and A. M. Botelho do Rego, *Thin Solid Films* 327–329, 252 (1998).

145. O. Pellegrino, M. Rei Vilar, G. Horowitz, F. Kouki, F. Garnier, J. D. Lopes da Silva, and A. M. Botelho do Rego, *Thin Solid Films* 327–329, 291 (1998).

146. A. M. Botelho do Rego, O. Pellegrino, J. D. Lopes da Silva, G. Horowitz, F. Kouki, F. Garnier, and M. Rei Vilar, *Synth. Met.* 101, 606 (1999).

147. D. Oeter, H.-J. Egelhaaf, Ch. Ziegler, D. Oelkrug, and W. Göpel, *J. Chem. Phys.* 101, 6344 (1994).

148. M. Rei Vilar, G. Horowitz, P. Lang, O. Pellegrino, and A. M. Botelho do Rego, *Adv. Mater. Opt. Electron.* 9, 211 (1999).

149. F. Kouki, Thèse d'Université, Univ. de Pierre et Marie Curie, Thiais, 1998.

150. See, for instance, M. Kawakasi and H. Inokuma, *J. Phys. Chem.* 103, 1233 (1999).

151. S. J. Kerber, J. J. Bruckner, K. Wozniak, S. Seal, S. Hardcastle, and T. L. Barr, *J. Vac. Sci. Technol. A* 14, 1314 (1996).

152. S. J. Gregg and K. S. W. Sing, "Adsorption, Surface Area and Porosity," p. 140. Academic Press, New York, 1978.

153. A. W. Adamson, "Physical Chemistry of Surfaces," 5th ed., p. 564. Wiley, New York, 1990.

154. R.C. Weast, "CRC Handbook of Chemistry and Physics," 78th ed., p. B-155. CRC, Press, Cleveland, 1997.

# 5. HIGH-PRESSURE SURFACE SCIENCE

Vladislav Domnich,

Department of Mechanical Engineering, University of Illinois at Chicago,
Chicago, Illinois, USA

Yury Gogotsi

Department of Materials Engineering, Drexel University, Philadelphia, Pennsylvania, USA

## 1 INTRODUCTION

In the majority of mechanical applications of materials, their surfaces experience contact with another material and take the external load before the bulk of the material is influenced. In some cases, surface interactions influence the bulk (e.g., propagation of cracks; dislocations or point defects from the surface in depth). In many cases, only the outermost surface layer is affected by the surface contact with no detectable changes in the bulk of the material. This is like a storm that is frightening and destructive on the ocean surface, but does not have any influence on deep-water life. We are primarily concerned in this review with that kind of interaction. The surface layer thickness affected by external mechanical forces ranges from nanometers to microns. Thus, in our case, the definition of "surface" is different from the one used by surface scientists, that is, physicists and chemists. We introduce here an engineering definition of **surface:** *the outermost layer of the material that can be influenced by physical and/or chemical interaction with other surfaces and/or the environment.* In this chapter, we consider only mechanical effects, but both mechanical and chemical interactions are possible and their synergy can lead to mechanochemical alteration of a material surface.

When we walk on the ground, our steps can densify soil, break up agglomerates, and leave footprints. In a similar way, any hard object can leave imprints on a material surface. The harder the counterbody material, the stronger its influence is on the material surface. In particular, when a hard indenter (e.g., diamond) touches the surface of another hard material (ceramic or semiconductor), very high pressures can be achieved under the indenter (Fig. 1) because there is no or minimal stress relaxation due to the plastic flow of the material.

At this point, it is important to define what we mean by "hardness." Hardness is the resistance of a material to penetration of a hard indenter. The softer the material, the deeper the indenter will penetrate its surface. The definition of hardness as "resistance to plastic deformation," which can be found in many textbooks, does not reflect the complexity of processes that occur in materials upon interaction

355

EXPERIMENTAL METHODS IN THE PHYSICAL SCIENCES
Vol. 38
ISBN 0-12-475985-8

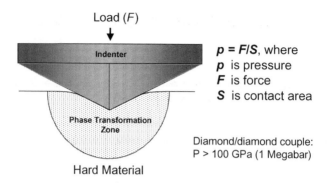

FIG. 1. Sharp indenter in contact with specimen surface.

with the indenter and therefore is incorrect to apply to many materials, including one of the most important materials today—silicon. This chapter will show that processes other than dislocation-induced ductility (plastic deformation) can be involved in the deformation of materials under an indenter. Recent data show that the hardness of many brittle materials depends on the stress (deformation) needed to initiate a phase transformation. As the contact area in the beginning of the penetration of the indenter into material is small, extremely high pressures can be achieved, as shown by the equation

$$p = F/A \qquad (1)$$

where $p$ is pressure, $F$ is force, and $A$ is contact area. These pressures can exceed the phase transformation pressure for many materials. Understanding and appreciating this fact can help researchers choose and/or optimize conditions of ductile-regime machining of materials, as well as give new insights into their surface properties. Information on the phase transformations in the surface layer of materials upon contact interactions is very important for understanding the mechanisms of wear, friction, and erosion. High shear stresses and flexibility of loading conditions allow one to drive phase transformation that cannot occur under hydrostatic stresses, or that would occur at much higher pressures.

In the following sections we will describe the phase transformations and amorphization that occur in many ceramics and semiconductors under contact loading, including that of indentation with hard indenters or scratching, grinding, milling, etc.

Contact loading is one of the most common mechanical impacts that materials can experience during processing or application. Examples are cutting, polishing, indentation testing, wear, friction, and erosion. This kind of loading has a very significant nonhydrostatic component of stress that may lead to dramatic changes in the materials structure, such as amorphization [1] and phase transformations

[2, 3]. Simultaneously, processes of plastic deformation, fracture and interactions with the environment, and counterbody can occur. The latter ones have been studied by mechanical engineers and tribologists, but the processes of phase transformations at the sharp contact have been investigated for only a few materials (primarily, semiconductors) and the data obtained so far can only be considered preliminary. One of the reasons for the lack of information may be the fact that the problem is at the interface between at least three scientific fields, that is, materials science, mechanics, and solid state physics. Thus, an interdisciplinary approach is required to solve this problem and understand how and why a nonhydrostatic (shear) stress in the two-body contact can drive phase transformations in materials.

## 1.1 Use of Indentation/Raman Tests as a Tool for High-Pressure Surface Studies

The recent observation of phase transformations under indentation using Raman microspectroscopy [4] opens a new area of research. It is known that the indentation of materials with diamond indenters creates high stresses (hydrostatic and deviatoric) under the indenter that can cause phase transformations [5–9]. It has also been shown that the hardness correlates with metallization pressures for a number of semiconductors (Fig. 2) [10]. However, the indentation technique has been used to study phase transformations only in Si and Ge [3, 5–7, 10–14] because of the difficulties associated with monitoring the process. Indirect monitoring of phase transformations during indentation through conductivity measurements [5, 6, 8] or sudden volume changes (known as "pop-out" events [9, 15]) occurring during unloading was used in most of the indentation experiments. However, the conductivity measurements can be used only to detect metallic-nonmetallic transitions. Pop-out events alone can suggest only that a transformation may have occurred if the volume change is large enough. Cracking around the impression and difficulties of analyzing the indentation force/displacement data may also lead to erroneous conclusions. Direct studies of the phase transformations induced by indentation have been conducted using transmission electron microscopy (TEM) [6, 13, 16–18]. However, TEM examination is possible only on quenched products and is complicated by the difficulty in handling small and stressed samples, and the instability of metastable phases during dimpling, ion thinning, and under the electron beam. Because Raman spectroscopy was used successfully to monitor phase transformation in conventional pressurization experiments [19–23] and to map internal stresses in materials [24–28], it was natural to use it in indentation experiments.

Raman spectroscopy studies of indentations have been conducted on a number of ceramics and semiconductors and new phases within and around indentations have been found in several materials. These preliminary results are in excellent

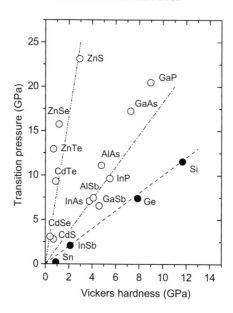

FIG. 2. Correlation of the metallization pressure as calculated from Hertzfeld's theory and the Vickers indentation hardness for tetrahedrally bonded semiconductors. The initial structures are cubic diamond and zincblende; the final structures are of the $\beta$-tin and rocksalt types. After Reference [10].

agreement with the predictions of Gilman [3, 10, 14]. The main hurdle was to demonstrate unambiguously that high-pressure metallization occurs in a variety of contact loading situations. The use of TEM and other techniques to support Gilman's ideas could supply only indirect evidence. Raman spectroscopy provided a tool to do this directly and even *in situ*. The feasibility of this technique has been clearly demonstrated on the most typical semiconductors and metastable phases in the hardness impressions were clearly shown. As these phases can be formed only via metallic Si and Ge, these experiments produced direct evidence of the metallization of semiconductors under contact loading.

In summary, the *advantages* of the indentation/Raman microspectroscopy tests include:

• fastest and most efficient of all methods available (1–5 min for the whole analysis);
• high shear stresses favor amorphization and phase transformations;
• simplicity of sample preparation and measurements;
• possibility of *in situ* studies;

- phase transformations upon loading and unloading, and reverse transformations upon heating;
- effects of the crystal orientation on the transformation can be studied;
- determination of both amorphous and crystalline phases;
- a very small amount of material is sufficient (e.g., a thin film of a few square micrometers in size);
- a variety of easily replaceable indenters allows the use of a wide range of loading conditions; and
- statistical treatment of numerous measurements minimizes errors and leads to reproducible results.

This technique has some *limitations*:

- difficulties with the quantification of the transformation conditions;
- experiments with soft materials require the use of thin films on hard substrates;
- scarce databanks of Raman data compared to those for the diffraction data;
- difficulties in calculation of structure parameters of new phases on the basis of the Raman spectra only; and
- metals cannot be studied, but a transition to metallic state can be detected.

Thus, the combination of indentation tests with Raman microspectroscopy provides a powerful and fast tool for monitoring of pressure-induced phase transformations in ceramics and semiconductors [4].

# 2 PHASE TRANSFORMATIONS IN MATERIALS UNDER STATIC CONTACT LOADING

It has been long recognized that when two extended surfaces are placed together, the actual contact occurs only at the tips of surface asperities [29]. The simplest model system for the scientific evaluation of all related phenomena would consist of a single hard asperity loaded statically against a softer half-space, which process can be easily scaled up to a well-known indentation (hardness) test. In general, however, even the indentation itself is too complicated for reliable theoretical modeling because of such possible concurrent processes as brittle macro- and microfracture, dislocation and defect formation, and structural transformations in the material beneath the indenter. All these processes impose uncertainties on the size and shape of the evolving elastic-plastic zone. Thus, some simplifying assumptions need to be introduced into the development of indentation theory.

## 2.1 Analytical Modeling

Stress distribution in indentation is largely affected by the indenter tip geometry, which is a vital factor in determining the boundary conditions for the field. The major types of indenter tips shown schematically in Figure 3 may be separated into two groups, viz. point-force (pyramidal and conical) and spherical indenters. Correspondingly, Boussinesq and Hertzian stress fields will describe point-force and spherical indentation in the case of purely elastic loading (Fig. 4). To account for possible elastic compliance of the indenter, a reduced elastic modulus $E_r$ is

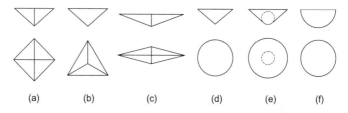

FIG. 3. Schematic of the various types of indenter tips used in indentation testing: (a) Vickers; (b) Berkovich; (c) Knoop; (d) conical; (e) Rockwell; and (f) spherical.

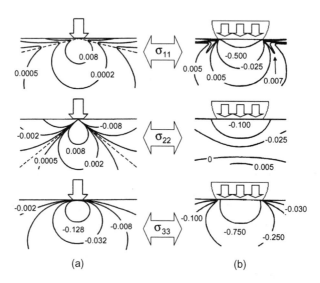

FIG. 4. Contours of principal normal stresses in (a) Boussinesq and (b) Hertzian fields, shown in the plane containing contact axis. After References [30, 31].

introduced as defined through the following relation [32]:

$$\frac{1}{E_r} = \frac{1 - v^2}{E} + \frac{1 - v_i^2}{E_i} \tag{2}$$

where $v_i$ and $E_i$ are Poisson's ratio and Young's modulus for the indenter, and $v$ and $E$ are the same parameters for the specimen.

Very high gradients in the stresses are expected around any sharp points or edges of an ideally elastic contact. As a result, nonlinear, inelastic deformation operates to relieve the stress concentration about the singularity by distributing the applied load over a non-zero contact area. This is illustrated in Figure 5 for the case of penetration of a regular tetrahedral pyramid into the half-space [33], which process is equivalent to Vickers indentation. For linearly elastic half-space, the contact pressures are singular along the edges of the pyramid (Fig. 5a), but the singularity is reduced when the nonlinearity of the half-space is presumed (Fig. 5b). Note also that the compressive radial stress $\sigma_{22}$ within the contact area (see Fig. 4) leads to the inward bending of the impression faces (dashed lines in Fig. 5).

Several models account for indentation-induced plastic deformation of the half-space material. Johnson [32, 34] considered the expansion of an incompressible hemispherical core of material subjected to an internal pressure and has derived an expression relating the mean pressure in the core $p_m$ to a combined parameter $E/Y \cot \psi$, where $E$ and $Y$ are elastic modulus and yield stress of the specimen, respectively, and $2\psi$ is the included angle of a conical indenter. According to

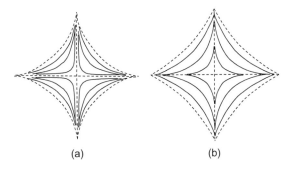

(a)                    (b)

FIG. 5. Contact areas (dashed lines) and contact stresses at the interface of the regular tetrahedral pyramid penetrating into the half-space. The maximum contact stresses are observed in the center of the indentation zone and decrease according to the stress isobars (solid lines). (a) Stress singularities in the purely elastic loading and (b) their reduced form when the nonlinearity of the half-space is taken into account. After Reference [33].

Johnson [32],

$$\frac{p_m}{Y} = \frac{2}{3}\left[1 + \ln\left(\frac{E/Y \cot \psi + 4(1 - 2\nu)}{6(1 - \nu)}\right)\right] \tag{3}$$

which equation applies to geometrically similar indentations, such as those made with a conical or pyramidal indenter, where the radius of the plastic zone increases at the same rate as the radius of the core. In the case of spherical indentation, Johnson [32] suggested that $\cot \psi$ in Eq. (3) be replaced with $a/R$ for $\psi \simeq \pi/2$, where $a$ is the radius of the contact area and $R$ is the radius of the indenter. However, such a procedure appears to invalidate the assumed condition for geometrical similarity. An empirical relationship similar to Eq. (3) was proposed recently for spherical indentation [35], and it may be used to predict the indentation stress-strain response over a wide range of indentation strains and for a wide range of material types.

Tanaka [36] proposed an indentation model that generalized Johnson's incompressible core model. In contrast to Johnson's treatment, the core in Tanaka's model is compressible and the term related to the plastic work dissipated to heat in the core is added:

$$\frac{p_m}{Y} = \frac{2}{3}\left(1 + \ln \frac{\sqrt{\pi} E/Y \cot \psi}{12(1 - \nu)}\right) + \alpha \tag{4}$$

where the material constant $\alpha$ varies from $1/3$ for ceramics to 1 for metals and polymers [36].

Accounting for the stress-induced phase transformations during indentation is possible in the following way. If the transformed layer is thin, the stresses under the indenter will not change noticeably after the transformation. If a metallic phase is formed, it can be modeled as a liquid and the distribution of stress $\sigma$ and pressure under the indenter $p = -\sigma$ will be uniform. It is easy to account for the viscosity of the metallic phase. For the case when the transformation zone differs from a thin and uniform film, Galanov et al. [37] suggested to modify the Tanaka's model to account for reversible phase transformations under a rigid indenter. The modified model leads to the following equation for the radial pressure $p_m$ acting on the core surface when the phase transformation characterized by volume change $\Delta$ and pressure $p_{pt}$ occurs in the core and in the plastic zone (i.e., $p_{pt} \leq p_m$, see Fig. 6a) [37]:

$$\frac{p_m}{Y} = \frac{2}{3}\left(1 + \ln \frac{\sqrt{\pi} E \cot \psi}{12Y(1 - \nu) + 4E\Delta \exp(-3p_{pt}/2Y)}\right) + \alpha \tag{5}$$

When the phase transformation occurs within the plastic core (i.e., $p_m < p_{pt} < HV$, see Fig. 6b; $HV$ is the Vickers hardness of the specimen), one needs to consider the transformation zone boundary (Fig. 6a)

$$z = (c - r)\cot \psi + t_{pt}, \quad r \leq c, \tag{6}$$

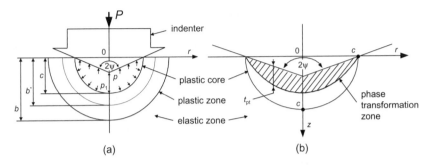

FIG. 6. Schematic representation of the indentation model of Galanov et al. [37] in spherical coordinates ($r \leq c$—plastic core; $c < r \leq b$—plastic zone; $r > b$—elastic zone; $r \leq b^*$—reversible phase transformations zone). (a) Phase transformation in the core and in the plastic zone. (b) Phase transformation within the plastic core.

where the thickness of the transformed layer $t_{pt}$ is derived from

$$t_{pt} = \left[ \sqrt{c^2 - r^2} - (c - r) \cot \psi \right] \xi \tag{7}$$

$$\xi = \frac{HV - p_{pt}}{HV - p_m} \tag{8}$$

Now, the radial pressure $p_m$ on the core surface is defined through the following equation [37]

$$\frac{p_m}{Y} = \frac{2}{3} \left( 1 + \ln \frac{E(\sqrt{\pi} \cot \psi - 4\xi \Delta)}{12Y(1 - v)} \right) + \alpha \tag{9}$$

The last equation is nonlinear with respect to $p_m$ ($\xi$ is the function of $p_m$) and must be solved numerically. Note also that Eqs. (5) and (9) transform into Tanaka's Eq. (4) when the volume change associated with the transformation is zero ($\Delta = 0$).

## 2.2 Depth-Sensing Indentation

Experimental values of the phase transformation pressures may be assessed through the depth-sensing (nano)indentation technique, which allows high-resolution *in situ* monitoring of the indenter displacement as a function of the applied load. Changes in a material's specific volume or mechanical properties during a phase transition are revealed as characteristic events in the load-displacement curve. The formation of a new phase under the indenter may result in the yield step ("pop-in") or the change in slope ("elbow") of the loading curve; a sudden displacement discontinuity ("pop-out") or an elbow in the unloading curve may be indicative of the reverse transition. In cyclic indentation, additional information can be extracted from the broad or asymmetric hysteresis loops or from the specific features in the reloading curves [38].

Once a specific event in the load-displacement curve has been associated with a particular phase transition, the pressure at which it occurs can be estimated by considering the elastoplastic behavior of the material under the indenter. For the point-force contact, the Sneddon's solution [39] to the problem of the penetration of an axisymmetric punch into an elastic half space predicts the following relation between the applied load $P$ and the indenter displacement $h$:

$$P = \alpha h^2 \tag{10}$$

where the parameter $\alpha$ is determined by indentation geometry and the mechanical properties of the specimen [39].

As described in the previous section, including plasticity in the modeling of indentation contact is a complex problem and analytical solutions are not easily obtained [32]. Fortunately, in most cases, at least the upper part of the unloading curve is elastic, leading to the following modified Sneddon's relation [40]:

$$P = \alpha(h - h_f)^m \tag{11}$$

where the residual plastic deformation $h_f$ (Fig. 7) is introduced to retrieve the elastic part of the indenter displacement during unloading.

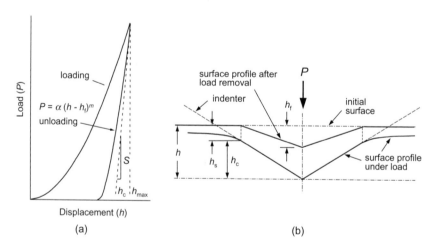

FIG. 7. (a) Load-displacement curve of a typical elastoplastic material and (b) the schematic of the indentation model of Oliver and Pharr [40]. $S$—contact stiffness; $h_c$—contact depth; $h_{max}$—indenter displacement at peak load; $h_f$—plastic deformation after load removal; $h_s$—displacement of the surface at the perimeter of the contact.

The power exponent $m$ in Eq. (11) needs special attention. Extensive nanoin-dentation studies performed by Oliver and Pharr [40] on materials with various mechanical properties implied that $m$ is a material constant and has the values in the range of 1.0 to 2.0. In fact, the routine analysis of nanoindentation data be-gins with the determination of the parameters $\alpha$ and $m$ for a particular experiment by numerically fitting the measured unloading curve. This suggests deviations from the Sneddon's solution ($m = 2$; Eq. (10)) in real cases. As shown by Sakai et al. [41], most commercially available indentation test systems (including that of [40]) are designed in such a way that the load frame compliance and mechanical contact clearances inevitably and significantly alter the load-displacement data. In contrast, for a "stiff" indentation system, the indenter displacement is indeed proportional to the square root of the applied load and the Sneddon's solution (Eq. (10)) is preserved in this case [41].

Oliver and Pharr [40] proposed a procedure to assess the average contact pres-sure at peak load $P_{max}$ from the experimental load-displacement curves by de-termining the projected contact area $A$, which is a function of the corresponding maximum contact depth $h_{c,max}$. In general, at a given load $P$, the contact depth $h_c$ can be determined as the difference between the indenter displacement $h$ and the surface deflection at the perimeter of indentation $h_s$ (see Fig. 7). Assuming that the elastic deflection of the sample surface $h_s$ is directly proportional to the square root of the indentation load $P$, Novikov et al. [42] developed a method to assess the instantaneous contact depth $h_c$ from the maximum surface deflection at peak load, provided $h_{s,max}$ has been determined by the Oliver-Pharr's technique. This allows estimation of the average contact pressure $p_m$ for both, loading (elasto-plastic) and unloading (elastic) segments simply as

$$p_m = \frac{P}{A(h_c)} \tag{12}$$

The quantification of contact pressures in spherical indentation is based on the indentation model of Field and Swain [43]. The procedure is similar to the point-force indentation and includes the evaluation of the contact radius at a given load ($a_c$) from its value at peak load ($a_{c,max}$). The mean contact pressure is then found as [43]

$$p_m = \frac{P}{\pi a_c^2} \tag{13}$$

## 2.3 Raman Tensometry

The values and spatial distribution of the residual stresses around indentations can be assessed by means of the Raman imaging technique. In general, mechani-cal strain may shift the frequencies of the Raman modes and lift their degeneracy.

For example, for silicon in the cubic diamond structure (Si-I phase), the frequencies of the three optical modes in the presence of strain, to terms linear in strain, can be obtained by solving the following secular equation [44, 45]:

$$\begin{vmatrix} p\varepsilon_{11} + q(\varepsilon_{22} + \varepsilon_{33}) - \lambda & 2r\varepsilon_{12} & 2r\varepsilon_{13} \\ 2r\varepsilon_{12} & p\varepsilon_{22} + q(\varepsilon_{11} + \varepsilon_{33}) - \lambda & 2r\varepsilon_{23} \\ 2r\varepsilon_{13} & 2r\varepsilon_{23} & p\varepsilon_{33} + q(\varepsilon_{11} + \varepsilon_{22}) - \lambda \end{vmatrix} = 0 \quad (14)$$

where $p$, $q$, and $r$ are material constants, the so-called phonon deformation potentials, and $\varepsilon_{ij}$'s are the strain tensor components. The shifts in the Raman band positions in the presence of strain can be calculated from the eigenvalues $\lambda_j$ of Eq. (14):

$$\lambda_j = \omega_j^2 - \omega_{j0}^2 \quad \text{or}$$

$$\Delta\omega_j \equiv \omega_j - \omega_{j0} \simeq \frac{\lambda_j}{2\omega_{j0}} \quad (15)$$

where $\omega_j$ and $\omega_{j0}$ are Raman frequencies of the mode $j$ ($j = 1, 2, 3$) in the presence and absence of stress, respectively. The values of the phonon deformation potentials have been determined for silicon in the uniaxially stressed state from the slopes of the experimental $\Delta\omega_j(\sigma)$ dependencies as $p = -1.85\omega_0^2$, $q = -2.31\omega_0^2$, and $r = -0.71\omega_0^2$ [46].

It can be easily shown that compressive uniaxial or biaxial stress results in the upshift of the Raman band to higher wavenumbers, whereas tensile stress decreases the Raman frequency in these cases [47]. However, in the complex stress field under the indenter, further complicated by the volumetric changes during possible phase transformations and the breakdown of constitutive equations due to macro- and microcracking, determination of the strain tensor components becomes a challenging task and the simplifying analytical models already discussed here need to be used.

## 2.4 Elemental Semiconductors

2.4.1 Silicon.    At atmospheric pressure, silicon has a cubic diamond structure (space group $Fd3m$) up to the melting temperature [48]. Conventionally, this phase is labeled Si-I. At elevated pressures, 11 other crystalline phases of Si have been identified by calculation and from pressure cell experiments [49]. Of interest for indentation studies are the phases referred to as Si-II ($\beta$-tin structure, space group $I4_1/amd$) [50–52], Si-III or $bc8$ (body-centered cubic structure with 8 atoms per unit cell, space group $Ia\overline{3}$) [52–54], Si-IV or $hd$ (hexagonal diamond structure, space group $P6_3/mmc$) [53, 55–60], Si-IX or $st12$ (tetragonal structure with 12 atoms per unit cell, space group $P4_222$) [61], and Si-XII or $r8$ (rhombohedral structure with 8 atoms per unit cell, space group $R\overline{3}$) [62–64]. In addition, amorphous silicon ($a$-Si) has often been observed within indentations [6, 16, 64, 65] and should be taken into consideration.

Pressurization experiments indicate that under nearly hydrostatic conditions, the Si-I → Si-II transition occurs in the pressure range of 9 to 16 GPa [51, 52, 66, 67]. This transition is not reversible: Si-II transforms to different metastable phases depending on pressure release conditions. In stepwise slow decompression, the first phase to form at 10-12 GPa is Si-XII [62, 63]. On further pressure release, the degree of rhombohedral distortion diminishes gradually, producing the mixture of Si-XII and Si-III, with the Si-XII persisting to ambient pressure (although in this case as only a minor component). The Si-III → Si-XII transition is fully reversible—recompression to 2.5 GPa results in a near complete Si-III transformation to Si-XII [62].

The tetragonal Si-IX phase was first obtained (supposedly) from a mixture of Si-I and Si-II after a rapid pressure release from 12 GPa [61]. Traces of Si-IX have also been reported in deposits on the Si-I substrate after thermal spraying experiments [68].

The lonsdaleite Si-IV phase can be obtained either from the metastable Si-III phase after heat treatment at 200–600 °C [53, 56, 57] or from Si-I after plastic deformation at elevated temperatures (350–700 °C) and under confining pressure [55, 58]. The Si-I → Si-IV transformation is closely related to deformation twinning and was described as a martensitic transformation taking place at twin-twin intersections or after secondary twinning [58–60].

The relative volumes of various Si phases as functions of pressure are shown in Figure 8. The Si-I → Si-II transformation is followed by ~20% densification of the material [51, 52, 71]. The equilibrium Si-III structure was found to be ~9% denser than Si-I [53, 54, 72], and ~2% less dense than Si-XII [63, 73]. Thus, upon slow decompression, the Si-II → Si-XII transition leads to ~9% volume expansion, with the residual ~2% being recovered during the gradual Si-XII → Si-III transformation at low pressures. No volume-pressure data is available for the Si-IX phase, formed upon rapid decompression from Si-II. However, the packing fraction for the $st$12 structure (0.385) was found to be only slightly higher than that for $bc$8 (0.372) [74], which suggests similar densities for both structures. Thus, we estimate an approximate 10% volume increase during the formation of Si-IX. Finally, the hexagonal diamond Si-IV phase has an atomic volume identical to that of cubic diamond Si-I [53].

The sequence of increasingly complex structures: cubic diamond–lonsdaleite–$bc$8–$st$12, has been used to study the development of the properties of amorphous silicon with increasing local disorder in the crystalline phase [75]. The $st$12 structure was chosen as a tractable model to simulate the effect of bond distortion in amorphous silicon; these two phases were found to have similar optical properties. In general, the presence of the oddfold-bonded rings in $st$12 and $r$8 greatly

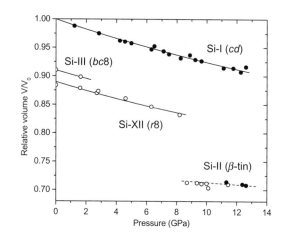

FIG. 8. Reduced volume of the high-pressure Si phases. Experimental points are from Reference [67] (Si-I and Si-II) and Reference [63] (Si-III and Si-XII). Filled and open circles correspond to increasing and decreasing pressure, respectively. The solid lines are fits to the 3rd-order Birch equation of state [69, 70] with the values of bulk modulus $B_0 = 108$ GPa and its first pressure derivative $B_0' = 4$ for Si-I; $B_0 = 117$ GPa for Si-III and $B_0 = 108$ GPa for Si-XII with $B_0'$ fixed at 5. Dashed line serves as a guide to the eye.

affects the electronic properties of these structures, which may have a variety of interesting applications [63, 75].

First attempts to explain the formation of a residual impression in Si accounted for the dislocation motion. Trefilov and Milman [76] experimentally measured the hardness of Si and Ge over a wide temperature range and observed two distinct regions in the hardness-temperature dependence (Fig. 9). The average hardness decreased $\approx 5\%$ up to $\sim 500\,°C$, whereas above this temperature, the curve shape changed abruptly and revealed a steep down slope. The authors quantitatively explained the high-temperature indentation behavior of Si by considering the dislocation glide in covalent crystals as a function of temperature and relating the resulting values of the flow stress to the hardness. The equation they obtained,

$$\sinh \frac{\beta H_V}{T} = A e^{U/kT} \tag{16}$$

allowed for an accurate fit of the experimental dependence of Vickers hardness $H_V$ on temperature $T$ in the high-temperature region by varying the dislocation activation energy $U$ and the parameters $A$ and $\beta$ (see Fig. 9).

For the low-temperature region, several dislocation-related hypotheses were proposed to explain the mechanism of plastic deformation. Trefilov and Milman [76] suggested that during indentation the theoretical shear strength was

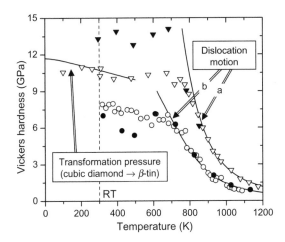

FIG. 9. Hardness vs temperature dependence of silicon: 10 mN loading, etched (110) surface [77] (filled triangles); 500 mN loading [78] (open circles); 2.3 N (for $T > 300$ K) and 1 N (for $T < 300$ K) loading, etched (111) surface [76] (open triangles); 10 N loading, ground surface [79] (filled circles). Solid lines are the calculated hardness due to dislocation glide [Eq. (16); (a) $U = 0.55$ eV, $A = 2.05 \cdot 10^{-4}$, $b = 44$ K/GPa; (b) $U = 0.51$ eV, $A = 8.5 \cdot 10^{-4}$, $b = 219$ K/GPa] and the cubic diamond $\rightarrow$ $\beta$-tin transformation pressure for Si (Reference [80], adjusted).

exceeded locally and dislocations arose as a result of the accommodation of the displacements due to block slip. Other considerations took into account the quantum properties of dislocations (tunneling effect) [81], or the generation of prismatic dislocation loops during the indenter intrusion with subsequent conservative climb [82]. Experimental data available at the time apparently confirmed the formation of dislocations during indentation of Si. Transmission electron microscopy (TEM) studies by Eremenko and Nikitenko [55] revealed the dislocation structure and defect formation in the vicinity of indentations made at temperatures of 20–700 °C. Hill and Rowcliffe [83] reported highly localized dislocation arrays around Vickers indentations made under 1 N load in (110) and (111) Si at both room temperature and 300 °C. They also observed the dislocation arrays to expand away from the indentation during annealing of indented samples at 600 °C. The authors' major conclusion for the mechanism of plastic deformation at low temperatures was similar to that made in Reference [76]. However, all the forementioned hypotheses were inconsistent with the observed behavior of either hardness or flow stress [84] in silicon below some characteristic temperature.

Gridneva et al. [5], and later Gerk and Tabor [85], noted the similarity of Si hardness as measured at low temperatures and the pressure of Si-I → Si-II transformation known from the high-pressure cell experiments [50, 51]. They suggested that the pressure-induced metallization occurs in semiconductors with cubic diamond structure during the indentation test, and that the phase transformation pressure determines the indentation hardness in this temperature range. Quite supportive of this assumption, the recent calculation of the phase boundary between the cubic diamond and $\beta$-tin structures in Si [80] reveals a slow decrease in the phase transformation pressure with increasing temperature (Fig. 9). This trend is in excellent agreement with the experimental hardness data.

Conductivity measurements performed by Gridneva et al. on Si wafers subjected to Vickers indentation showed a decrease in electrical resistance of the material beneath the indenter. The authors assumed the formation of a thin metallic layer between the indenter and the specimen, and calculated the thickness of this layer ($\sim$0.05 $\mu$m) to be independent of the applied load [5].

These results were further confirmed by using different experimental setups. Gupta and Ruoff [86] reported an abrupt drop in the resistance of Si during slow indentation with diamond spheres (Fig. 10). The corresponding pressure threshold was dependent upon the crystallographic orientation of the sample and varied from 8 GPa for loading in the [111] direction to 12 GPa for the [100] direction. Both the pressures and the resistance changes implied that the Si-I $\rightarrow$ Si-II transformation occurs during indentation of Si similar to the high-pressure cell results [50, 51]. The variation in the transformation pressure in this case may be attributed to the higher degree of deviatoric loading when uniaxial pressure is

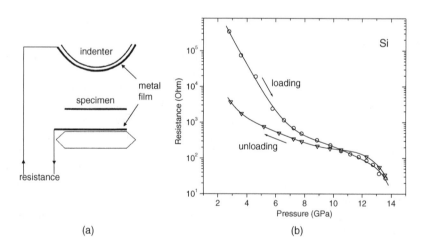

(a)                                        (b)

FIG. 10. Resistance of silicon during slow indentation in the [100] direction: (a) experimental setup and (b) resistance vs pressure curve. The initial large change in resistance is due to the gradual change in contact area. Abrupt drop in resistance at $\sim$12 GPa is an indication of the transformation of silicon under loading into a metallic phase. After Reference [86].

applied along the [111] direction, which facilitates the transformation in accordance with Gilman's predictions [87]. Finally, *in situ* conductivity measurements by Clarke et al. [6] (Vickers and Knoop indentations; (100), (110), and (111) orientations) and Pharr et al. [8] (Berkovich indentation; (110) orientation), showed that for sharp tips, the transformation to the electrically conducting form of silicon occurs instantaneously as the load is applied. This is due to the much smaller contact areas as compared to indentation with spherical tips, which results in pressures sufficiently high to start a transformation at the initial stages of indentation.

Although indicative of the semiconductor-metal transition, the conductivity data alone cannot be accepted as rigid proof of the formation of Si-II during indentation. Unfortunately, due to obvious complications with the experimental setup, no *in situ* indentation diffraction data are available as of today. On the other hand, a number of papers provide indirect evidence of the Si-I → Si-II transformation during indentation based on the post-indentation characterization techniques. We will start with a discussion of the electron microscopy results.

The TEM studies by Clarke et al. [6] revealed amorphous silicon at the bottom of Vickers and Knoop indentations made at room temperature. The results were reproducible for crystal orientations of (100), (110), and (111), and with applied loads between 100 and 500 mN. Among possible formation paths of the amorphous silicon during indentation, the authors proposed that: (i) the high-pressure Si-II phase transforms to the metastable amorphous form of silicon in the unloading stage because it cannot transform fast enough to another crystalline state; or (ii) Si-I transforms to *a*-Si directly on loading as the local pressure exceeds the metastable extension of the liquidus curve in the *P-T* phase diagram, and *a*-Si persists on unloading because of insufficient thermal energy to allow rearrangement back to the diamond cubic form.

A characteristic deformation feature is often observed in the electron microscopy studies of Si indentations [7, 64]. Figure 11 shows the scanning electron microscope (SEM) image of a Vickers indentation in (111) Si made at room temperature with a 1-N loading. The micrograph clearly reveals a thin layer adjacent to the impression faces, apparently plastically extruded during the indentation process. Such extrusions are possible only if a thin layer of highly plastic material is squeezed between the diamond tip and the relatively hard surrounding silicon, which suggests a very high ductility of the extruded material and provides more evidence for the indentation-induced phase transformation to the metallic state.

Similar extrusions were revealed by TEM in the Berkovich indentations made in (100) Si at ultralow loadings (~10 mN) [16]. Although isolated diffraction patterns were not obtained from the extruded material due to the presence of underlying crystal, the amorphous nature of this material was inferred from the absence of any crystalline diffraction and tilting experiments. Further, at these experimental conditions (below the cracking threshold), no evidence of dislocation activity or other mechanisms of plastic deformation operating outside the clearly demar-

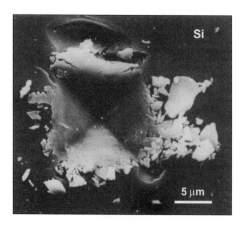

FIG. 11. Scanning electron micrograph (SEM) of a 1-N Vickers indentation in (111) silicon revealing plastically extruded material [64].

cated transformation zone was found. The impression consisted of an amorphous core with an adjacent region of plastically extruded material and a layer of polycrystalline silicon at the near-surface transformation interface. As for the origin of the polysilicon layer, the authors [16] suggested a possible shear fragmentation or interface recrystallization.

Extensive TEM studies by Page et al. [65] delivered all previous low-temperature electron microscopy results to the consistent view that: (i) silicon becomes amorphous in response to the high contact stresses under a hardness indenter; and (ii) limited dislocation arrays are generated around the deformed volume at contact loads exceeding some threshold value. The authors also argued that the dislocation arrays might occur as a means of accommodating the displacements from the densification transformation, rather than as a primary response to the indenter intrusion.

We should emphasize that the formation of amorphous silicon in the preceding references was confirmed using electron diffraction and diffraction contrast methods. As pointed out by Suzuki and Ohmura [77], the material appeared amorphous only on a diffraction scale; it might as well be polycrystalline of very fine grain or nanocrystalline. Another inconsistency was the persistent absence of any traces of $a$-Si after decompression from Si-II in the related high-pressure cell experiments [52, 53, 61, 62].

To clarify these issues, Wu et al. performed both plain-view [88] and cross-sectional [89] high-resolution electron microscopy (HREM) studies on a large array of Vickers indentations made in (110) and (001) Si with loadings of 10 and 50 mN at room temperature. Their results confirmed that the indentation region consisted of $a$-Si; moreover, they revealed numerous clusters showing lattice character inside the amorphous region near the periphery of the indentations. The cubic diamond lattice was distorted at the interface and continuously transited into an amorphous structure. Based on these observations, the authors suggested that the formation of $a$-Si might be induced by lattice distortion in the loading stage rather than the transformation from a high-pressure metallic phase during unloading. This assumption does not contradict the conductivity data because the amorphous silicon also exhibits a reversible semiconductor-metal transition at $\sim$10 GPa, observed both in the high-pressure cell experiments [90] and during indentation [6].

Interestingly, the metastable crystalline Si phases (Si-III and Si-XII) routinely obtained in the high-pressure cells upon depressurization from Si-II have been for a long time escaping detection in the TEM studies of indentations. We envisage two reasons for this. Firstly, as will be shown later, unloading rates play a crucial role in the transformation path of the material beneath the indenter, with fast unloading (which was typically used in the previous TEM studies) leading to the formation of amorphous silicon. Secondly, the metastable phases are probably unstable and decompose into the amorphous state during preparation of the indented samples for TEM (polishing, dimpling, and ion thinning), and/or under the electron beam. Indeed, a more careful TEM characterization of the indentations in Si reported in two very recent papers [91, 92] that appeared during the preparation of this review, confirmed that the metastable crystalline phases of Si can be found within the residual impressions using TEM. Plain view TEM analysis by Mann et al. [91] showed that the indentation transition path in Si is scale-dependant, with the residual impression consisting mostly of Si-III in the "small" indents ($<$600 nm) and mostly of amorphous material in the "large" indents ($>$1 $\mu$m). However, the loading rates for both types of indentations were not given and the statistical analysis of the reproducibility of the results was not performed. In contrast, the cross-sectional TEM analysis by Bradby et al. [92] emphasized the importance of the unloading rates and revealed the formation of Si-XII upon slow unloading as opposed to the formation of amorphous silicon when the fast unloading rates were used. We note here that most commercially available hardness testers can operate only at relatively fast loading and unloading rates, which is probably inhibiting the kinetically controlled Si-II – Si-XII transformation. The kinetics of phase transitions in Si will be discussed in detail below.

Taking into account all aforementioned factors, it is clear that the TEM analysis of indentations may be in many cases nontrivial and does not necessarily produce reliable and reproducible results. In this regard, Raman microspectroscopy serves as a unique tool for a fast nondestructive characterization of indentations [4]. We begin with a brief review of the vibrational properties of the relevant Si phases that will facilitate a proper interpretation of the Raman spectra taken from the residual impressions on silicon.

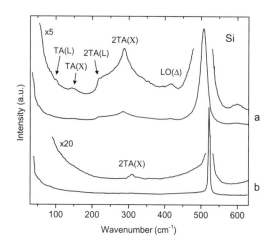

FIG. 12. First-order Raman scattering in (a) Si nanocrystals and (b) bulk silicon at room temperature [95].

In the absence of stress, the first-order [93] Raman spectrum of Si-I exhibits a single line at 520 cm$^{-1}$ [94], which corresponds to the light scattered by a triply degenerate optical phonon in the center of the Brillouin zone (Fig. 12b). Finite size of the crystalline grains (i.e., nano- or cryptocrystalline material) has a twofold influence on the Raman spectrum. The relaxation of the $k = 0$ selection rule due to quantum confinement makes possible the first-order Raman scattering by the phonons in other than $\Gamma$ high-symmetry points of the Brillouin zone, e.g., $X$ and $\Delta$. As a result, the Raman spectrum of nanocrystalline Si-I has a characteristic 2TA(X) overtone peak (at $\sim$300 cm$^{-1}$) of an enhanced intensity as compared to the bulk silicon (Fig. 12a). Another consequence of the quantum confinement is the distorted shape of the major Si-I band. This problem is normally tackled with the phonon confinement model [96, 97], which takes into account the contribution to Raman scattering from the phonons in the vicinity of the $\Gamma$ point. The model predicts the shape of the Raman line based on the optical phonon dispersion relations and the average size and shape of the crystallites. The resultant Raman line is asymmetric and is displaced to the lower wavenumbers (Fig. 12a).

Several theoretical and experimental studies assess the vibrational properties of the high-pressure phases of silicon. A group-theoretical analysis of lattice vibrations in the $\beta$-tin structure has been made by Chen [98]. In the vicinity of the $\Gamma$ point, the optical modes consist of one longitudinal optical (LO) branch and at higher frequencies of a doubly degenerate transverse optical (TO) branch, both of which are Raman active. Zone-center phonon frequencies of Si-II have been calculated as a function of pressure using the *ab initio* pseudopotential method

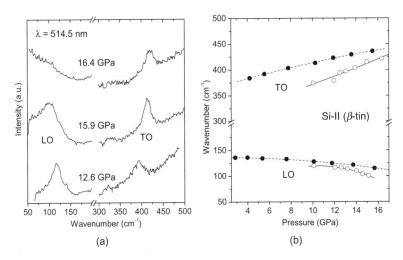

FIG. 13. (a) Experimentally determined Raman spectra of Si-II and (b) their dependence on pressure (open circles) [101] together with theoretically predicted phonon frequencies for the TO and LO modes of Si-II (filled circles) [100]. The lines serve as guides to the eye.

[99, 100], and showed good agreement with the experimental high-pressure results [101] (Fig. 13). Interestingly, the LO mode "softens" (shifts to the lower wavenumbers) with increasing pressure, which results from bond bending [102] associated with the transition of Si-II into a denser phase with a simple hexagonal lattice [99].

Group-theoretical considerations predict three Raman-active modes for the lonsdaleite structure [56, 103]. Theoretical analysis of the vibrational modes in Si-IV, based on the Raman Hamiltonian and the model of a tetrahedrally bonded solid [104], shows that two of these modes are degenerate, thus suggesting a spectrum with only two distinct lines (at 500 and 520 $cm^{-1}$) [105]. The reported Raman spectra assigned to Si-IV (the crystal structure was reportedly confirmed by the diffraction techniques), exhibit a broad asymmetric band centered at $\sim$510 $cm^{-1}$ [56, 106] (Fig. 14), which cannot be readily deconvolved into separate peaks. Note that the spectrum in Figure 14a was obtained from a sample containing 40% Si-I phase, which may explain the appearance of the 520 $cm^{-1}$ line in this case. A striking similarity of the spectra in Figure 12b and Figure 14 questions the credibility of the results of [56, 106]. The Raman spectra of Si-IV appear indistinguishable from those of nanocrystalline Si-I. Moreover, when applied to the calculation of the vibrational properties of Si-III, the method of [104] did not prove very successful [107]. Based on these considerations, we cannot

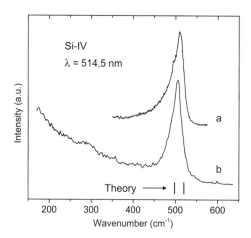

FIG. 14. Raman spectrum assigned to Si-IV: (a) From Reference [56]. Sample composition 60% Si-IV, 40% Si-I; effective sample temperature 700 K. (b) From Reference [106]. Presumably pure Si-IV; effective sample temperature 450 K. The vertical lines show the theoretical predictions for the Raman-active modes in Si-IV [105].

accept the Raman spectrum of Si-IV as established. More discussion on this topic will be given in the foregoing text.

The *ab initio* calculation of Piltz et al. [63] predicts seven Raman-active modes for the $bc8$ structure and eight modes for the $r8$ structure. As the $r8$ structure is considered to be a rhombohedral distortion of $bc8$ [62], the two phases are expected to have similar Raman spectra. Indeed, the calculated zone-center phonon frequencies of $r8$ and $bc8$ differ by only $\pm 10\%$ [63], which hampers the separation of these phases based on the experimentally measured Raman spectra. Pressurization experiments [108, 109] may also serve as a guide to the proper identification of Si-III and Si-XII. Figure 15 shows the Raman spectra of Si recorded at various pressures during depressurization from Si-II and the corresponding frequency-pressure plot. The Raman spectrum at 5.9 GPa (Fig. 15a) should be assigned to Si-XII because only this phase was shown to exist in this pressure range [62]. On the other hand, Si-III is predominant at ambient pressure [62] and thus the spectrum at 0 GPa (Fig. 15a) is due mostly to this phase. Analysis of Figure 15 suggests that most Raman lines indeed represent the overlapping vibrational modes of $r8$ and $bc8$; however, the lines at 182, 375, and 445 cm$^{-1}$ definitely belong to the Si-XII phase only.

The experimental Raman spectra of Si-IX have not been reported. Theoretical calculations of the vibrational properties of a similar $st12$ structure in germanium [107, 110] predict 21 Raman-active modes. Thus, observation of a large number

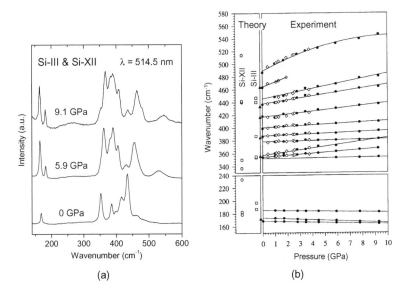

FIG. 15. (a) Raman spectra of Si obtained during slow unloading from Si-II. (b) Pressure shift of Raman-active modes in Si during unloading (filled circles) and reloading (open diamonds) [109]. Filled triangles represent the Raman bands of Si-III at ambient pressure from Reference [108]; open hexagons and squares are theoretically calculated lines of Si-XII and Si-III, respectively [63].

of unidentified lines in the Raman spectrum of silicon would suggest the presence of Si-IX.

It is accepted that the features in the Raman spectra of amorphous solids resemble the vibrational densities of states (VDOS) of their crystalline counterparts [111]. The Raman spectrum of amorphous silicon is characterized by four broad bands around 160, 300, 390, and 470 $cm^{-1}$ [112, 113]. They correspond to the features in the vibrational density of states of $a$-Si and are referred to as TA-, LA-, LO-, and TO-like bands, respectively [114]. In general, the TA/TO intensity ratio, their linewidths, and frequency positions depend on the method of preparation, deposition conditions, and the degree of structural disorder [115] (Fig. 16).

When applied to the study of phase transformations during indentation, Raman spectroscopy drastically extends the electron microscopy results. Raman spectra indeed reveal amorphous silicon within the indentation area [64, 116]; however, they also indicate that the formation of $a$-Si strongly depends on the experimental conditions [64]. Figure 17 shows the Raman spectra obtained from various points around Rockwell indentations in silicon. The original Si-I phase is observed outside the contact area, which is confirmed by a single Raman line at 520 $cm^{-1}$ (Fig. 17a). In contrast, the Raman spectra taken from the indentation area look completely different. The spectrum in Figure 17b, assigned to Si-IV by the authors [64], is in fact more characteristic of the nanocrystalline cubic diamond silicon, perhaps a mixture of the Si-I nanocrystals and Si-IV. Figure 17c,d suggests the presence of Si-III and Si-XII within indentations. The two spectra are similar in terms of band positions; however, the relative intensities of the same

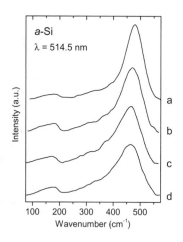

FIG. 16. First-order Raman spectra for (a) mostly ordered through (d) mostly disordered *a*-Si [113].

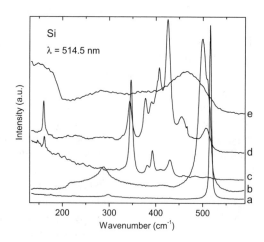

FIG. 17. Raman spectra taken from Rockwell indentations in silicon. (a) Pristine material outside the contact area. (b)–(d) Various points within the indentation area, slow unloading. (e) Indentation area, fast unloading. Data from Reference [64].

bands (i.e., $I_{430}/I_{350}$) are very different for (c) and (d). The laser annealing of the sample in the region showing a (c) type of Raman spectrum led to a continuous increase in the $I_{430}/I_{350}$ ratio and a gradual transformation of (c) into (d) [64]. Based on the intensity ratio considerations, the authors proposed that the Raman spectrum in Figure 17c originates from the Si-XII phase, which then transforms into Si-III (Fig. 17d) under the laser beam. Although the characteristic lines of Si-XII (at 182, 375, and 445 cm$^{-1}$) are not pronounced in the spectrum of Figure 17c, the general tendency of increasing the $I_{430}/I_{350}$ ratio with the increase in Si-III content (see Fig. 15a) is quite supportive of this assumption.

When fast unloading rates (>1 mm/min) were applied, only amorphous silicon was observed within indentations (Fig. 17e) [64]. Note also that in all cases the original line at 520 cm$^{-1}$ has vanished, indicating that no Si-I remains within the contact area after the microhardness test. Similar results were obtained for Vickers indentations [64]. Thus, in contrast to previous electron microscopy studies, Raman microspectroscopy studies have confirmed the presence of the Si-III and Si-XII phases in silicon indentations, providing additional evidence of a Si-I → Si-II transformation under the indenter.

Theoretical calculations [72, 117] show that the Si-III phase has a slightly higher enthalpy than Si-I and Si-II at all pressures (Fig. 18). The Si-III phase is metastable at ambient conditions and is expected to transform into Si-I with time. However, such transformation has not been observed experimentally. Biswas

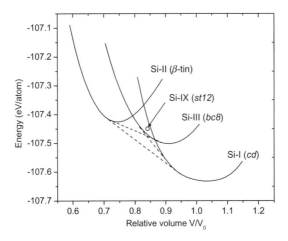

FIG. 18. Energy of silicon in the relaxed $bc8$, cubic diamond, and $\beta$-tin phases as a function of reduced volume. The $st12$ energy is shown at one volume. The dashed common tangent lines show that the stable phases are Si-I and Si-II; Si-III and Si-IX have a slightly higher enthalpy and are thus metastable. Data from Ref. [72].

et al. [74] argued that because the *cd* and *bc*8 structures are topologically very different, a possible Si-III → Si-I transition by shear distortion would require breaking and reforming of as many as one-fourth of all tetrahedral bonds, which implies a high activation energy barrier between Si-I and Si-III and may explain why the Si-III phase persists at ambient pressure. Fewer bonds need to be broken if a similar transition path (shear distortion) is followed between the Si-III and Si-IV phases [74]. Still, the Si-III → Si-IV transformation has not been observed at ambient conditions, but the formation of Si-IV has been confirmed in the experiments when the Si-III samples were only moderately heated [53, 56, 57].

Hexagonal diamond Si-IV is also known to form directly from the cubic diamond Si-I when the indentation test is performed at elevated temperatures [55, 58]. Using the TEM analysis of indentations made in Si single crystals under different experimental conditions, Eremenko and Nikitenko [55] established the temperature range of 350–650 °C for the Si-I → Si-IV transformation. The Si-IV phase appeared as ribbons or platelets embedded in Si-I around the indentation contact area, with a $\{511\}_{cd}$ habit plane and the following orientation relationship:

$$(0\bar{1}1)_{cd}\|(0001)_{hd};\ [011]_{cd}\|[\bar{1}2\bar{1}0]_{hd}$$

These results found further confirmation in the HREM studies by Pirouz et al. [58] and the mechanism for the formation of Si-IV by double twinning in Si-I was established [58, 59]. The Si-IV phase is not thermodynamically stable with respect to Si-I; its localization in the narrow ribbons around indentations is related to the stress relaxation during twin interactions [58]. It is still unclear if the Si-I → Si-IV transformation occurs within the indentation contact area as well.

Our recent Raman microspectroscopy studies of the temperature effect on phase stability in Si indentations revealed the formation of an unidentified phase of silicon at elevated temperatures [118]. The Raman spectrum of this phase (designated here as Si-XIII) has three characteristic lines at 200, 330, and 475 cm$^{-1}$ (Fig. 19), which do not match any of the established spectra of various Si phases (Figs. 12–16). The new phase is observed both in indentations made at elevated temperatures (150–250 °C), and in the same temperature range during heating of the samples with indentations made at room temperature. In the latter case, the Si-XII → Si-III → Si-XIII transformation can be monitored by Raman spectroscopy *in situ* as the temperature increases. The similarities in the formation process of Si-XIII and Si-IV strongly suggest the equivalence of the two phases; however, at this time, the association of the new phase with the hexagonal diamond silicon is inhibited by the rather low temperatures of its formation and a completely different Raman spectrum (provided the published spectra of Si-IV (Fig. 14) indeed represent this phase). Further diffraction studies may help to clarify this issue.

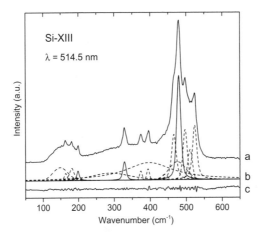

FIG. 19. (a) Raman spectrum of the new Si phase observed under experimental conditions similar to those of Si-IV. (b) Peak deconvolution reveals three characteristic lines at 200, 330, and 475 cm$^{-1}$ that cannot be assigned to Si-I, Si-III, Si-XII or $a$-Si. (c) The difference between the observed and calculated profiles.

In addition to Si-XIII, the Raman spectrum in Figure 19 clearly shows the presence of amorphous silicon and the traces of Si-III and Si-XII within indentation. With increasing temperature, the lines associated with Si-XII disappear and the lines of Si-XIII become predominant in the Raman spectrum. This process is accompanied by partial amorphization of the metastable silicon phases. In fact, we always observe the formation of amorphous silicon as an intermediate step before the final structural recovery to Si-I [118]. At relatively fast heating rates, the first traces of $a$-Si appear at ~150 °C as revealed by Raman microspectroscopy. In the temperature range of 150–250 °C, $a$-Si coexists with Si-XIII in indentations, with a continuous decrease in the relative amount of Si-XIII as the temperature goes up. At the same time, the Si-I line appears in the Raman spectrum of silicon indentations >150–200 °C, and its relative intensity increases with further heating until the $a$-Si bands completely vanish at ~600 °C. The whole process depends strongly on annealing time and starts at lower temperatures when the heating rates are slow.

In Figure 20, we present a schematic of the phase transformations occurring in silicon under static contact loading and during subsequent annealing. The initial and final phase is always Si-I, which is the only thermodynamically stable phase at ambient conditions. Annealing greatly facilitates a material's recovery to the cubic diamond structure; a sufficient time span is expected to have the same effect at room temperature. We did not include the Si-IV phase into the loading stage of Figure 20 because its formation in the narrow ribbons around indentations is

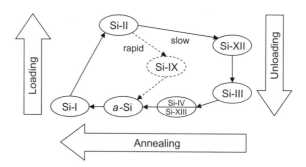

FIG. 20.   Phase transformations in silicon under contact loading.

related to the defect formation and interactions, which topic is outside the scope of this discussion. We should note that the tetragonal Si-IX phase, whose formation is expected under rapid unloading in analogy to the high-pressure cell results [61], has not been observed to date in indentations. Among the other silicon phases, Si-IX has the structure closest to the amorphous silicon [75] and thus is believed to precede the formation of $a$-Si at faster unloading rates. However, monitoring of Si-IX may be complicated by its presumed low stability during the external probe characterization (a similar $st12$ structure in germanium has been shown to transform rapidly into a mixture of amorphous and cubic diamond structures under the laser beam [109, 119]).

    Additional information on the phase transformations in Si during indentation can be obtained from the experimental load-displacement curves. Following the pioneering work of Pethica et al. [120], several groups extensively applied the nanoindentation technique to the analysis of silicon behavior under static contact loading. Pharr [12] first reported a sharp discontinuity in the Si unloading curve (Fig. 21a) and established a load threshold (5–20 mN) below which the pop-out was no longer observed in indentation experiments with a Berkovich tip. At peak loads below the threshold value, an elbow appeared in the unloading curve and subsequent reloading revealed the broad hysteresis loops, suggesting the irreversible Si-I → Si-II transition accompanied with a volume change of 20% (Fig. 21b). Pharr argued that the discontinuity in the unloading curve was induced by formation of deep lateral cracks around indentations [12]. However, TEM studies by Page et al. [65] of indentations made in silicon to similar loads showed no evidence of cracking, either lateral or radial/median. In general, TEM results suggested that the appearance of both pop-out and elbow in the unloading segment was due to the undensification or relaxation of some portion of the amorphized material previously densified in the loading stage [121]. As no specific features were observed in the loading curve when the Berkovich indenter was used, the transformation apparently started immediately after the sharp tip had

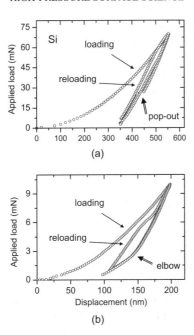

FIG. 21. Characteristic events observed in the depth-sensing indentation of silicon during unloading: (a) pop-out; and (b) elbow. Indentations were made at the loading rate of 3 mN/s and maximum load of 50 mN.

contacted the material's surface and occurred continuously throughout the entire loading segment, in agreement with the conductivity results already discussed here [8, 12].

On the other hand, indentation with spherical tips is characterized by contact pressures significantly lower than those achieved in Berkovich indentation at the same load, and is thus expected to reveal signs of the Si-I → Si-II transformation onset in the loading segment. Indeed, Weppelmann et al. [9, 15] and Williams et al. [122] observed discontinuities in the loading curves in addition to the pop-outs during unloading in their studies of Si nanoindentations made with spherical-tipped indenters. Again, no evidence of cracking was found based on the surface interferometry results and on the analysis of a material's elastic compliance during loading and unloading [9]. Using Hertzian contact mechanics and accounting for the material's elastoplastic behavior during indentation with spherical tips (Eq. (13)), Weppelmann et al. calculated the pressures at which the discontinuities occurred in the load-displacement curves of silicon as 11.8 GPa on loading and 7.5–9.1 GPa on unloading [9]. These values are very close to the Si-I → Si-II [52] and Si-II → Si-XII [63] transformation pressures, respectively, obtained under quasihydrostatic conditions in the high-pressure cells. Such similarities in the pressure values imply that the discontinuities in the loading and unloading curves are indeed related to the high-pressure phase transitions and not to the dislocation nucleation (on loading) or the crack formation or instant undensification of amorphous material (on unloading) as suggested elsewhere [12, 121]. This is further supported by a recent direct observation of the Si-XII and Si-III phases in silicon nanoindentations using Raman spectroscopy [38] and TEM [91, 92].

Finally, Domnich et al. [123] performed Raman microanalysis on a large set of Berkovich nanoindentations made in (111) and (100) silicon at peak loads of 30–50 mN and at the loading/unloading rates of 1–3 mN/s. At these experimental conditions, either a pop-out or an elbow in the unloading curve was observed with approximately the same probability. In most cases, the Raman spectra of indentations with a pop-out in the unloading curve showed the presence of the Si-III and Si-XII phases only, while the spectra of indentations with an elbow in the unloading curve revealed only amorphous silicon [123]. A few indentations that showed a mixed response in the unloading curve (an elbow followed by a pop-out), produced Raman spectra with the lines of both $a$-Si and the metastable Si-III and Si-XII phases. Based on these observations, the authors suggested that the pop-out during unloading is a consequence of a Si-II $\rightarrow$ Si-XII transformation, accompanied by a sudden volume release leading to the uplift of material surrounding the indenter, whereas an elbow appears in the unloading curve as a result of the material's expansion during slow amorphization of the metallic Si-II phase.

The contact pressures in the point-force indentation can be quantified using the technique of Novikov et al. [42] (Eq. (12)). This method was successfully applied to retrieve the $p_m(h_s)$ dependencies from the load-displacement data of Berkovich nanoindentation experiments [38, 42] and the pressures of $\sim$12 GPa for the Si-I $\rightarrow$ Si-II transformation on (re)loading [38] and 5–8 GPa for the Si-II $\rightarrow$ Si-XII transformation on unloading [123] were obtained, again favorably correlating with the high-pressure cell results. The amorphization onset during unloading, revealing itself as a kink in the $p_m(h_s)$ dependence (Fig. 22), has been determined to occur at pressures slightly higher than 4 GPa [123]. The step in the reloading curve at 8 GPa (Fig. 22) suggests the terminal stage of the reverse transition to the high-pressure Si-II phase.

The Raman imaging provides information on the residual stress fields around indentations. Assuming purely elastic loading, Lucazeau and Abello [116] calculated the stress components around Vickers indentation as functions of the radius to indentation center $R$. The calculated hydrostatic pressure $\sigma(R) = -1/3[\sigma_1(R) + \sigma_2(R) + \sigma_3(R)]$ was then compared with the experimentally measured shift of the main Raman band of Si-I from its equilibrium position [$\Delta\omega(R)$]. The experimental pressures were found to be significantly higher than the theoretical ones at intermediate distances from the indentation center, which implied

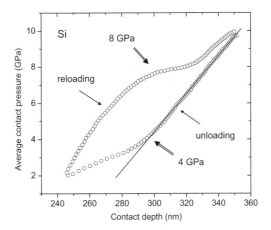

FIG. 22. Hysteresis loop in the cyclic nanoindentation of Si (maximum load 70 mN) in the average contact pressure vs contact depth coordinates.

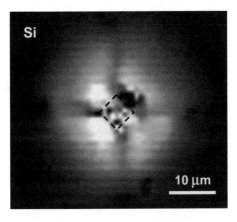

FIG. 23. Raman stress map around a 1-N Vickers indentation in (111) silicon. The residual indentation contact area is shown with dashed lines. Brighter regions correspond to the higher compressive stresses.

deviations from the purely elastic case considered in the theory. In general, the Raman maps around Vickers indentations reveal a compressive stress field [124, 125] in agreement with the calculated stress isobars [32]. Radial cracks emerging from the impression corners contribute to stress relaxation and result in a characteristic flower-like stress map (Fig. 23).

2.4.2 Germanium. Similar to silicon, germanium transforms from the semiconducting cubic diamond phase (Ge-I, space group $Fd3m$) into a metallic phase with $\beta$-tin structure (Ge-II, space group $I4_1/amd$) at elevated pressures [50, 51]. Under quasihydrostatic conditions, this transition begins at about 10 GPa, with the mixture of Si-I and Si-II persisting to higher pressures [126]. The presence of shear stresses apparently lowers the transformation onset (the Ge-I $\rightarrow$ Ge-II transformation pressures of 6–7 GPa have been reported for samples compressed without a pressurizing medium [127]) and increases the pressures for transformation completion [126, 127]. Upon decompression, transformation into a metastable phase with a structure similar (but not identical) to that of Si-IX is observed in most experiments at $\sim$7.5 GPa. This phase has a simple tetragonal structure with 12 atoms in the unit cell (space group $P4_32_12$) and is usually referred to as Ge-III or $st12$ [54, 126, 128]. At ambient conditions, Ge-III partially transforms back to the cubic diamond structure [128, 129], with reportedly 80% of Ge-III remaining after two months at room temperature [129].

Another metastable phase with the structure identical to that of Si-III, referred to as Ge-IV or $bc8$, has been obtained on pressure release from metallic Ge-II at low temperatures [130], and at room temperature by using amorphous Ge as the starting material (prior to first compression) [131]. Additionally, Nelmes et al. [129] reported a reproducible formation of Ge-IV after rapid unloading from metallic Ge-II at room temperature. Due to very high unloading rates (<1 s for unloading from 14 GPa to ambient pressure), no data is available for the Ge-II $\rightarrow$ Ge-IV transformation pressure. Germanium-IV is very unstable at ambient conditions and rapidly transforms into a hexagonal diamond structure (Ge-V, space group $P6_3/mmc$) [129, 130]. In regard to this transition, Ge-IV is very similar to Si-III, which is also known to transform into the hexagonal diamond Si-IV phase. However, Si-III is remarkably stable at room temperature (a "lifetime" of more that 100 yr was reported [57]) so the observable Si-III $\rightarrow$ Si-IV transition occurs only at elevated temperatures.

The possibility of formation of the $r8$ structure during unloading from Ge-II has also been discussed, based on the analogies between germanium and silicon and the well-established $\beta$-tin $\rightarrow$ $r8$ $\rightarrow$ $bc8$ transformation path upon slow decompression of Si. Lyapin et al. [132] reported the existence of a new phase after low-temperature decompression from Ge-II near 150 K, with the X-ray pattern and crystallographic density similar to those of the $r8$ modification of silicon (Si-XII). However, the result was not reproducible and the conditions for preparation of $r8$ germanium have not been specified.

Figure 24 shows the relative volumes vs pressure dependencies of various Ge phases [126]. The Ge-I $\rightarrow$ Ge-II transformation is accompanied by $\sim$19% volume decrease, whereas on pressure release, there is a 9% volume recovery during the Ge-II $\rightarrow$ Ge-III transformation. No data are available for the Ge-IV phase. However, the packing fraction of the $bc8$ structure was shown to be only slightly less than that for $st12$ [74], which implies that the Murnaghan curve for Ge-IV would be located next to the Ge-III curve in Figure 24. As in the case of silicon, the

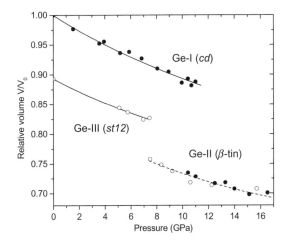

FIG. 24. Reduced volume of the high-pressure Ge phases. Filled and open circles correspond to increasing and decreasing pressure, respectively. The solid lines are fits to the
3rd order Birch equation of state [69, 70] with the values of bulk modulus $B_0 = 70$ GPa
($B_0'$ fixed at 4) for Ge-I; $B_0 = 76$ GPa ($B_0'$ fixed at 5) for Ge-III. Dashed line serves as a
guide to the eye. Data from Reference [126].

volume of the lonsdaleite phase (Ge-V) is identical to that of the cubic diamond
phase (Ge-I).

The possibility of phase transformation during indentation of germanium was
first suggested by Gridneva et al. [5] based on the experimentally determined
Ge hardness as a function of temperature. Similarly to silicon (see Fig. 9), two
distinct regions are observed in the hardness vs temperature dependence of Ge
(Fig. 25). Above the characteristic temperature (450 K for Ge), the hardness behavior may be accurately described by the temperature-dependent flow stress due
to dislocation glide (Eq. (16)), whereas at low temperatures, hardness is essentially temperature-independent and has the values close to the Ge-I → Ge-II
phase transformation pressure under nonhydrostatic loading [126, 127]. Again,
as for silicon, the calculated phase boundary between the cubic diamond and $\beta$-
tin structures in Ge [80] is similar to the experimentally measured hardness at low
temperatures and follows the same trend (Fig. 25).

*In situ* conductivity measurements by Clarke et al. [6] (Vickers and Knoop
indenters) and by Pharr et al. [8] (Berkovich indenter) showed a drop in Ge resistance during indentation with sharp tips and thus provided the first experimental support to the notion of indentation-induced metallization in germanium. The
TEM analysis of the indented Ge samples by Clarke et al. [6] revealed amorphized material within the indentations, similar to the results obtained on silicon.

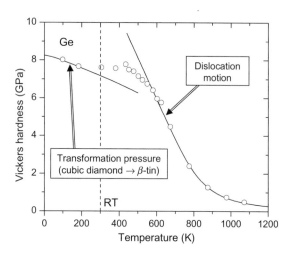

FIG. 25. Hardness vs temperature dependence of germanium: 2.3 N (for $T > 300$ K) and 1 N (for $T < 300$ K) loading, etched (111) surface [76] (open circles). Lines represent the calculated hardness due to dislocation glide (Eq. (16); $U = 0.53$ eV, $A = 0.43 \cdot 10^{-3}$, $b = 323$ K/GPa) and the cubic diamond $\rightarrow$ $\beta$-tin transformation pressure for Ge [80].

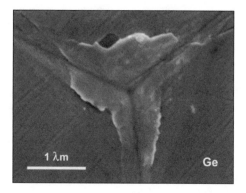

FIG. 26. The SEM image of a typical nanoindentation in Ge (Berkovich tip; 70 mN loading; (111) surface) revealing plastically extruded material.

The plastically extruded layer around indentations is observed more often in Ge than in Si [38, 121]. Such a layer (Fig. 26) is believed to represent a squeezed-out ductile metallic phase that forms under the indenter during the loading stage. In general, although much less effort was put into the indentation studies of Ge as compared to Si, it is acknowledged in the literature that the behavior of both group IV semiconductors is very similar.

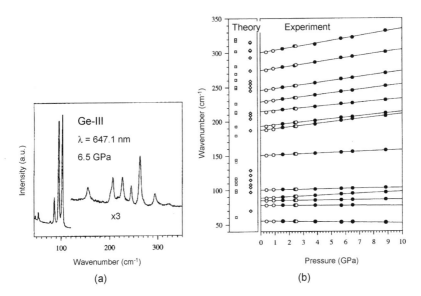

FIG. 27. (a) Raman spectra of Ge obtained during unloading from Ge-II at 6.5 GPa. (b) Pressure shift of Raman-active modes in Ge during unloading (filled circles) and reloading (open circles) [109]. Theoretically calculated lines of Ge-III are represented by open diamonds [107] and squares [110].

The formation of nonequilibrium phases during unloading from metallic Ge-II has been confirmed by means of Raman spectroscopy [107–109]. Due to similarities in the crystal structure of silicon and germanium, most features inherent to the Raman spectra of various polymorphs of Si are present in the corresponding spectra of Ge. This is particularly true for the cubic diamond and lonsdaleite phases, as well as for nanocrystalline and amorphous structures of Ge, which are characterized by Raman spectra identical to those of Si, but displaced to lower wavenumbers due to the weaker interatomic forces in Ge. However, correct assignment of the Raman modes arising from the Ge-III and Ge-IV phases is inhibited by the high instability of these phases under the laser beam [109, 119] and also by a complicated phonon structure of Ge-III (*st*12) near the center of the Brillouin zone, giving rise to as many as 21 Raman-active modes [107, 110] and thus hindering the bands of Ge-IV. Experimentally, 13 lines (apart from the lines of Ge-I and Ge-V) have been reported to appear in the Raman spectrum of germanium upon decompression from Ge-II [109]. Figure 27 shows the pressure dependence of these lines on unloading from 10 GPa and subsequent reloading. Following Olijnyk and Jephcoat [109], we tentatively assign all the lines in Figure 27 to Ge-III because this phase is expected to form under these experimental conditions (slow decompression from Ge-II).

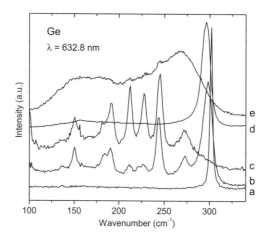

FIG. 28. Raman spectra taken from hardness indentations in germanium. (a) Pristine material outside the contact area. (b, c) Indentation area, slow unloading. (d, e) Indentation area, fast unloading. Data from Reference [134].

Raman microspectroscopy analysis of hardness indentations in Ge [133, 134] revealed the formation of different Ge structures depending on the unloading conditions. Upon fast unloading (typically beyond 1 mm/min), the bands of amorphous (Fig. 28e) or nanocrystalline Ge (Fig. 28d) were observed in various points within the indentation area. When slower unloading rates were applied, a number of new peaks appeared in the Raman spectra of indentations, indicating formation of the metastable Ge-III phase [134] and thus implying indentation-induced metallization. Based on the band intensity variations, Gogotsi et al. [133] assigned the lines at 210 and 227 $cm^{-1}$ to Ge-IV, which does not seem justified in view of the later *in situ* high-pressure results of Olijnyk and Jephcoat [109] (Fig. 27). In addition, the formation of Ge-IV is rather expected at fast unloading rates during decompression from Ge-II [129]. Taking into account the remarkable instability of Ge-IV and its rapid transformation into a lonsdaleite Ge-V phase at ambient conditions (further enhanced by the laser beam), as well as the similarities in the Raman spectra of Ge-V and nanocrystalline Ge-I, the formation of Ge-IV in hardness indentations at fast unloading rates may be argued assuming the Raman spectrum in Figure 28d belongs at least partially to the lonsdaleite Ge-V phase.

In Figure 29, we present a sequence of phase transitions occurring in Ge under contact loading. The loading part of Figure 29 is identical to that of Figure 20 for Si, viz. the formation of the metallic phase with $\beta$-tin structure under contact stresses of 8–10 GPa with high deviatoric components. Unloading part is a reverse copy of Si: at slow unloading, Ge-II transforms into a rather stable *st*12 phase, whereas at fast unloading, there is a formation of the *bc*8 phase, which is

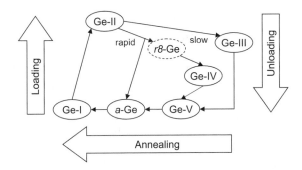

FIG. 29. Phase transformations in germanium under contact loading.

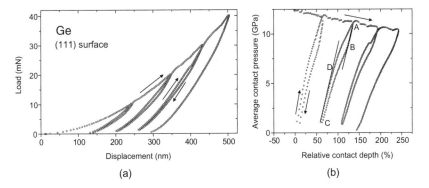

FIG. 30. (a) Load-displacement and (b) ACP vs relative contact depth curves for Ge obtained in cyclic nanoindentation at the loading/unloading rate of 1 mN/s. Arrows indicate the loading direction. After Reference [38].

very unstable and transforms rapidly into a hexagonal diamond structure at ambient conditions. We also included the hypothetic $r8$-Ge phase in the unloading stage, the existence of which in Ge has not to date been unambiguously confirmed. However, the formation of $r8$-Ge as an intermediate high-pressure phase between Ge-II and Ge-IV is thought very plausible, and the experimental observation of the $r8$-Ge is probably inhibited by a rather narrow pressure range of its existence under high depressurization rates.

Nanoindentation load-displacement curves of Ge are not as characteristic as in Si. The only feature discussed in the literature is the formation of numerous small discontinuities ("pop-ins") in the upper portion of the loading curve (Fig. 30a). As Ge is known to be very prone to radial cracking, it has been argued that the pop-ins occur as a result of the discontinuous propagation of radial cracks [135]. Another explanation of loading discontinuities associated the pop-ins with the nucleation of dislocation slips [121].

The seemingly featureless unloading curves of Ge might imply that the material behaves purely elastically during unloading, thus contradicting the assumption of the pressure-induced metallization in the loading stage. However, a close examination of the unloading curves in Ge shows that a unique power law relation (Eq. (11)) does not hold throughout the whole unloading segment and the material's mechanical properties indeed change during unloading. Gogotsi et al. [38] used the technique described in Reference [42] to assess average contact pressures (ACP) during indentation and replotted their cyclic load-displacement data (Fig. 30a) into the ACP vs relative contact depth curves (Fig. 30b). The resultant hysteresis loops clearly revealed elbows in the unloading/reloading curves similar to those observed in Si (compare Fig. 30b and Fig. 22). The nanoindentation behavior of Ge thus can be explained as follows—elastic unloading of the metallic Ge-II phase in the AB range (Fig. 30b) is followed by a gradual amorphization that begins in point B and leads to a faster decrease of the indentation depth with decreasing pressure until the transformation is complete in point C. On reloading, the reverse process takes place, with the formation of Ge-II from $a$-Ge beginning in point D. From Figure 30b, the pressure for the Ge-II $\leftrightarrow$ $a$-Ge transformation may be estimated as $\sim$8 GPa.

Only very rarely are the traces of the $st12$ phase observed in Ge nanoindentations [38]. In view of the Si nanoindentation results, where only the elbow in the unloading curve and never a pop-out (associated with the Si-II → Si-XII transformation [91, 123]) is observed below some load threshold ($\sim$10 mN for Si) [12, 121], we conjecture that a minimum volume of material may be necessary for the abrupt transformation from the metallic into a metastable phase to occur. The maximum loads reported for Ge nanoindentation experiments did not exceed 120 mN [8], and perhaps the lattice constraints in this case inhibit the abrupt Ge-II → Ge-III transformation. On the other hand, at higher loads ($>1$ N), the formation of metastable phases in Ge indentations has been experimentally confirmed (see Fig. 28).

## 2.5 Compound Semiconductors

Following the pioneering work of Jamieson [51], traditional understanding of structural systematics in elemental and binary semiconductors was based on the notion of pressure-induced transitions to the high-symmetry structures of increasing coordination, from covalent fourfold (cubic diamond, zincblende, wurtzite) to metallic quasisixfold ($\beta$-tin). In the octet semiconductors with increased ionicity ($>0.45$), this transformation path included an intermediate sixfold-coordinated rocksalt (NaCl) structure, with the zincblende (wurtzite) → rocksalt transformation pressure continuously decreasing until the rocksalt structure becomes thermodynamically stable at ambient conditions in the mostly ionic compounds (Fig. 31) [136, 137]. The universality of the zincblende → rocksalt → $\beta$-tin tran-

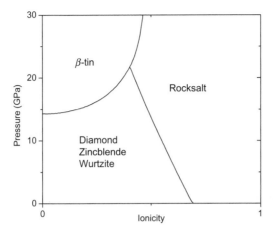

FIG. 31. Traditional phase diagram for a prototypical octet semiconductor with an atomic volume corresponding to GaAs [136, 137].

sition sequence in most octet semiconductors was apparently confirmed by extended experimental [138] and theoretical work [139–142]. Moreover, in contrast to the elementary Ge and Si, which were known to form the metastable tetrahedral phases upon depressurization from the $\beta$-tin phase (see Section 2.4), the transitions in binary semiconductors appeared to be fully reversible (with hysteresis), and the diatomic analogs of the $bc8$ or $st12$ structures were not observed experimentally [143].

Recent advances in X-ray diffraction techniques due to the use of synchrotron radiation sources and image plate detectors have led to results that completely change the established structural systematics of octet semiconductors. The high-pressure phases found in most binary semiconductors turned out to be orthorhombic distortions of the dense $\beta$-tin or rocksalt structures, with larger unit cells and lower symmetries [144–147], ranging from the site-disordered $Imma$ structure in GaSb and site-ordered $Immm$ in InSb to site-disordered $Cmcm$ in GaP and the site-ordered $Cmcm$-like structure in several other III-V and II-VI semiconductors of higher ionicity [147, 148]. Inclusion of the newly discovered phases in the traditional density-functional calculations showed only partial agreement with the experiment, predicting, for instance, that the $\beta$-tin phase is thermodynamically stable in GaP and GaAs [149–151].

The next breakthrough in understanding of structural systematics in octet semiconductors was made by Ozoliņš and Zunger [148], who pointed out that the standard theoretical approach is somewhat deficient in the sense that it can predict only *static* stability of a set of preselected structures and does not account for their *dynamical* properties. In contrast, the calculations of Ozoliņš and Zunger, based on the linear response phonon model [148], showed that both the rocksalt structure in elementary semiconductors and the $\beta$-tin structure in binary semiconductors are characterized by the dynamical phonon instabilities in their TA[00$\xi$] and LO[00$\xi$] branches, respectively. The authors proposed that the universal absence of the NaCl ($\beta$-tin) phases in covalent (ionic) semiconductors is caused by dynamical rather than static instabilities of these phases.

Another intermediate structure that has to be accounted for in the studies of pressure-induced phase transformations in compound semiconductors is the cinnabar structure (space group $P3_121$). This structure, found in highly ionic HgO and HgS at ambient conditions and for a long time known to precede the formation of NaCl structure in mercury chalcogenides [152], has recently been reported to exist at elevated pressures also in CdTe, ZnTe [144, 146] and GaAs [153]. The complete transition path in compound semiconductors may be thus viewed as the following sequence: zincblende $\rightarrow$ wurtzite $\rightarrow$ cinnabar $\rightarrow$ rocksalt $\rightarrow$ *Cmcm* $\rightarrow$ *Imma*. Not necessarily all the links should be present in a particular binary semiconductor; moreover, the general trend is toward reducing the transformation path with increase in ionicity, ending up with a mere cinnabar $\rightarrow$ rocksalt sequence in the highly ionic HgO and HgS.

Until recently, the dense tetrahedral structures similar to the $bc8$ and $st12$ phases in Si and Ge have not been experimentally observed in binary semiconductors. The theoretical calculations were very confusing—they posited that a diatomic equivalent of the $bc8$ structure, denoted $sc16$ after Crain et al. [143] (who erroneously assumed that the diatomic analog of $bc8$ would require breaking inversion symmetry, leading to a double-sized simple cubic unit cell, space group $P2_13$; as shown later by Mujica et al. [149], $sc16$ does contain a center of inversion and the true symmetry is $Pa3$ [154]), has its narrow pressure range of stability and, in fact, should replace the cinnabar structure in the transformation sequence of binary semiconductors [155, 156]. Crain et al. [143] proposed that the transition to $sc16$ is inhibited for kinetic reasons, the problem arising from the formation of unlike-atom bonds in compound semiconductors in contrast to monatomic Si and Ge. Indeed, in their recent high-pressure studies of GaAs, McMahon et al. [154] observed the formation of the $sc16$ phase during heating of GaAs samples while holding the pressure at 13–14.5 GPa. Moreover, the $sc16$ phase is retained after cooling and decompressing to ambient pressure [154], which fact strongly supports the assumption of a large influence of kinetic effects on the stability of $sc16$.

Having reviewed the current state of knowledge in the field of high-pressure phase transformations in octet semiconductors, we now proceed to the discussion of how the forementioned phase transitions may affect the behavior of these compounds during contact loading. Among the group of binary semiconductors, GaAs has by far the greatest technological importance and as a result has attracted the most attention in experimental studies. We will thus restrict our discussion to GaAs and outline more general trends on an example of InSb.

**2.5.1  Gallium Arsenide.**   At ambient conditions, GaAs exists in zincblende structure (space group $F\bar{4}3m$), commonly referred to as GaAs-I. The phase transitions sequence in GaAs has been clarified recently by McMahon and Nelmes using the angle-dispersive X-ray diffraction technique through the high-pressure diamond anvil cells [145, 153, 154]. Under quasihydrostatic loading, GaAs-I transforms into a structure with *Cmcm* symmetry at 17–24.5 GPa [153]. Traditional notation assigned the GaAs-II and GaAs-III labels to the orthorhombic *Pmm2* and *Imm2* structures, respectively [157]. However, new diffraction data suggest that the true structure of both *Pmm2* and *Imm2* is *Cmcm* [145, 146], which is also supported by the most recent computational results [149, 158]. In this regard, the GaAs-II label should be assigned to the *Cmcm*-GaAs, which seems to be the only stable structure that exists in GaAs above 24 GPa [146].

The zincblende → *Cmcm* transition in GaAs is not fully reversible: although some portion of the material in the *Cmcm* state transforms back to zincblende structure directly at 11–8 GPa, an intermediate phase having the cinnabar structure (space group $P3_121$) is also observed ≈12–10 GPa [153]. Cinnabar-GaAs is not likely to be a true equilibrium phase [155, 156]; rather, its formation is induced by kinetic factors that inhibit the formation in this pressure range of a thermodynamically stable *sc*16 structure (space group *Pa3*) at room temperature. This finds further confirmation in the recent high-temperature experiments of McMahon et al. [154] (see previous section). Once transformed into the *sc*16 phase, GaAs retains this structure to ambient conditions for the same kinetic reasons that inhibit its formation from *Cmcm* [154].

The indentation behavior of GaAs has been studied by means of various experimental techniques. The results of depth-sensing indentation indicate that a pop-in event similar to that observed in Si during indentation with spherical tips [122] (in the case of Si, the pop-in correlates with the transition from cubic diamond to $\beta$-tin structure, see Section 2.4), is sometimes present in the loading curves of GaAs as well [121, 122, 159]. Its apparently random variation with load (the critical loads varying by a factor of two were reported in the same sets of experiments [159]) suggests that the pop-in in GaAs is induced by a process of a statistical nature, such as crack or defect formation. As the SEM analysis of indentations in GaAs revealing pop-in in the loading curve showed no evidence of cracking [122] and the TEM observations clearly indicated that this phenomenon is not related to microtwinning [159], it is now accepted that the occurrence of a pop-in in the load-displacement curves of GaAs is due to the nucleation of dislocation slip.

A more typical load-displacement curve of GaAs (Fig. 32a), however, is fairly featureless. As clearly seen from Figure 32b, there are no deviations from the power law relation (Eq. (11)) during unloading; the slope of the unloading curve in

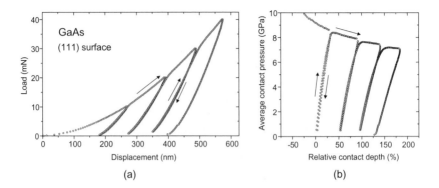

FIG. 32. Typical (a) load-displacement and (b) ACP vs relative contact depth curves for GaAs. Arrows indicate the loading direction. After Reference [38].

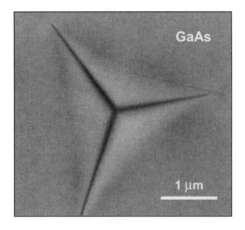

FIG. 33. The SEM image of a typical nanoindentation in GaAs (Berkovich tip; 50 mN loading; (111) surface). After Reference [38].

each cycle almost coincides with the slope of the unloading curve of the previous cycle; and the unloading curve and the elastic portion of the reloading curve do overlap for each cycle [38]. The SEM micrograph of a typical nanoindentation in GaAs also does not have any features that could be assigned to a pressure-induced phase transformation (Fig. 33). In the indentation with spherical tips, character-ized by lower values of contact pressures as compared to the point-tip indenters, replotting of the loading curve in terms of average contact pressure reveals a catas-trophic plastic deformation leading to a large increase in penetration depth and re-duction in the registered hardness values from 12.5–16 to $\sim$8 GPa [122] (Fig. 34). Further loading leads to a slight work-hardening of the indented material [122].

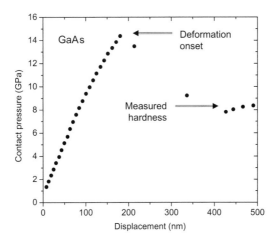

FIG. 34. ACP vs indenter displacement curve for GaAs obtained in indentation with spherical tip. The catastrophic plastic deformation starting at ~200 nm leads to the drop in hardness value from 15 to 8 GPa. Data from Reference [122].

In fact, this behavior is typical for the ductile metals and may be suggestive of dislocation-induced plasticity.

This is further supported by the TEM results. Clarke et al. [6] observed no signs of amorphized material in the Vickers indentations made in GaAs, as opposed to Si and Ge. The indentation area in GaAs was reportedly heavily dislocated [6]. A more dedicated analysis by Le Bourhis and Patriarche [159] revealed a similar picture—the dislocation density, very high in the center of the indentation contact area, was reduced at the edges of the plastic zone, where perfect dislocations as well as stacking faults could be observed. Again, TEM analysis of indentations in GaAs revealed no signs of phase transformations or densification of material during the hardness test [159].

The Raman microspectroscopy studies of indentations in GaAs did not provide direct evidence of the indentation-induced phase transformations [133, 134]. No new peaks were observed in the spectra from indentations other than the TO and LO bands of the zincblende structure (Fig. 35a). However, the relative intensities of the two bands vary significantly between the spectra obtained from the pristine surface and those obtained from microhardness impressions (Fig. 35). The decrease in frequency of the two bands in addition to the greatly reduced intensity of the LO band relative to the TO band, are suggestive of the presence of nanocrystalline material within the indentations as opposed to the original GaAs-I single crystal [95, 160]. Similar Raman spectra were observed from the GaAs samples composed of 5–10 nm disoriented crystallites by Besson et al. [160] af-

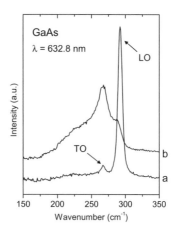

FIG. 35. (a) Raman spectra from the polished surface and (b) Vickers indentation in GaAs. Data from Reference [134].

ter decompression from the high-pressure orthorhombic *Cmcm* phase. There is also good correlation with the Raman spectra of GaAs nanoparticles of the size on the order of 10 nm [95] (Fig. 36a). The broad features extending from ~170 to 300 cm$^{-1}$ in the spectrum of Figure 35b may be attributed to the presence of partially amorphized material in the indentation area [161].

Consequent annealing of the indentation area with a laser beam reveals the emergence of two additional lines at lower frequencies (Fig. 36c), which were thought to be the possible signs of yet another heat-induced transformation at ambient pressure [162]. However, a direct comparison with the Raman spectrum of nanocrystalline GaAs-I (Fig. 36b) strongly suggests that the indentation area in Figure 36c consists of zincblende nanocrystallites of the size on the order of 5 nm. The overall red-shift of the Raman bands in Figure 36c may be attributed to the effect of heating. Emergence of the new lines is due to the relaxation of the selection rules in nanocrystalline material and correlates well with the one-phonon density of states for bulk GaAs [163]. The corresponding phonon modes [95] are shown by arrows in Figure 36b.

It is instructive to include in Figure 36 the Raman spectrum registered by Venkateswaran et al. [164] at ambient conditions after depressurizing the GaAs thin-film single crystals from 21 GPa. The authors assigned this spectrum to the cinnabar phase of GaAs based on the rough correlation between the two major lines at 248 and 261 cm$^{-1}$ and the mass-scaled cinnabar-HgS frequencies. Moreover, in the paper establishing the stability range for cinnabar structure in GaAs, McMahon and Nelmes [153] argued that the observation of Venkateswaran et al. [164] might imply a different stability range of cinnabar-GaAs in the thin

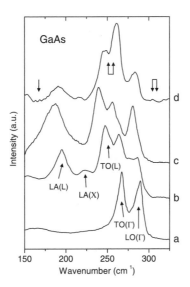

FIG. 36. (a) Raman spectra of the nanocrystalline zincblende GaAs, crystallite size ~10 nm [95]. (b) The same, crystallite size ~4 nm (arrows denote the high-symmetry phonon modes). (c) Laser-annealed indentations in GaAs [162]. (d) Presumably cinnabar GaAs phase registered at ambient pressure after decompression from 21 GPa (arrows denote the mass-scaled cinnabar-HgS frequencies) [164].

films as opposed to the bulk samples. In our opinion, a better correlation exists between the spectrum of Venkateswaran et al. [164] (Fig. 36d) and that of zincblende nanocrystallites (Fig. 36b), which is clearly indicated in Figure 36 by arrows denoting various modes of the mass-scaled cinnabar-HgS and the nanocrystalline GaAs-I. The differences in Raman response of the samples depressurized from 17 GPa and those decompressed from 21 GPa reported by Venkateswaran et al. [164] can be easily attributed to various characteristic sizes of nanocrystals in the recovered GaAs-I phase.

In summary, the results of Raman microspectroscopy analysis of indentations in GaAs suggest that the indented material consists of the disordered nanocrystals, the characteristic dimensions of which are further reduced with subsequent heating. This may be indicative of a pressure-induced transformation into a metallic *Cmcm* phase. Another possibility is that the shape of the Raman bands in Figure 35b is induced by extreme deformation during loading, leading to significant lattice damage. The nanocrystalline structure forms only after subsequent heating, in a way similar to the case of silicon (Section 2.4.1). Such an explanation brings the results of Raman microspectroscopy in accord with nanoindentation and electron microscopy results, both of which indicate dislocation-induced plas-

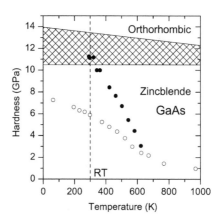

FIG. 37. Hardness vs temperature dependence of GaAs: (111) surface, load unknown (open circles) [5]; (100) surface, 0.1 N load [169] (filled circles). Crosshatched area is the phase boundary between the zincblende GaAs-I and orthorhombic GaAs-II phases with the following meaning: above it, GaAs-I is thermodynamically unstable; below it, GaAs-II is unstable [160].

ticity. As can be seen from Figure 37, in the case of GaAs the athermal plateau in the hardness vs temperature dependence is not reached at room temperature (compare with Fig. 9 for Si and Fig. 25 for Ge), which implies that the dislocation activity is still high at ambient conditions and also that the indentation-induced metallization may be expected at lower temperatures in GaAs. We will show in Section 3.1 that the scratches in GaAs produced at low temperatures indeed support this assumption.

2.5.2 Indium Antimonide. Among the binary semiconductors, InSb has received extensive attention in the experimental and theoretical studies due to the complex transition mechanisms involved in its high-pressure behavior [165]. The traditional phase diagram in InSb [166] was substantially revised recently [167, 168]. Two high-pressure phases of InSb are the site-ordered orthorhombic Cmcm and Immm structures, the transition between which proceeds via an intermediate site-disordered orthorhombic phase with Imma symmetry [167]. At room temperature, the transition sequence in InSb depends on the pressurization rate. At slow loading, zincblende structure transforms into Cmcm at ~2 GPa, whereas rapid loading rates lead to a direct zincblende → Immm transition at ~3 GPa [167]. The zincblende ↔ Cmcm transformation seems to be reversible and the phases other than zincblende have not been observed at ambient conditions in InSb.

Although the values of metallization pressures in InSb are fairly low, there is no evidence of pressure-induced transformations to metallic phases during indentation. The hardness vs temperature dependence in InSb [5, 169] is similar to that

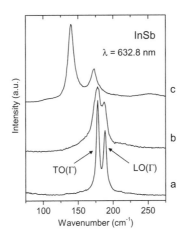

FIG. 38. Raman spectra from Vickers indentations in InSb: (a) polished surface; (b) indented area, slow unloading; (c) the same, fast unloading. Data from Reference [134].

of GaAs (Fig. 37), with the values of hardness and metallization pressure [168] downshifted accordingly. The load-displacement curves of InSb are characterized by a more plastic response with the lower percentage of elastic recovery as compared to GaAs, and are similarly featureless. Thus, the behavior of InSb during indentation at room temperature is determined by high activity of dislocations and (by analogy with GaAs) the pressure-induced metallization in InSb is expected at lower temperatures.

Interestingly, Raman microspectroscopy suggests that the indentation response of InSb is to a great extent dependent on the unloading rate [134] (Fig. 38). Slow unloading rates (0.01 mm/min) lead to Raman spectra with distorted TO and LO bands of the zincblende structure (Fig. 38b), very similar to the corresponding spectra of GaAs. In contrast, rapid unloading yields a new line of an enhanced intensity at $\sim 145$ cm$^{-1}$ and leads to a vanishing LO($\Gamma$) mode (Fig. 38c). The proposed explanation based on the formation of a wurtzite phase in InSb [134] is arguable bacause the wurtzite structure is characteristic of more ionic octet semiconductors [148] and has never been observed in the highly covalent InSb (with an ionicity index of 0.294 [170]). Although a definite conclusion cannot be made before the diffraction pattern from an indentation characterized by the spectrum in Figure 38c is obtained, by comparison with GaAs we conjecture that the enhancement of the low-frequency features in the Raman spectra of rapidly decompressed InSb indentations is related to lattice disorder and probably scales with the size of nanocrystallites. A direct comparison with the phonon dispersion curves in InSb should be suggestive. The formation of nanocrystalline material in InSb indentations after rapid unloading is further supported by the observation that a similar band at $\sim150$ cm$^{-1}$ is emerging in the Raman spectra as that in Figure 38b after laser-beam annealing [162].

## 2.6 Diamond

At ambient conditions, the thermodynamically stable phase of carbon is graphite (space group $P6_3/mmc$ [171]), which consists of stacked sheets with the threefold ($sp^2$) coordinated carbon atoms within the layers arranged in a two-dimensional network of regular hexagons. Diamond (space group $Fd3m$) is a high-pressure polymorph of carbon consisting of the three-dimensional regular network of fourfold ($sp^3$) coordinated atoms. From thermodynamical considerations, diamond is metastable at ambient conditions, but its spontaneous transition to graphite is inhibited by extremely slow kinetics. Another polymorph of carbon that has been experimentally observed is the so-called hexagonal diamond (lonsdaleite, space group $P6_3/mmc$), which is not to be confused with graphite because in lonsdaleite structure all atoms are fourfold-coordinated [172].

Diamond is the hardest among the known materials and shows enormous stability under static compression. Achievement of the pressure limit for diamond has challenged the scientific community for several decades. Similarly to silicon and germanium, diamond is expected to undergo a transition to a metallic state at sufficiently high loads. In terms of electronic band structure, metallization would mean overlapping of the electronic wavefunctions and the closure of the gap between the valence and conduction bands, leading to delocalization of the valence electrons (Mott transition, see Reference [173]). The bandgap of diamond was calculated to vanish under hydrostatic compression at 760 GPa based on a dielectric model and <900 GPa based on extrapolation of measurements of the diamond absorption edge from 168 to 405 GPa [174]. Theoretical calculations [175] indicate that the bandgap of diamond actually increases under purely hydrostatic compression and only a combination of hydrostatic and [100] uniaxial strains (which is exactly the case in the high-pressure anvil cells [176, 177]) leads to the overall positive pressure coefficient of the bandgap. Theoretical estimations for the Mott transition pressure in diamond under uniaxial [100] compression predict the values of 290 and 400 GPa depending on the calculation model used [175, 178, 179].

Possible candidates for the high-pressure metallic phases of carbon include a number of hypothetic structures. Simple cubic ($sc$), bcc, $fcc$, $hcp$ and $\beta$-tin structures have been initially included in the $ab$ $initio$ density-functional calculations of Yin and Cohen [180], which list was later augmented by the $bc8$ [117] and $r8$ structures [181], mostly by analogy with silicon. The total-energy considerations suggest that, among the structures considered, diamond will transform to the $r8$ structure at a pressure of 500 GPa [181]. At this pressure, the calculated electronic band structure of $r8$-C features a contact between the valence and conduction bands at the Z point of the Brillouin zone [181], suggesting that $r8$-C is semimetallic. However, owing to consistent 30–50% underestimation of the bandgaps in semiconductors and insulators in the density-functional calculations based on the local-density approximation (LDA) for the electronic structure, it is almost certain that the $r8$ structure of carbon is semiconducting. A true metallic phase of carbon with simple cubic structure ($sc1$) was calculated to be thermodynamically stable with respect to diamond >1.9 TPa [182]. Finally, the most recent constant-pressure molecular dynamic simulation of the behavior of diamond at terapascal pressures suggests its spontaneous transformation into a sixfold-coordinated metallic phase ($sc4$) in this pressure range [183].

A single experimental account for the possible metallization of diamond dates back to the works of Vereshchagin et al. [184, 185] in the early 1970s. The authors reported a reversible sudden decrease (of the order of $10^7$ $\Omega$) in the resistance of diamond powder subjected to quasihydrostatic loading at a surprisingly low pressure of 100 GPa. At fixed load, a similar resistance drop was observed at elevated temperatures [185], which allowed Vereshchagin et al. to establish the pressure-temperature boundary between diamond and a metallic carbon phase [185]. However, the details of the experimental set-up have never been given [184, 186], and the results of Vereshchagin et al. are commonly regarded with a high degree of disbelief. This is only aggravated by the fact that the most advanced high-pressure equipment failed to drive diamond to its limits, and diamond showed enormous stability under the highest static pressures that the modern high-pressure technology could achieve ($\sim$450 GPa) [187, 188].

It may be argued [3] that the goal hardly attainable in the experiments using high-pressure diamond anvil cells could be more easily achieved in as simple an experiment as a conventional hardness test. The well-documented indentation size effect (ISE) [189] reveals itself in the following relation between the Meyer hardness $HM$ (equivalent to the mean contact pressure) and the applied load $P$ [190]:

$$HM = \text{const} \cdot (P)^{1-2/n} \qquad (17)$$

where the material-dependent parameter $n$ has values <2 [191]. As a consequence, the mean contact pressures during indentation are very high at low loads in the initial stages of the test. Direct comparison of the calculated pressure necessary to close the bandgap in diamond under [001] uniaxial compression (Fig. 39a) with the plot of the values of diamond hardness measured at various nominal loads for loading in the [001] direction (Fig. 39b) suggests that the metallization pressures might be achieved at the initial stages of indentation with the point-force tips.

*In situ* optical microscopy observation of the indentation process demonstrated that diamond becomes nontransparent to visible light in the loaded zone [196] (Fig. 40; white regions show the reflection of light). This is consistent with a narrowing optical window in a diamond anvil that was observed at much higher

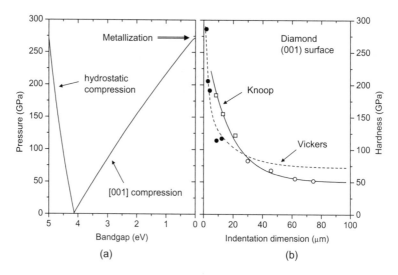

FIG. 39. (a) Calculated bandgap for diamond as a function of applied pressure for the case of hydrostatic compression and for compression along the fourfold axis of diamond (combined shear and hydrostatic) [175] and (b) the measured hardness values of diamond in the indentation in [001] direction [192] (experimental data from References [193] (filled circles), [194] (open squares), and [195] (open circles)).

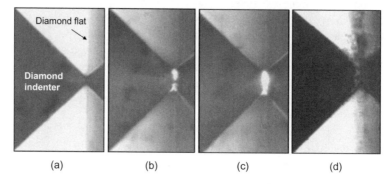

FIG. 40. Optical micrographs of the diamond indenter conical tip in contact with the (111) surface of a synthetic diamond single crystal: (a) prior to loading; (b, c) on increasing load during indentation; and (d) after unloading. After Reference [196].

pressures under quasihydrostatic compression [174, 197]. Although microfracture and a double refraction of light can also be responsible for this optical effect, a visible deformation of the tip (Fig. 41b) is a good argument in favor of a high ductility of diamond under contact loading. The size of reflecting area increases

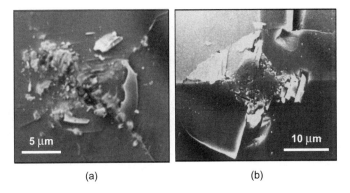

(a)                                      (b)

FIG. 41. (a) SEM micrographs of a 30-N Vickers impression in diamond [196] and (b) the corresponding deformed indenter tip [162].

with pressure, and it noticeably deforms before the indenter collapses and the pressure drops at the same load [196] (Fig. 40). Reportedly, only sufficiently sharp indenters produced the picture shown in Figure 40; otherwise, brittle fracture of the indenter was observed [196].

Scanning electron microscopy revealed the formation of debris around the indentation contact area in diamond [196] (Fig. 41a). This correlates with the behavior of silicon and germanium under contact loading, where the formation of plastic extrusions around indentations is believed to be indicative of the pressure-induced metallization (see Section 2.4). The formation of ductile extrusions was reported along the edges of the Vickers impression in diamond and around the deformed top of the diamond indenter [196] (Fig. 41), suggesting that similar transformations occurred in both the indenter and the crystal.

Raman spectroscopy analysis demonstrated dramatic broadening and weakening of the triply degenerate $(LO(\Gamma) + 2TO(\Gamma))$ mode of diamond at 1332 cm$^{-1}$ after indentation, as well as increasing background and appearance of several additional new bands in the Raman spectrum [4, 196]. The most typical spectra from various points in the indentation area are shown in Figure 42. All spectra in Figure 42b–d feature two broad bands at $\sim$1330 and 1570–1600 cm$^{-1}$ in addition to the original line of cubic diamond at 1332 cm$^{-1}$. The band at 1570–1600 cm$^{-1}$ correlates with the so-called G-band of graphite, commonly assigned to the C=C stretching in this $sp^2$ polymorph of carbon [103]. The broad band at $\sim$1330 cm$^{-1}$ is similar to the so-called D-band observed in the small graphite crystallites. This band was assigned to the coupled electron-phonon scattering event based on the correlations between its position and the electronic band structure and phonon dispersion curves of graphite [198]. In general, a combination of G- and D-bands of variable relative intensities is characteristic to the threefold-coordinated carbon

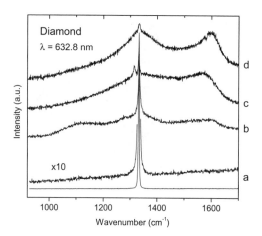

FIG. 42. Typical Raman spectra obtained from the Vickers impression in diamond: (a) unaffected pristine surface; and (b)–(d) various points in the indentation area. Data from References [4] and [196].

structures. Thus, the presence of G- and D-bands in the Raman spectra of indentations in diamond confirms the formation of $sp^2$ carbon during the hardness test [199]. It should be noted that the scattering cross section of graphite is some 50 times higher than that of diamond [200] and hence the G- and D-bands dominate the Raman spectrum even at small contents of $sp^2$ carbon. In the extreme case, the signal from $sp^2$ carbon covers completely the diamond band (Fig. 42d). As a result, the quantitative evaluation of the amount of graphitic carbon in the hardness impressions in diamond is impossible. Nevertheless, the phase distribution around hardness indentations can be monitored qualitatively by mapping the normalized Raman intensities of the line of cubic diamond and the G-band of graphite [199] (Fig. 43).

The spectrum in Figure 42c has an additional band at 1315 cm$^{-1}$. The band at this position has been reported as the only observable feature in the Raman spectra of hexagonal diamond powder prepared by a shock loading technique [103]. However, the authors did not provide the details of a complementary sample characterization or the degree of powder dispersity [103]. In contrast, Solin and Kobliska [201] observed only one single Raman band at 1332 cm$^{-1}$ for a 40–100 Å powder with a known mixture of 60% cubic diamond and 40% lonsdaleite phase. Group-theoretical considerations predict three Raman-active modes for the lonsdaleite structure [103]. The considerations based on the folded Brillouin zone concept [201] suggest that a particular shape of the phonon dispersion curves for the cubic diamond leads to triple degeneracy of the hexagonal diamond modes [202], which could help explain the single broad Raman band instead of a pair of sep-

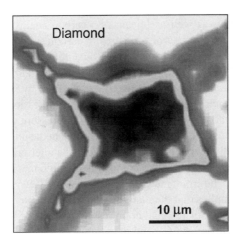

FIG. 43.  Distribution of the diamond phase around a Vickers impression in diamond based on the Raman intensity mapping. Bright area outside the contact area corresponds to the pristine diamond surface unaffected by indentation. Minimum amount of diamond phase remains in the center of indentation (dark area). After Reference [199].

arate Raman bands found for the mixtures of lonsdaleite and cubic diamond in Reference [201] and probably in Reference [103]. Thus, it is not possible to assign unambiguously the Raman band at 1315 cm$^{-1}$ (Fig. 42c) to hexagonal diamond. The simultaneous emergence of two bands at 1315 and 1322 cm$^{-1}$ in the hardness impressions in diamond (Fig. 42c) may also be attributed to a complex stress state in a particular sample area, leading to splitting of the cubic diamond TO and LO bands and the downshift of the corresponding transverse Raman mode.

The broad band at ~1110 cm$^{-1}$ (Fig. 42b) has also been repeatedly observed in the Raman spectra of indentations in diamond. It was initially assigned to amorphous diamond [4, 196]. However, both the experimental Raman spectra of $sp^3$-bonded amorphous films [203] and the calculated vibrational density of states of an amorphous carbon network constrained to be fully $sp^3$-bonded [204] show a strong peak at 1200 cm$^{-1}$. In contrast, the vibrational density of states of tetrahedral amorphous carbon ($ta$-C) features a broad peak centered at 1100 cm$^{-1}$ [205], which is quite similar to that observed in Figure 42b. The $ta$-C networks should not be confused with amorphous diamond because they contain some portion of aromatically bonded $sp^2$ carbon.

In addition to the Raman bands shown in Figure 42 and already discussed here, several weak but rather sharp lines were occasionally observed in the Raman spectra from both the sample indentation area and the deformed surface of the indenter [196]. These could be attributed to various hexagonal and rhombohedral diamond polytypes [202] or to the new structures built out of both $sp^3$- and $sp^2$-

bonded carbon atoms, as well as linear ($sp^1$) carbon chains. The possibility of formation of the dense tetrahedral structures of carbon similar to the $r8$, $bc8$ and $st12$ phases of Si and Ge should not be ruled out either.

The preceding facts are in agreement with the hypothesis that diamond experiences a phase transition upon contact loading. The question remains as to whether the appearance of the new phases after decompression supplies evidence for transition to metallic carbon or to another high-pressure phase. Graphite and tetrahedral amorphous carbon were probably formed in the unloading stage. Their formation upon loading would contradict the Le Chatelier principle because of a larger specific volume of graphite as compared to diamond, and atomic force microscopy reveals that the entire indentation area is slightly elevated above the crystal surface, which implies that the material under the indenter expands. This fact, along with the presence of residual compressive stresses of $\approx$10–15 GPa (measured from the upshift of the G-band of graphite) on the sample surface [196], excludes the possibility of surface graphitization as the mechanism of formation of $sp^2$-coordinated carbon in the indentation area. Thus, the explanation for the observed effects can be given by assuming the transformation of diamond into a high-pressure phase and a subsequent graphitization or amorphization under decompression.

## 2.7 Ceramics

### 2.7.1 Boron Carbide.
Boron carbide has a complex structure related to the simpler structure of $\alpha$-rhombohedral boron (space group $R\bar{3}m$), which consists of one $B_{12}$ icosahedron per unit cell positioned at eight vertices of a rhombohedron [206]. Specific to boron carbide is a linear three-atom chain in the center of the rhombohedron oriented along the threefold axis. The structure can also be described in terms of a hexagonal lattice based on a nonprimitive unit cell, in which case the $c$-axis of the hexagonal lattice corresponds to the [111] rhombohedral direction, which is along the three-atom chain axis. For a nominal stoichiometry of $B_4C$, there would be one $B_{11}C$ icosahedron and one CBC chain per unit cell. According to the standard phase diagram for boron carbide [207], this is the only phase known, although the actual homogeneity range extends from stoichiometries of $B_{4.03}C$ to $B_{10.24}C$ with the additional boron accommodated within the three-atom chain so that the icosahedra remain $B_{11}C$ for all but the most boron-rich compositions.

The high-pressure behavior of boron carbide was studied solely due to the presumption that it belonged to the class of the so-called inverted molecular solids [208], which is revealed through the higher compressibility of $B_{11}C$ icosahedra than that of space between them [209]. However, the inverted molecular behavior of boron carbide has been disputed recently based on the *ab initio* density-functional calculation of its vibrational properties [210]. Not much is known about the high-pressure behavior of boron carbide. Under quasihydrostatic pressure, the diffraction pattern of the powder sample was consistent with the $R\bar{3}m$ structure up to 11 GPa, and no extra lines were reported [209]. We are not aware of any other experimental studies of boron carbide to pressures >11 GPa, or of the theoretical investigation of phase stability in boron carbide at elevated pressures. Thus, the $R\bar{3}m$ structure observed at ambient conditions apparently remains the single phase of boron carbide ever reported.

   Polycrystalline boron carbide has a hardness of about 40 GPa at room temperature [211, 212], third only to diamond and cubic boron nitride. There is a linear increase in hardness with the carbon content, in the phase homogeneity range [212]. Nanoindentation studies on the (111) surface of boron carbide single crystal with the nominal stoichiometry $B_{4.3}C$ give even higher hardness values of 45 GPa. This suggests that the contact pressures during indentation of boron carbide are very high, and the possibility of a phase transition as a mechanism for accommodation of extreme deformations should not be eliminated. We will now proceed to a discussion of the results of a complex characterization of nanoindentations in $B_{4.3}C$ single crystal based on our recent work.

   The SEM images of the Berkovich indentations in the (111) surface of $B_{4.3}C$ reveal the presence of discrete deformation bands within the indentation contact area, which apparently follow the sample crystallography (Fig. 44). Similar features have been observed previously in $TiB_2$ and $Al_2O_3$ [213] and discussed in detail in connection with the indentation size effect by Bull et al. [214]. It was proposed [214] that the yielding beneath the indenter occurs nonuniformly in some hard materials and the discrete dislocation slip-steps are generated to re-

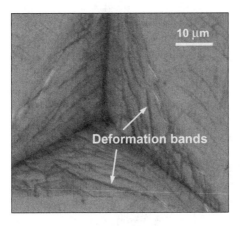

FIG. 44. The SEM image of a typical nanoindentation in boron carbide (Berkovich tip; 150 mN loading; (111) surface).

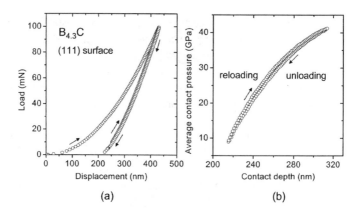

FIG. 45. (a) Load-displacement and (b) ACP vs contact depth plot for the indentation of boron carbide single crystal.

lieve the energy stored in the elastically bent material around the periphery of the indenter-substrate contact area. This is in contrast to the more plastic materials where the yielding beneath the indenter is uniform and continuous (compare to GaAs, Fig. 33). Thus, an observation of the deformation bands in $B_{4.3}C$ may be suggestive of a dislocation slip as the primary response to indenter intrusion. Obviously, Figure 44 does not readily show the signs of phase transformation or amorphization that might have occurred during indentation.

Nanoindentation load-displacement curves (Fig. 45a) and ACP vs contact depth curves (Fig. 45b) of the $B_{4.3}C$ single crystal are similar to those of GaAs in terms of the absence of any specific features and a purely elastic behavior during unloading (as follows from the coincidence of the unloading and reloading curves in Figure 45), though with a higher degree of elastic recovery. In fact, this may be the case of a completely irreversible phase transformation in the loading stage.

Dramatic structural changes in the $B_{4.3}C$ single crystal under contact loading are implied by Raman spectroscopy (Fig. 46). The Raman bands associated with numerous (and not entirely understood at this time; see References [210, 215–217]) vibrational modes of the pristine boron carbide (Fig. 46a) either disappear or are completely overwhelmed by a set of new high-frequency bands, the most intense of which is at $\sim 1330$ cm$^{-1}$ (Fig. 46b). Varying the laser wavelength rules out a possible assignment of these new high frequency bands to a resonant scattering coupled with the excitation energy (fluorescence, etc.). There is some correlation between the shift of the most prominent band at $\sim 1330$ cm$^{-1}$ with excitation wavelength and a similar dependence of the D-band position in polycrystalline graphite [198], which may imply disproportionation of boron carbide during indentation to form either amorphous carbon or a microcrystalline

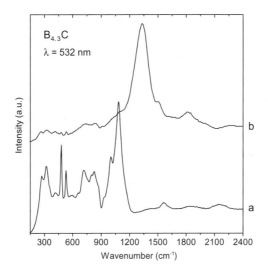

FIG. 46. Raman spectra from (a) the polished surface and (b) a 5 N Vickers indentation in the (111) surface of boron carbide single crystal.

form of graphite, with the latter possibly incorporating some boron. However, this explanation seems to be in conflict with the: (i) energy-dispersive spectroscopy (EDS) results, showing no alteration in the boron : carbon ratio in the hardness impressions in $B_{4.3}C$ as compared to the pristine surface unaffected by the indenter (neither does EDS reveal oxygen in the indentation contact area, which rules out surface oxidation as well); and (ii) complete absence of the graphitic G-band at 1580 $cm^{-1}$ in the spectrum of Figure 46b (the G-band is always present in all carbon structures involving $sp^2$ bonding). These forementioned considerations suggest that the pressure-induced phase transformation in boron carbide during indentation is still possible. However, additional characterization of the material in the vicinity of indentations by diffraction techniques will be required to establish definitely whether a new phase has been produced in indentation of boron carbide.

2.7.2 Silicon Carbide.    More than 100 crystallographic modifications of SiC have been identified at atmospheric pressure [218]. All SiC polytypes are characterized by identical tetrahedrally coordinated bonding between Si and C atoms but a different stacking sequence of the Si-C double layers along the [111] or [0001] direction. The polytypes are classified into cubic ($C$), hexagonal ($H$), or rhombohedral ($R$) arrangements of $n$ double layers within a unit cell. The most common SiC polytypes are $3C$ (zincblende structure, space group $F\bar{4}3m$), $6H$, $4H$, and $2H$ (wurtzite structure, space group $P6_3mc$). The hexagonal and cubic modifications of silicon carbide are also referred to as $\alpha$-SiC and $\beta$-SiC, respectively.

*Ab initio* density-functional calculations restricted to the configuration space of zincblende, rocksalt, and $\beta$-tin structures predicted a phase transition in $3C$ SiC from zincblende to the rocksalt structure at 66–67 GPa [219, 220]. Similar high-pressure behavior was also predicted for $2H$ and $4H$ SiC [220]. Additionally, the semiconducting rocksalt phase was shown to be energetically preferred to the metallic $\beta$-tin phase of SiC at all pressures [219]. According to Karch et al. [220], the rocksalt structure in SiC has a weakly pronounced semimetallic character due to lowering of the conduction band below the Fermi level in the K point of the Brillouin zone. Again, taking into account the consistent underestimation of the bandgap in the density-functional LDA calculations, the semiconducting nature of the rocksalt phase of SiC is not to be questioned.

Experimentally, a phase transition to a rocksalt-type structure was observed in $3C$ SiC under a static pressure of $\sim 100$ GPa by X-ray diffraction (XRD) studies [221]. This transformation was followed by an abrupt volume reduction of $\sim 20\%$. Among the hexagonal SiC polytypes, only $6H$ has received extensive attention in the high-pressure studies. In the same high-pressure XRD studies found in Reference [221], $6H$ SiC was found to be stable to 95 GPa with emergence of extra peaks attributed to the premonition of a phase transition at this pressure. Shock compression data on $6H$ SiC indicate a phase transition to a sixfold-coordinated phase (most likely rocksalt), which starts around 100 GPa [222, 223] and completes at 137 GPa [223].

The hardness of SiC is on the order of 25 GPa, which is significantly lower than the 100 GPa pressure corresponding to the transition to rocksalt structure upon compression. Nonetheless, the hardness vs temperature dependence in SiC is very similar to those of silicon and germanium, featuring an athermal region below $\approx 600\,^{\circ}$C [79, 224]. Extending the analogy with Si and Ge, the presence of this plateau should be suggestive of the transformation to a new phase during loading. However, in the case of SiC there is no correlation between the low-temperature hardness and the phase transformation pressure. An argument that deformation and not stress is a major driving force in indentation loading as opposed to the quasihydrostatic compression in the high pressure cells [3] does not seem applicable here because the transition pressure of 100 GPa was also obtained in the shock experiments, which themselves are characterized by extreme deformations. Thus, a mechanism other than a pressure-induced metallization should be considered for the explanation of the indentation behavior of SiC at low temperatures. This is further confirmed by extensive experimental data.

The nanoindentation load-displacement curves of SiC reveal a displacement discontinuity in the loading curve [65, 121] (Fig. 47), similar to the pop-in event often observed at low loads in GaAs (see Section 2.5.1). Below the loads of pop-in, the silicon carbide behavior is purely elastic, as indicated by overlapping loading and unloading curves in the initial stage of the indentation in Figure 47. The SEM images of nanoindentations in (0001) $6H$ SiC clearly show characteristic

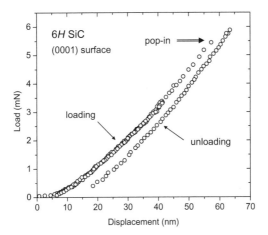

FIG. 47. Typical nanoindentation load-displacement curve of SiC revealing displacement discontinuity in the loading curve and purely elastic behavior below the pop-in onset. Data from Reference [65].

dislocation slip steps throughout the impression outline [213]. Similarly, TEM analysis of indentations in (0001) $4H$ SiC [225] and (0001) $6H$ SiC [213] reveals dislocated material within and around the indentation contact area. At the loads corresponding to the pop-in event, only a few dislocation loops were observed, whereas upon further loading, the density of dislocations was increasing and the radial cracks emerged around indentations [213]. By contrast to Si, high-resolution electron diffraction showed no evidence for amorphized material anywhere in the contact-affected zone [213]. Thus, the experimental evidence provided by electron microscopy indicates that the formation of a pop-in in the loading curves of SiC is linked to the nucleation of dislocation loops at the critical resolved shear stress during indentation [65], and more generally, the overall indentation deformation behavior of SiC at room temperature is dominated by dislocation slip rather than any sort of densification (e.g., transformation to a crystalline metallic phase or pressure-induced amorphization).

Raman microspectroscopy studies indicate that the spectra from hardness impressions in SiC and those from the pristine surface outside the indentation area are significantly different [4, 134]. This is illustrated in Figure 48 for a polycrystalline chemical-vapor deposited (CVD) $3C$ SiC film. The results for a single crystal $2H$ polytype of SiC are essentially the same [134], except for the extra line at 770 cm$^{-1}$ in the Raman spectrum of pristine $2H$ SiC, related to the splitting of the TO($\Gamma$) modes in hexagonal $2H$ as compared to the cubic $3C$ SiC [226]. This indicates that the deformation mechanism during indentation of SiC is independent of its microstructure prior to loading. Comparison of a typical spectrum

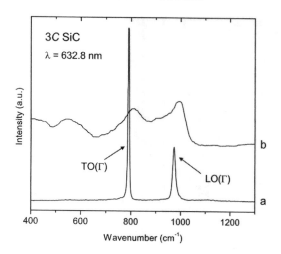

FIG. 48. The Raman spectra from (a) the surface of polycrystalline CVD 3*C* SiC and (b) the center of a 10 N Vickers impression in the same material. Data from Reference [4].

from indentations in SiC (Fig. 48b) with the vibrational densities of states of both 3*C* and 2*H* polytypes of SiC [226], suggests the assignment of the broad bands at 980, 810, 550 cm$^{-1}$ and the band <500 cm$^{-1}$ (Fig. 48b) to the LO-, TO-, LA- and TA-like modes of amorphous SiC, respectively (the upshift in the band positions is due to residual compressive stresses in the indentation area).

The presence of amorphous material in the hardness impressions in SiC revealed by Raman microspectroscopy does not closely correlate with the TEM observations; however, the scale effect might play an important role here—the indentation loads in Reference [213] (TEM) did not exceed 250 mN, whereas the loads in References [4, 134] (Raman spectroscopy) were on the order of 10 N. Since no transition to a high-pressure phase is expected in SiC at the contact pressures that might be achieved during indentation, the mechanism of amorphization in silicon carbide should be different from that in silicon. By analogy with a similar process in $\alpha$-quartz [22], the mechanism of amorphization by shearing or distortion of the crystal structure during loading was proposed [4].

2.7.3 Quartz and Silica Glass.    At low pressures, $SiO_2$ (silica) can exist in a variety of crystalline modifications that all share the same structural unit, a $SiO_4$ tetrahedron [227]. The fourfold coordination of silicons within the tetrahedra results from the strong $sp^3$ bonding to the oxygens. Linking of rather rigid $SiO_4$ units is typically realized through the soft and easily deformable Si−O−Si bonds, which allows a large number of the corner-sharing packing sequences. At ambient conditions, the thermodynamically stable phase is $\alpha$-quartz (space group $P3_221$), which consists of the helix chains of $SiO_4$ tetrahedra aligned along the $c$-axis of the crystal and interconnected to two other chains at each tetrahedron [227]. Coesite (space group $C2/c$) and stishovite (space group $P4/mmm$) are the high-pressure polymorphs of $SiO_2$, with stability ranges of 3–7 and >7 GPa, respectively [227]. At room temperature, however, the transformation of $\alpha$-quartz to both coesite and stishovite is kinetically inhibited.

Similarly to its crystalline counterparts, the amorphous phase of $SiO_2$ (silica glass) is composed of a three-dimensional network of slightly distorted $SiO_4$ tetrahedra interconnected at their vertices [227]. The disorder in this structure originates mainly from the variations in the bond and torsion angles describing the relative orientations of the tetrahedra, and only to a minor extent from the distortion of the tetrahedra themselves. The distribution of $O-Si-O$ bond angles and lengths in the $SiO_4$ tetrahedra is very narrow, which gives silica glass its highly pronounced short range order. The structural similarities between the amorphous and crystalline forms of $SiO_2$ play an important role in the high-pressure transformation mechanism in $\alpha$-quartz, which is essentially different from that observed in semiconductors.

Following Hemley's [228] early observation of amorphous material in the $\alpha$-quartz samples quenched from above 30 GPa, Kingsma et al. [22] performed *in situ* Raman spectroscopy studies of $\alpha$-quartz powder under quasihydrostatic loading and reported abrupt changes in the Raman spectra at ~21 GPa, with the emergence of rather broad bands associated with noncrystalline material on further compression. A careful examination of the quenched samples by TEM revealed the presence of an amorphous phase in the parent crystalline structure with a characteristic lamellar habit, apparently grown at planar defect sites. For the samples recovered from the highest pressures (>30 GPa), a complete amorphization with no sign of residual crystallinity was observed. The authors proposed the amorphization mechanism that involved compression-induced lattice distortions producing irreversible strains and a high density of planar defects in $\alpha$-quartz at high pressures, formation of amorphous lamellae inducing large heterogeneous stresses at the interfaces between the crystalline and amorphous phases, and the shear-assisted growth and coalescence of the amorphous lamellae [22].

Subsequent X-ray diffraction studies [229] revealed that the high-pressure amorphization of $\alpha$-quartz is preceded by the formation of another crystalline phase of $SiO_2$ ("quartz II") at 21 GPa, and that exactly this transition corresponds to the abrupt changes in the Raman spectra of $\alpha$-quartz reported in Reference [22]. At further compression of the samples beyond 21 GPa, broader and weaker bands were observed, indicating a gradual amorphization of $SiO_2$ [229], again in accordance with the previous Raman spectroscopy results [22]. Moreover, in the later XRD experiments on $\alpha$-quartz without a pressure-transmitting medium (i.e., under highly nonhydrostatic conditions close to those of contact loading) [230], the formation of yet another, highly coordinated crystalline phase of $SiO_2$ ("quartz III") was observed around 17 GPa. At least one amorphous and two crystalline phases coexisted in the pressure range of 21 to 43 GPa, with the diffraction pattern from the quartz II disappearing at 43 GPa and the quartz III persisting up to 213 GPa [230]. Both Raman spectra and XRD patterns indicated that the new crystalline phases are neither coesite nor stishovite, although the structure of quartz III was possibly similar to stishovite. Another observation was that all high-pressure phases (amorphous as well as crystalline) could be quenched to ambient conditions from a maximum pressure of 43 GPa [230].

These experimental observations prompted extensive theoretical research to identify possible candidates for the high-pressure polymorphs of $SiO_2$ and establish the transformation path upon compression of $\alpha$-quartz. In particular, molecular-dynamics (MD) simulations by Tse et al. [231] suggested a fourfold-coordinated triclinic structure for quartz II, with the angle between the original $a$- and $b$-axes in the $\alpha$-quartz unit cell essentially retained but the cell underwent a shearing and tilting of the axes originally perpendicular to the basal plane in the trigonal cell. The agreement between the calculated XRD pattern [231] and the observed one [229] proved very promising. Furthermore, Ovsyuk and Goryainov [232] pointed out that the internal parameters of the $SiO_4$ tetrahedra also needed to be taken into account in MD calculations. Specifically, the authors showed that the twist angle (deviation from a 90° angle between two opposite symmetrical tetrahedron edges) rather than the tilt angle (which determines relative orientations of the pairs of $SiO_4$ tetrahedra) is responsible for the structural instabilities in $\alpha$-quartz at 21 GPa and thus stimulates the transition to a triclinic phase with subsequent amorphization [232]. For the structure of quartz III, the MD simulations of Wentzcovitch et al. [233] suggested a $SiO_2$ phase with two-thirds of silicons in sixfold-coordinated sites and the remaining one-third in five-fold sites, which produced an XRD pattern favorably correlating with the experimentally observed one [229]. Finally, Teter et al. [234] noted that near amorphization pressure, a large class of energetically competitive phases could be generated from *hcp* arrays of oxygen with silicon occupying one-half of the octahedral sites.

Silica glass has also been shown to experience structural changes upon compression. In his *in situ* Brillouin scattering experiments on $a$-$SiO_2$ up to 17 GPa, Grimsditch [235] observed continuous irreversible changes in the elastic properties of silica glass >9 GPa, which were consistent with a $\sim$20% densification. Raman spectroscopy studies on the quenched samples [235, 236] suggested that the new phase was also amorphous. Hemley et al. [21] performed *in situ* Raman scattering experiments on $a$-$SiO_2$ up to $\sim$40 GPa and recorded the gradual changes in the Raman spectra at intermediate pressures. This is illustrated in Figure 49. The Raman spectrum at ambient pressure (0.1 MPa in Fig. 49) resembles the vibrational density of states of $a$-$SiO_2$ [237]. A strong diffuse band at $\sim$440 $cm^{-1}$ is associated with symmetrical Si$-$O$-$Si stretching modes principally involving motion of the oxygen atom [238–240]. Continuous reduction of this band upon compression to 8 GPa (Fig. 49) indicates a significant increase in intermediate-

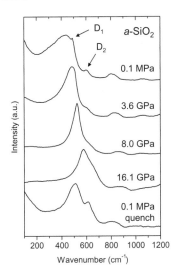

FIG. 49. Raman spectra of silica glass upon quasihydrostatic compression. The "defect" bands are marked $D_1$ and $D_2$. Data from Reference [21].

range order, specifically, a narrowing and shifting of the broad distribution of the $Si-O-Si$ bond angles that is characteristic of the structure of $a$-$SiO_2$ at ambient pressure [21]. Also of interest are two rather sharp lines at 490 and 605 $cm^{-1}$, commonly referred to as the "defect" bands, $D_1$ and $D_2$, respectively. These are associated with symmetric stretching modes of small-ring configurations in the three-dimensional random network of $SiO_4$ tetrahedra [240, 241]. It is evident from Figure 49 that the intensities of $D_1$ and (to a lesser degree) $D_2$ are enhanced on compression >8 GPa, and these lines (although displaced to higher frequencies) are predominant in the spectrum of the quenched sample. Such changes may arise from the shift in the ring statistics, with a possible formation of two-membered rings or edge-sharing tetrahedra [21].

The identical Raman spectra were obtained from Vickers indentations in silica glass [134], suggesting similar structural changes in $a$-$SiO_2$ during indentation and during quasihydrostatic compression in the diamond anvil cell. The SEM observations of Vickers impressions in amorphous silica by Kurkjian et al. [242] did not reveal any details within the indentation contact area that could be associated with shear flow. Thus, it seems very likely that the indentation behavior of $a$-$SiO_2$ is determined mainly by the gradual transformation into another amorphous phase with a narrower distribution in the $Si-O-Si$ bond angles and smaller ring configurations in the three-dimensional network of $SiO_4$ tetrahedra. The transformation is irreversible and the second amorphous phase of $SiO_2$ is retained in the indentations after load removal.

In the case of indentations in $\alpha$-quartz, Kurkjian et al. [242] reported the brittle-fracture behavior at both liquid nitrogen and room temperatures. A more detailed, combined SEM and TEM analysis of Vickers indentations in $\alpha$-quartz by Ferguson et al. [243] revealed an intensely deformed region fringing the indenter impression, with cryptocrystalline microstructure of an average size of ∼0.5 $\mu$m caused by a large mismatch between different blocks formed by intersecting fractures. Adjacent to many fractures, narrow zones of amorphous material were observed. It was conjectured [243] that the glass phase of $SiO_2$ develops in the early stages of the loading cycle as a way of accommodating the shearing associated with the forward motion of the indenter. However, no structural information on the indentation core was provided.

The Raman spectra taken from the center of the hardness impressions in $\alpha$-quartz [4, 134] reveal significant structural changes after indentation (Fig. 50). The disappearance of the weak sharp bands of the $E$-symmetry [228] is well documented in the high-pressure research and has been shown to correlate with the formation of quartz II at 21 GPa [229]. The three remaining bands at 205, 352, and 463 $cm^{-1}$ are of the $A_1$-symmetry and persist in the Raman spectra (although weakened and broadened) after the $\alpha$-quartz $\rightarrow$ quartz II transformation has started [22]. The band at 485 $cm^{-1}$ in Figure 41b favorably correlates with the $D_1$ band of amorphous silica (see Fig. 49) and may be indicative of the indentation-induced amorphization of $\alpha$-quartz. Of particular interest though is a pronounced band at 245 $cm^{-1}$ in the Raman spectrum of a Vickers indentation in $\alpha$-quartz (Fig. 50b). It is definitely not related to amorphous silica, either in

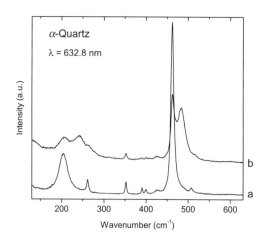

FIG. 50. Raman spectra from (a) the polished surface and (b) a 2 N Vickers indentation in $\alpha$-quartz. Data from Reference [134].

its ambient or densified form (see Fig. 49; also Reference [237]). Accidentally, the position of this band corresponds to the $B_{1g}$-mode of stishovite [228], which was being considered the first possible candidate for the quartz III phase observed at 17 GPa under highly nonhydrostatic compression of $\alpha$-quartz [230]. It is only regretful that the Raman spectrum of quartz III has never been recorded; however, taking into account the sharpness of the 485 cm$^{-1}$ line (Fig. 50b) (not typical for the noncrystalline material), we do not rule out the possibility that the bands at 245 and 485 cm$^{-1}$ are indicative of the presence of quartz III in the Vickers impressions in $\alpha$-quartz and thus of the indentation-induced $\alpha$-quartz $\rightarrow$ quartz III transformation.

**2.7.4 Zirconia.** At ambient conditions, the stable phase of $ZrO_2$ (zirconia) has a monoclinic ($m$-$ZrO_2$) baddeleyite structure (space group $P2_1/c$) [244]. The pressure-temperature phase diagram for pure zirconia is given in Figure 51a. Upon pressure increase, there is a transformation into a centrosymmetric orthorhombic phase with the $Pbca$ symmetry [247], which occurs at 3.5 GPa at room temperature. At ambient pressure, the monoclinic phase is stable up to $\sim$1150 °C, when it transforms to a tetragonal phase ($t$-$ZrO_2$) with the $P4_2/nmc$ symmetry [207]. At $\sim$2380 °C, a cubic phase ($c$-$ZrO_2$) with the fluorite structure (space group $Fm\overline{3}m$) forms, and its stability range extends up to the melting point at 2710 °C [207]. The displacive monoclinic-to-tetragonal phase transformation is

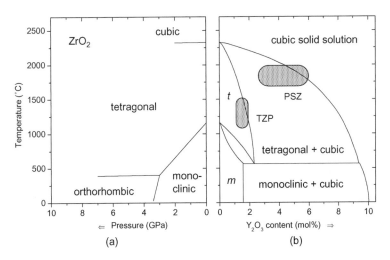

FIG. 51. (a) Temperature-pressure phase diagram of pure $ZrO_2$ [244] and (b) the zirconia-rich section of a $ZrO_2$-$Y_2O_3$ phase diagram [245]. Shaded areas indicate the ranges of compositions for partially stabilized zirconia (PSZ) and tetragonal zirconia polycrystals (TZP) [246].

accompanied by a ~4% volume reduction [244], which is reflected as a negative slope of the $m/t$ phase boundary in Figure 51a. There is little or no volume change associated with the tetragonal-to-monoclinic transformation [244] so that the slope of the $t/c$ boundary is close to zero (Fig. 51a).

The range of technological applications of pure zirconia is very limited, which is attributed to the tetragonal-to-monoclinic transformation on cooling. If not accommodated, the sudden volume changes accompanying this transformation can result in catastrophic fracture and, hence, structural unreliability of fabricated components [246]. To retain the tetragonal phase under ambient conditions, doping with stabilizing oxides (e.g., $Y_2O_3$) is used, combined with high cooling rates from sintering and solution treatment temperatures [246]. For illustration, we present the phase diagram for the $ZrO_2$-$Y_2O_3$ system in Figure 51b. As indicated in Figure 51b, the two most commonly used types of zirconia-based engineering ceramics are the partially stabilized zirconia (PSZ), generally consisting of a $c$-$ZrO_2$ matrix with a dispersion of tetragonal precipitates, and the tetragonal zirconia polycrystals (TZP), where the matrix grains are stabilized to a single-phase tetragonal form at room temperature [246] (Fig. 51b).

As evident from Figure 51b, the tetragonal phase in zirconia-based ceramics is metastable at room temperature, and its spontaneous transformation to the monoclinic phase is mainly inhibited by the lattice constraints to the associated volume expansion. Under an applied elastic stress, however, the high activation energy for the $t \rightarrow m$ transformation can be overcome, leading to the formation of monoclinic precipitates in the tetragonal matrix that create compressive strain fields in the matrix as a way of accommodation for the transformation shape change. This mechanism is known as transformation toughening or transformation plasticity [248] and reveals itself through the greatly improved fracture toughness of zirconia-based ceramics as compared to pure $ZrO_2$.

As a stress-assisted process, the $t \rightarrow m$ transformation inevitably accompanies indentation of zirconia-based ceramics. The most common way to monitor the formation of the martensitic phase in the hardness impressions in $ZrO_2$ is by using Raman microspectroscopy (e.g., References [249–251]). The vibrational modes of the cubic, tetragonal, and monoclinic phases of zirconia are well understood [252, 253] and the experimental Raman spectra of the three phases have been recorded [254] (see Fig. 52). The sharp differences in the spectral features of monoclinic (the lines at 178 and 189 $cm^{-1}$ in Fig. 52a) and tetragonal phase (the lines at 145 and 258 $cm^{-1}$ in Fig. 52b) are commonly used for estimation of the extent of transformation (e.g., References [249, 251]). This is illustrated in Figure 53 for a 10 N Vickers indentation in TZP ceramic. The bright rims around the residual impression in Figure 53b correspond to the maximum content of $m$-$ZrO_2$.

The volume changes associated with the $t \rightarrow m$ transformation around hardness impressions in zirconia-based ceramics have been monitored and quantified

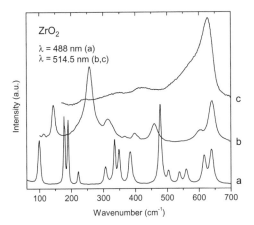

FIG. 52. Raman spectra of various phases in zirconia-based ceramics: (a) monoclinic (pure $ZrO_2$) [255]; (b) tetragonal ($ZrO_2$ + 2 mol% $Y_2O_3$) [256]; and (c) cubic ($ZrO_2$ + 9.5 mol% $Y_2O_3$) [257]. Gradual broadening of the spectral features upon increasing content of $Y_2O_3$ is due to the lattice mismatch caused by the differences in size between Zr and Y atoms.

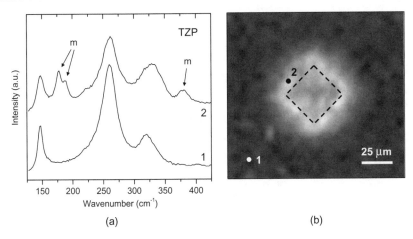

FIG. 53. (a) Raman spectra and (b) a phase distribution map from a 10-N Vickers indentation in TZP ceramic ($ZrO_2$ + 3 mol% $Y_2O_3$). Points of analysis are indicated by 1 and 2. Bright area in (b) corresponds to higher amount of the monoclinic phase. Dashed line delineates the residual impression.

with Tolansky interferometry [249] and, more recently, with atomic force microscopy [258]. These observations are consistent with the results of Raman spectroscopy, revealing an uplift of the surface surrounding the residual impressions. On further annealing, both the surface recovery and the decrease in intensity of the Raman bands associated with the monoclinic phase are observed [249], indicating a reverse $m \rightarrow t$ transformation in accordance with the phase diagrams for $ZrO_2$ and its alloys (Fig. 51).

A combined scanning acoustic microscopy (SAM), scanning electron microscopy (SEM), and Raman microspectroscopy investigation of spherical and Vickers indentations in yttria-stabilized TZP ceramics by Suganuma [251] revealed a general picture of indentation-induced damage in these materials. This is shown schematically in Figure 54. A cylindrical plastic zone beneath the indentation is characterized by a microcracked core surrounded by a transformation zone with a high content of martensitic phase. The microcrack density distribution is different for spherical and Vickers indenters, as revealed in Figure 54, which is due to the higher shear stresses in the case of point-tip indenters. The

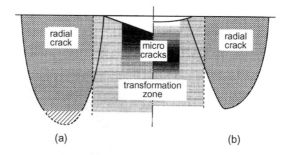

FIG. 54. Schematic of indentation damage for (a) Vickers and (b) spherical indentations in zirconia-based ceramics. Dark areas correspond to higher microcrack density. Dashed lines show the elastic-plastic boundary. After Reference [251].

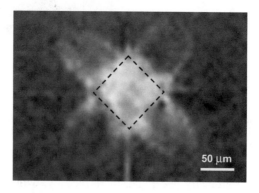

FIG. 55. Distribution of the monoclinic phase around a 20 N Vickers indentation in PSZ crystal ($ZrO_2 + 3$ mol% $Y_2O_3$) based on the Raman intensity mapping. Bright area corresponds to higher content of $m$-$ZrO_2$. Dashed line delineates the residual impression.

kidney-shaped radial cracks formed during indentation are very similar for both types of indenters, indicating the indenter-insensitive residual stress field [251]. We should add to this analysis that the $t \rightarrow m$ transformation in zirconia-based ceramics occurs in all areas subjected to stress, and the formation of a monoclinic phase (although to a lesser degree) is observed along the radial and lateral cracks as well (Fig. 55).

# 3 PHASE TRANSFORMATIONS IN MATERIALS UNDER DYNAMIC CONTACT LOADING

## 3.1 Scratching and Ultraprecision Cutting

Now consider a single hard asperity loaded statically and then made to translate across the sample surface at some (steady) speed. Frictional tractions in this case restrain mutual tangential displacements at the contact and the resultant distribution of tangential forces acts on the specimen in the direction of motion of the asperity [259]. Exact solutions for the complete stress fields in sliding interface between spherical indenter and elastic half-space have been given by Hamilton and Goodman [260], for the case of complete slip. Figure 56 shows the plots of the greatest principal stress acting in plane of contact of an elastic specimen with sliding sphere, for the coefficients of kinetic friction for (a) $f = 0.1$, and (b) $f = 0.5$. Also shown are trajectories of the lesser principal stresses, starting from the point of maximum tension in the field. As evident from Figure 56, an increase in the sliding friction is accompanied by an enhancement of tension behind the indenter and corresponding suppression ahead of it. Similar tendencies

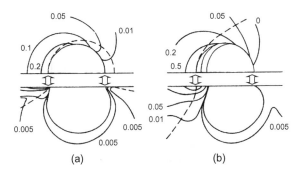

(a)                    (b)

FIG. 56. Plain and side views of contours of greatest principal stress in elastic half-space in contact with sliding sphere (marked with arrows) for the friction coefficients (a) $f = 0.1$ and (b) $f = 0.5$. Motion from left to right, unit of stress $p_0$. Dashed lines denote the trajectories of the lesser principal stresses, starting from place of maximum tensile stress in the specimen. After Reference [261].

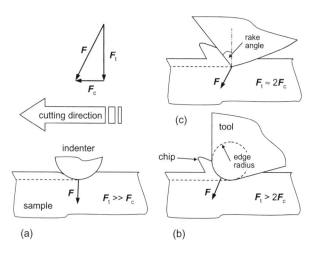

FIG. 57. Schematics of (a) the indentation sliding and (b, c) ultraprecision cutting. $F_t$—
thrust force; $F_c$—cutting force; $F$—resultant force acting on the sample.

are observed for the reduction in stress gradient below the trailing edge and for
deviation from axial symmetry of the stress trajectory patterns.

The two operations involving dynamical contact between a single asperity and
the specimen are scratching—that is, sliding of the hard indenter/stylus across the
sample surface (Fig. 57a), and ultraprecision cutting—that is, removal of work-
piece material using an inclined tool with a large edge radius relative to the depth
of cut (Fig. 57b), or a sharp tool with a negative rake angle (Fig. 57c). In ultraprecision cutting, the clockwise rotation of the resultant force vector $F$ leads to the
formation of chips ahead of the tool rather than debris alongside the grooves, typical for indentation sliding. In both cases, however, the material may be removed
in a ductile regime that involves plastic flow (severely sheared chips or debris),
without introducing fracture damage (propagation and intersection of cracks) into
the finished surface, which is intuitively expected for brittle semiconductors and
ceramics.

The possibility of cutting brittle materials in a ductile regime was originally
proposed by Blake and Scattergood [262] over a decade ago. The authors investigated single-point diamond turning (Fig. 58) of silicon and germanium in order to
specify the experimental conditions that would lead to material removal by plastic
flow. Tool rake angle (Fig. 57c) appeared to be the most important factor: critical
chip thickness for transition from brittle to ductile regime increased by an order
of magnitude (up to ~200 nm) for rake angles ranging from 0 to −30°. Further
characterization of Ge chip topographies using SEM [263] showed that fracture
damage was manifested as a frayed topology along the thicker portion of the chip,

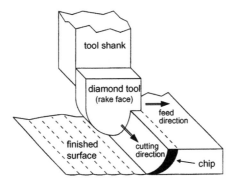

FIG. 58. Single point diamond turning using a round-nose tool.

beginning at the ductile-to-brittle transition point at the tool nose. In contrast, the chip appeared intact from the thinner end, from the tool center (Fig. 58) up to the initiation of the frayed topology, thus providing the width of that portion of the chip for which material removal mechanism by plastic shear flow (ductile regime) was operative. The results were consistent with previous observations [262] and once again suggested the maximum thickness for a plastically deformed layer of $\sim$200 nm, which corresponded to a tool rake angle of $-30\,^{\circ}$.

Transmission electron microscopy studies of the grooves made in the polished silicon wafers ((001) and (111) orientations) with conical [264] and pyramidal [17] diamond indenters showed the presence of a thin layer of amorphous material along the grooves, with a lightly dislocated/median-cracked region just beneath the amorphous layer. The amorphized zone was apparently equal to the contact area, and under given experimental conditions (10–20 mN maximum loads), it extended to depths of 150–200 nm below the sample surface. Similar TEM analysis of the cutting chips and a finished surface of the (110) germanium wafer subjected to ultraprecision cutting [18] also revealed the amorphous nature of the produced chips, this time with inclusions of microcrystalline fragments that were believed to originate from pitting in areas of the wafer with the highest resolved tensile stresses on cleavage planes. The morphological character of the chips showed evidence of high ductility and no evidence of molten flow, thus excluding high-temperature melting from the list of possible plasticity mechanisms.

Tanikella et al. [265] performed Raman microspectroscopy analysis of the grooves made in (100) Si with a Vickers indenter, below the cracking threshold of $\sim$50 mN (verified by the absence of acoustic emission). Within the groove, the material appeared amorphous, whereas the spectra taken from some debris along the groove flanks showed an additional asymmetric peak, suggesting the existence of nanocrystalline Si-I domains within the less dense amorphous phase, in agreement with previous TEM results [18].

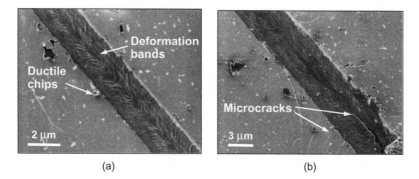

FIG. 59. The SEM images of scratches produced in (111) Si by a sharp Vickers indenter. Silicon exhibits (a) ductility on a microscale and (b) brittle fracture at the depth of cut beyond the critical value. Scratching direction from left to right.

A comprehensive study of Si response to scratching has been performed by Gogotsi et al. [266]. A single point diamond turning machine was used to make grooves in (111) silicon wafers at room temperature. Both sharp (Vickers) and blunt (Rockwell) diamond indenters were used for scratching. The depth of cut was increased gradually to monitor the transition from ductile to brittle regime of material removal. Post-scratching phase and stress analysis was done by means of Raman microspectroscopy, and the grooves morphology was characterized by scanning electron microscopy, atomic force microscopy, and optical profilometry.

Figure 59 shows the SEM images of a scratch produced using the Vickers (sharp) indenter. The periodic features within the groove (Fig. 59a) mark the intermittent shear bands formed ahead of the indenter [265] and suggest a ductile deformation mode. Beyond some critical deformation value, the formation of microcracks starts (Fig. 59b) as a way to release the elastic energy stored in the lattice. For the experimental conditions of Reference [266], the critical depth of cut corresponding to microcracking onset constituted ~250 nm for both sharp and blunt indenters. The grooves produced using a Rockwell (blunt) indenter exhibited even more interesting morphological features. The AFM image in Figure 60 reveals three to five microridges within the groove. The direction of the ridges changes along the scratch: ridges 1 and 2 deviate at a certain position, and then two new ridges 3 and 4 form. This phenomenon was not a result of tip irregularities, because in that case the pattern would have remained unchanged along the full length of the groove. However, if the plastic flow is accommodated via pressure-induced transformation into a metallic silicon phase (Si-II), then the shape of the deformation bands is dependent on the local conditions for transformation, producing complex groove morphologies similar to those in Figure 60.

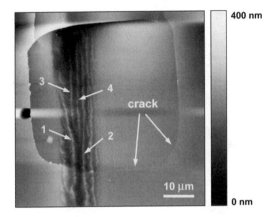

FIG. 60. The AFM image of the scratch produced on Si by the spherical tool. Micro-ridges 3 and 4 were formed between ridges 1 and 2. The orientation and number of ridges change along the scratch. Scratching direction from bottom to top.

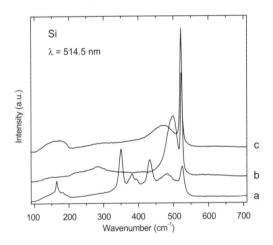

FIG. 61. Typical Raman spectra from various points in the scratches made in silicon: (a, c) contact area within the groove; and (b) debris along the groove flanks.

The Si-I → Si-II transformation during scratching is further confirmed by Raman microspectroscopy. The typical spectra taken from the areas within the grooves show the presence of the $r8$ (Si-XII) and $bc8$ (Si-III) polymorphs of silicon (Fig. 61a). The debris/chips produce Raman spectra with a "double-peak" feature (Fig. 61b), indicative of the hexagonal Si-IV phase or nanocrystalline material. The beginning portions of the grooves (at small depths of cut) are mostly amorphous (Fig. 61c). This perfectly correlates with the observations of static

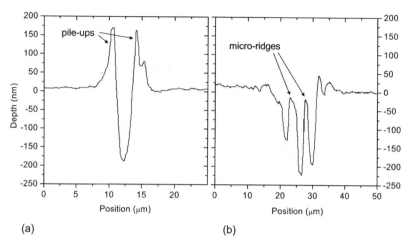

FIG. 62. Surface profiles of scratches in Si obtained using an optical profilometer. A high ratio of height of pile-up to depth proves the significant ductility of silicon under the tool. (a) Pyramidal tool; depth/width = 0.05; height of pile-up/depth = 0.89. (b) Spherical tool; depth/width = 0.016; height of pile-up/depth = 0.24.

contact loading (Section 2.4.1), when rapid unloading from the metallic Si-II phase was leading to the formation of amorphous silicon, whereas slow unloading produced the mixture of the Si-III and Si-XII phases (see Fig. 20). In the scratch experiments of Reference [266], the cutting speed was kept constant. At small depths of cut, the residual stresses were relatively small. This was leading to a faster stress release than at large depths, when significant residual stresses and the constraint of surrounding material resulted in a slow reverse transformation of Si-II. Therefore, $a$-Si was found in the beginning of the groove at small depths of cut, and Si-III and Si-XII at larger depths of cut.

Silicon response to scratching in the ductile regime is thus envisaged in the following fashion. Highly localized stresses underneath the tool lead to the formation of the metallic Si-II phase, which deforms by plastic flow and subsequently transforms into $a$-Si or a mixture of Si-III and Si-XII behind the tool. These reverse phase transformations, accompanied with a ~10% volume increase, are at least partly responsible for the complex groove morphology after the load is removed. The amount of Si polymorphs in the groove made with a spherical indenter was much higher compared to their amount in the groove made with a pyramidal indenter, which is probably due to the fact that most of the transformed material was displaced (squeezed out) to the flanks of the groove by indenter edges in the latter case. This is in agreement with optical profilometry results that revealed much higher pile-ups on both sides of the groove after scratching with a pyramidal tool (Fig. 62). No debris or chips were produced using a blunt tool: most

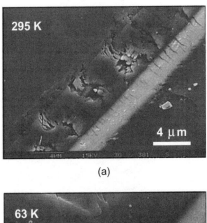

(a)

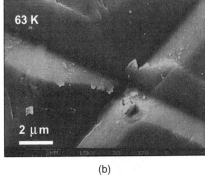

(b)

FIG. 63. The SEM images of scratches made in GaAs (a) at room temperature and (b) at liquid nitrogen temperature. The change from brittle to ductile behavior during contact loading with temperature decrease is evident. After Reference [133].

of the material was compressed and transformed to denser polymorphs in this case.

A similar deformation mechanism is expected for germanium. For GaAs, the pressure-induced phase transformation during ultraprecision cutting was inferred [267] from the high values of residual strains in the finished surface, as measured by Raman microspectroscopy. However, as discussed in Section 2.5.1, the pressure-induced transformation in GaAs under contact loading is not likely to occur at room temperature; it is rather expected at low temperatures based on the analysis of the experimental hardness vs temperature dependence for this semiconductor. Indeed, the SEM image of a scratch made in the GaAs wafer at liquid nitrogen temperature (Fig. 63b) clearly reveals ductile deformation behavior in contrast to purely brittle fracture in a scratch made at room temperature (Fig. 63a). This fact further supports the notion that the pressure-induced phase

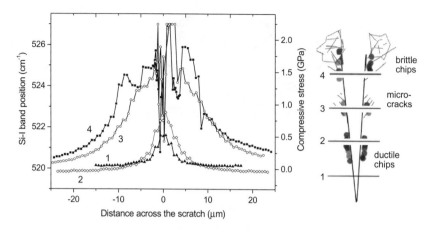

FIG. 64. Raman line scans across the groove in Si showing the increase in residual compressive stress with the depth of cut.

transformation is a common deformation mechanism in semiconductors under contact loading, in an appropriate temperature range.

The residual stresses in the grooves have also been assessed by means of Raman microspectroscopy [266, 268]. Figure 64 shows the stress distribution across the scratch made in silicon with a pyramidal tool by gradually increasing the contact depth [266]. Four different cross sections were selected, each representing a characteristic portion of the groove. Scan 1 in Figure 64 was acquired from the beginning part of the groove, at minimal depths of cut preceding the formation of ductile debris and chips. Scan 2 corresponds to the region of ductile chipping along the groove, whereas scans 3 and 4 both represent the fractured groove, in the areas of microcracking and brittle chipping, accordingly. Clearly, residual stresses increase with the depth of cut. However, above the critical depth of cut, the formation of cracks leads to local strain relaxation and as a result to strong stress variation across the groove (scans 3 and 4 in Fig. 64). At the same time, the highest residual stresses exist in the vicinity of pile-ups along the groove flanks.

## 3.2 Machining

Independently of a particular type of machining operation used for brittle materials, the process involves contact loading of multiple abrasive particles against the work surface of the specimen. Each of the particles may be regarded as a single asperity in dynamical contact with the specimen surface, in a way similar to indentation sliding and single-point cutting discussed in the preceding section. The tool rake angle together with the effective depth of cut turned out to be the key factors in attaining ductile regime cutting of brittle materials [262]; the same parameters are expected to be determinative in the selection of machining conditions that would inhibit brittle fracture in work materials. In the process of machining, although one cannot identify a definite rake angle as it is unknown and varies continuously during the process due to wear and friability (self-sharpening action) of the abrasive, it is generally accepted that the tool presents a large negative rake as high as –60 ° [269] and the radius of the tool edge has an action similar to that of an indenter. However, it is necessary to understand that during machining, the work surface of the specimen is in contact with numerous abrasive particles *at the same moment* and those cause different contact pressures and produce different depths of cut due to their different shapes and sizes. Thus, phase transformations, plastic deformation, and brittle fracture may occur simultaneously, and the particular machining conditions determine which of the preceding processes is predominant.

3.2.1 Silicon.  The production of silicon wafers with near-perfect surface quality requires a series of manufacturing processes [270]. Once a wafer is sliced from a bulk Si ingot it undergoes lapping, and/or grinding. Both processes introduce a certain amount of lattice damage in near-surface wafer regions. Typically, this damage is removed by chemical etching. New developments in grinding technology aim at skipping the costly and environmentally critical etching step. In such a case, the final polishing step would follow after high-quality grinding. Wafer edges also need special processing including grinding, etching, and polishing. In order to create external gettering capability, wafers can be intentionally damaged on the wafer backside. In such a case, quartz particles can be used as abrasives. Depending on the final device application, further wafer manufacturing steps may follow after polishing such as epilayer deposition or annealing treatment. In any case it is of utmost importance to assure a high surface quality of the wafer frontside. This requires consistent control and improvement of the sequential processing steps as described in the foregoing text.

Following the pioneering research on ductile-regime grinding of brittle materials [271], much effort has been made toward understanding the mechanism of ductility as well as characterizing machined surfaces of silicon [272–276]. High-resolution TEM analysis of nanomachined Si surfaces [274] suggested that dislocation networks and slip planes in Si-I accommodate plastic flow; no evidence of metastable Si phases or amorphous material in the subsurface zone was found. Early Raman microspectroscopy results [273] indicated the presence of nanocrystalline material in the vicinity of the wafer surface, with the size of the crystallites ranging from 3 to >30 nm depending on the machining (lapping) conditions. Detailed Raman microspectroscopy studies of Si wafers subjected to various machining operations (viz., lapping, chemical etching, quartz back-damaging, edge grinding, and dicing), as well as of the wear debris collected during dicing of Si wafers, were performed recently by Gogotsi et al. [275].

The Raman spectra taken from various points of machined Si surfaces fall into the three categories presented in Figure 61, although with the characteristic features less pronounced than in the case of scratching. Nevertheless, a mixture of Si-III and Si-XII phases, Si-IV or nanocrystalline Si, and amorphous material are all found in machined Si surfaces. Figure 65a shows a typical optical image of the rough wafer surface after lapping. The amount of transformed material in the wafer surface was estimated from the intensity ratio between the signal in the range of $330$–$515$ cm$^{-1}$ and the signal in the range of $516$–$525$ cm$^{-1}$ (Si-I line). Figure 65b shows the distribution of the transformed phases in the wafer surface of Figure 65a. Areas where lapping resulted in material removal via brittle fracture show a predominant presence of Si-I (dark areas in Fig. 65b,c); the lighter areas in Figure 65b correspond to the transformed material. However, after chemical etching, the surface is pristine without residual stresses and does not display any significant intensity of phases other than Si-I (Fig. 65c). Hence, sufficient etching removes the damaged layer completely.

A nonuniform stress distribution exists on the surface of quartz back-damaged Si wafers (Fig. 66). The surface is predominantly under residual compression, with significant variations across most areas [275]. However, no evidence of the phases other than Si-I is revealed by Raman microspectroscopy in this case.

Examination of the cut surface of diced Si wafers (Fig. 67) shows that the differences in the pressure applied to different sides of the cut and/or asymmetry of the cutting edge of the diamond saw lead to different machining conditions on each side of the cut. One of the cut surfaces (Fig. 67a) exhibits a relatively small amount of new phases, mostly amorphous silicon (Fig. 67c). In contrast, on the other side of the cut (Fig. 67d), the metastable Si-III and Si-XII phases are found in large quantities compared to Si-I (Fig. 67f). This clearly demonstrates that phase composition of the wafer surface can be controlled by changing the machining parameters. Similar results were obtained on Si wafers [277] subjected to two different regimes of edge grinding, and the magnitude and uniformity of residual stresses across the wafer edge correlated with the amount of metastable Si-III and Si-XII phases in the ground wafer edge.

These results show that the mechanical operations involving high contact pressure between the tool and the machined Si surface may lead to pressure-induced phase transformation. As Si-III and Si-XII can be formed only via metallic Si-II (see Section 2.4.1), the presence of these phases in the machined surface suggests the formation of metallic silicon beneath the tool in the machining processes under study. Thus, metallization and further transformation into various polymorphs of Si occur along with brittle fracture and plastic deformation. Slicing from an ingot is the first fabrication process that results in heavy mechanical damage. The lapping process involves high-pressure contact between abrasive grains and the Si surface. The bulk of the layers damaged during slicing would be removed here in the lapping process. Nevertheless, many regions of transformed phases and resid-

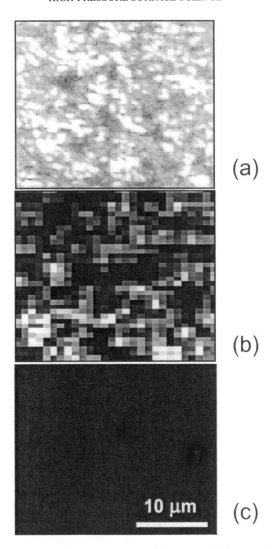

FIG. 65. Raman maps of the surface of a lapped Si wafer [275]. (a) Optical micrograph. (b) The ratio of transformed Si (amorphous material and metastable phases) to pristine Si-I. (c) The same, after the wafer surface was chemically etched. Here, all of the transformed phases are removed. Brighter areas in (b) and (c) correspond to higher content of transformed material.

ual stress persist and a new transformation layer may be produced during lapping. When the wafer edge is ground by a diamond tool, again, the surface is exposed to high pressure. As a result, many transformed phases were found along the edge.

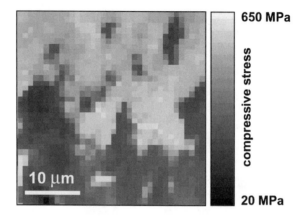

FIG. 66. Raman map of residual stress distribution on the surface of Si wafer that had been back-damaged with quartz grit. After Reference [275].

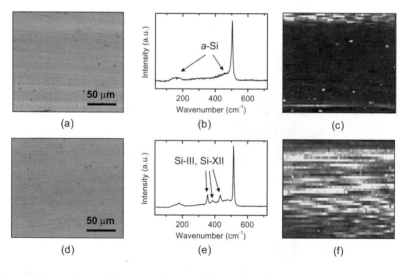

FIG. 67. Two sides of the cut surfaces of diced Si wafers: (a, d) optical images; (b, e) typical Raman spectra; and (c, f) intensity ratio maps, $I_{330–515}/I_{516–525}$. After Reference [275].

However, chemical etching removes the layers of transformed phases and residual stress, both in the top surface and edge surface.

Brittle fracture leaves damages to the underlying layers as deep as 5–10 $\mu$m due to crack propagation (Fig. 68a). In the case of metallic ductile removal, the underlying layers do not have fractures because the ductile, metallized Si lay-

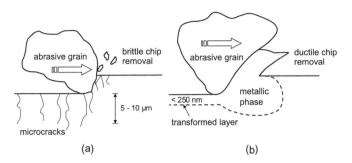

FIG. 68. Schematics of the two regimes of machining: (a) brittle fracture; and (b) ductile removal of material via pressure-induced metallization in the surface.

ers are scraped off (Fig. 68b). When the metallization occurs on the Si surface induced by high contact pressure, the metallic layers are $<1$ $\mu$m deep and the remaining transformed layer is only 100–200 nm in thickness. Therefore, a substantial amount of time and chemicals could be saved during subsequent etching to remove damaged layers if the depth of etching is reduced from 25 $\mu$m to, for example, 1–2 $\mu$m. The major hurdle in development of this approach is in the techniques of monitoring and controlling the metallization of silicon, and whether it is feasible to establish the machining conditions so as to remain within the ductile regime mediated by metallization without microfractures.

Such tasks are feasible with Raman microspectroscopy. Optimization of machining conditions in this case can be done in the following way. After applying the different conditions of a given tool operation (feed rate, spindle speed, applied load, etc.), the finished surface is scanned with a Raman spectrometer to yield maps of stress and phase composition with their relative ratios. Mapping will continue after etching away series of surface layers to determine the depth of damage. Finally, the maps will be compared to determine the optimum level of ductile machining condition that yields the most metallization in the absence of fracture. This approach may help to select the optimum machining conditions, as well as minimize etching depth and time.

# 4 CONCLUSIONS

Interaction between two material surfaces in a real environment is a complex process that may involve material fracture, deformation, mechanochemical interactions, and phase transformations. These processes must be considered together because of the existing synergy between them. None of the current classical fields of research, such as tribology or contact mechanics, address the complexity of the topic. Thus, a new area of research is emerging, which we suggest calling "high-pressure surface science."

In particular, phase transformations under contact loading need a more detailed investigation. Both static and dynamic interactions between hard surfaces may result in phase transformations. Hydrostatic and deviatoric stresses must be taken into account and phase transformations in contact loading can be described as deformation-induced transformations. At the same time, the transformation pressures for silicon obtained in indentation tests are in good agreement with the results from high-pressure cell experiments, which utilize hydrostatic loading.

Phase transformations in semiconductors, including pressure-induced metallization, have been described in this review. Currently, only silicon has been studied thoroughly enough. However, even for this simple elemental semiconductor, not all issues concerning identification of new phases, transformations mechanisms, and transformation paths have been resolved.

Mechanisms of phase transformations in ceramics can be different from those in semiconductors, but pressure- or deformation-induced amorphization has been observed for both classes of materials. Even the hardest material known— diamond—experiences a phase transformation under contact load. Changes in structure and density have also been reported for amorphous materials, such as silica glass. Recent studies suggest indentation-induced phase transformations in intermetallic compounds and thus other classes of materials may show a similar behavior as well.

Raman microspectroscopy is the fastest and most powerful tool for analysis of phase transformations in contact loading. It can additionally provide information on residual stresses and/or chemical changes in the surface layers. However, limited databases of Raman spectra and difficulties with the interpretation of Raman spectra, as well as low accuracy of existing predictive tools for calculations of Raman spectra of solids, make it necessary to complement Raman data with electron or X-ray diffraction studies. Fourier-transform infrared microspectroscopy is another technique that can provide useful information on structural and compositional changes in the surface layer.

Studies of phase transformations induced by contact loading may help to better understand such a basic phenomenon as materials hardness. Information on the phase transformations in the surface layer of materials upon contact interactions is very important for understanding the mechanisms of wear and contact damage that occur in many industrial processes. Currently, only mechanical or tribochemical processes are considered responsible for damage of sliding parts. Similar problems exist in electronics, where local pressures of up to 10 GPa were measured under a spreading resistance probe. Better understanding of interactions between the tool and the machined material may lead to improved machining techniques and help to identify regimes for ductile machining of brittle materials. Studies of phase transformations in the surface layer of materials resulting from contact interactions can give a new insight into these technologically important problems.

## Acknowledgments

Our research in this area was supported by the National Science Foundation, grant Nos. DMR-9874955 and DMI-9813257. V. Domnich was also supported by a UIC Fellowship and Dean's Scholar Award.

Y. Gogotsi is thankful to his former students and postdoctoral scientists at the University of Illinois at Chicago including C. Baek, M. Gardner, M. Rosenberg, G. Zhou, and Dr. A. Kovalchenko, as well as to his collaborators Dr. A. Kailer and Prof. K.G. Nickel of the University of Tübingen, Germany; Prof. M. Trenary of UIC; Dr. F. Kirscht of Mitsubishi Silicon America, Salem, Oregon; Dr. S. Dub of the Institute of Superhard Materials, Kiev, Ukraine; Dr. B.A. Galanov and Dr. O.N. Grigoriev of the Institute for Materials Science, Kiev, Ukraine; and Ms. L. Riester and Dr. K. Breder of the Oak Ridge National Laboratory. Helpful discussions with Prof. J.J. Gilman of the University of California at Los Angeles and Prof. Y.V. Milman of the Institute for Materials Science, Kiev, Ukraine, are greatly appreciated.

## References

1. V. V. Brazhkin and A. G. Lyapin, *High Pressure Res.* 15, 9 (1996).
2. R. W. Cahn, *Nature* 357, 645 (1992).
3. J. J. Gilman, *Czech J. Phys.* 45, 913 (1995).
4. Y. G. Gogotsi, A. Kailer, and K. G. Nickel, *Materials Research Innovations* 1, 3 (1997).
5. I. V. Gridneva, Y. V. Milman, and V. I. Trefilov, *Phys. Stat. Sol.* (a) 9, 177 (1972).
6. D. R. Clarke, M. C. Kroll, P. D. Kirchner, R. F. Cook, and B. J. Hockey, *Phys. Rev. Lett.* 60, 2156 (1988).
7. G. M. Pharr, W. C. Oliver, and D. S. Harding, *J. Mater. Res.* 6, 1129 (1991).
8. G. M. Pharr, W. C. Oliver, R. F. Cook, P. D. Kirchner, M. C. Kroll, T. R. Dinger, and D. R. Clarke, *J. Mater. Res.* 7, 961 (1992).
9. E. R. Weppelmann, J. S. Field, and M. V. Swain, *J. Mater. Res.* 8, 830 (1993).
10. J. J. Gilman, *Mat. Res. Soc. Symp. Proc.* 276, 191 (1992).
11. G. M. Pharr, W. C. Oliver, and D. R. Clarke, *Scripta Metall.* 23, 1949 (1989).
12. G. M. Pharr, *Mat. Res. Soc. Symp. Proc.* 239, 301 (1992).
13. J. C. Morris and D. L. Callahan, in "Microstructure of Materials" (K. M. Krishnan, Ed.), p. 104, San Francisco Press, San Francisco, 1992.
14. J. J. Gilman, *J. Mater. Res.* 7, 535 (1992).
15. E. R. Weppelmann, J. S. Field, and M. V. Swain, *J. Mater. Sci.* 30, 2455 (1995).
16. D. L. Callahan and J. C. Morris, *J. Mater. Res.* 7, 1614 (1992).
17. J. C. Morris and D. L. Callahan, *J. Mater. Res.* 9, 2907 (1994).

*18.* J. C. Morris, D. L. Callahan, J. Kulik, J. A. Patten, and R. O. Scattergood, *J. Am. Ceram. Soc.* 78, 2015 (1995).

*19.* A. Jayaraman, *Rev. Modern Phys.* 55, 65 (1983).

*20.* J. R. Ferraro, "Vibrational Spectroscopy at High External Pressures: The Diamond Anvil Cell." Academic Press, Orlando, 1984.

*21.* R. J. Hemley, H. K. Mao, P. M. Bell, and B. O. Mysen, *Phys. Rev. Lett.* 57, 747 (1986).

*22.* K. J. Kingma, C. Meade, R. J. Hemley, H. K. Mao, and D. R. Veblen, *Science* 259, 666 (1993).

*23.* S. Roberts and I. Beattie, in "Microprobe Techniques in the Earth Sciences" (P. J. Potts, J. F. Bowles, S. J. B. Reed, and M. R. Cave, Eds.), p. 387, Chapman & Hall, London, 1995.

*24.* S. K. Sharma, H. K. Mao, P. M. Bell, and J. A. Xu, *J. Raman Spectrosc.* 16, 350 (1985).

*25.* R. G. Sparks and M. A. Raesler, *Prec. Eng.* 10, 191 (1988).

*26.* H. Shen and F. Pollack, *J. Appl. Phys.* 64, 3233 (1988).

*27.* J. F. Di Gregorio and T. E. Furtak, *J. Am. Ceram. Soc.* 75, 1854 (1992).

*28.* G. Lucazeau and L. Abello, *Analusis* 23, 301 (1995).

*29.* D. Tabor, *Philos. Mag.* A 75, 1207 (1996).

*30.* B. R. Lawn and M. V. Swain, *J. Mater. Sci.* 10, 113 (1975).

*31.* B. R. Lawn, *J. Appl. Phys.* 39, 4828 (1968).

*32.* K. L. Johnson, "Contact Mechanics." Cambridge University Press, Cambridge, 1985.

*33.* B. A. Galanov, *Soviet Applied Mechanics* (Prikladnaya Mekhanika) 18, 711 (1983).

*34.* K. L. Johnson, *J. Mech. Phys. Solids* 18, 115 (1970).

*35.* A. C. Fischer-Cripps, *J. Mater. Sci.* 32, 727 (1997).

*36.* K. Tanaka, *J. Mater. Sci.* 22, 1501 (1987).

*37.* B. A. Galanov, O. N. Grigor'ev, and Y. G. Gogotsi, *Mat. Res. Soc. Symp. Proc.* 481, 249 (1998).

*38.* Y. G. Gogotsi, V. Domnich, S. N. Dub, A. Kailer, and K. G. Nickel, *J. Mater. Res.* 15, 871 (2000).

*39.* I. N. Sneddon, *Int. J. Engng. Sci.* 3, 47 (1965).

*40.* W. C. Oliver and G. M. Pharr, *J. Mater. Res.* 7, 1564 (1992).

*41.* M. Sakai, S. Shimizu, and T. Ishikawa, *J. Mater. Res.* 14, 1471 (1999).

*42.* N. V. Novikov, S. N. Dub, Y. V. Milman, I. V. Gridneva, and S. I. Chugunova, *Journal of Superhard Materials* (Sverkhtverdye Materialy) 18, 32 (1996).

*43.* J. S. Field and M. V. Swain, *J. Mater. Res.* 8, 297 (1993).

*44.* E. Anastassakis, A. Pinczuk, and E. Burstein, *Solid State Commun.* 8, 133 (1970).

*45.* S. Ganesan, A. A. Maradudin, and J. Oitmaa, *Ann. Phys.* 56, 556 (1970).

*46.* E. Anastassakis, A. Cantarero, and M. Cardona, *Phys. Rev.* B 41, 7529 (1990).

*47.* I. De Wolf, *Semicond. Sci. Tech.* 11, 139 (1996).

*48.* J. F. Cannon, *J. Phys. Chem. Ref. Data* 3, 781 (1974).

*49.* A. George, in "Properties of Crystalline Silicon" (R. Hull, Ed.), p. 104, INSPEC, the Institution of Electrical Engineers, London, 1999.

*50.* S. Minomura and H. G. Drickamer, *J. Phys. Chem. Solids* 23, 451 (1962).

*51.* J. C. Jamieson, *Science* 139, 762 (1963).

52. J. Z. Hu, L. D. Merkle, C. S. Menoni, and I. L. Spain, *Phys. Rev.* B 34, 4679 (1986).
53. R. H. Wentorf and J. S. Kasper, *Science* 139, 338 (1963).
54. J. S. Kasper and S. H. Richards, *Acta Crystall.* 17, 752 (1964).
55. V. G. Eremenko and V. I. Nikitenko, *Phys. Stat. Sol.* (a) 14, 317 (1972).
56. R. J. Kobliska and S. A. Solin, *Phys. Rev.* B 8, 3799 (1973).
57. J. M. Besson, E. H. Mokhtari, J. Gonzalez, and G. Weill, *Phys. Rev. Lett.* 59, 473 (1987).
58. P. Pirouz, R. Chaim, U. Dahmen, and K. H. Westmacott, *Acta Metall. Mater.* 38, 313 (1990).
59. U. Dahmen, K. H. Westmacott, P. Pirouz, and R. Chaim, *Acta Metall. Mater.* 38, 323 (1990).
60. P. Pirouz, U. Dahmen, K. H. Westmacott, and R. Chaim, *Acta Metall. Mater.* 38, 329 (1990).
61. X. S. Zhao, F. Buehler, J. R. Sites, and I. L. Spain, *Solid State Commun.* 59, 678 (1986).
62. J. Crain, G. J. Ackland, J. R. Maclean, R. O. Piltz, P. D. Hatton, and G. S. Pawley, *Phys. Rev.* B 50, 13043 (1994).
63. R. O. Piltz, J. R. Maclean, S. J. Clark, G. J. Ackland, P. D. Hatton, and J. Crain, *Phys. Rev.* B 52, 4072 (1995).
64. A. Kailer, Y. G. Gogotsi, and K. G. Nickel, *J. Appl. Phys.* 81, 3057 (1997).
65. T. F. Page, W. C. Oliver, and C. J. McHargue, *J. Mater. Res.* 7, 450 (1992).
66. H. Olijnyk, S. K. Sikka, and W. B. Holzaffel, *Phys. Lett.* A 103, 137 (1984).
67. J. Z. Hu and I. L. Spain, *Solid State Commun.* 51, 263 (1984).
68. R. Goswami, S. Sampath, H. Herman, and J. B. Parise, *J. Mater. Res.* 14, 3489 (1999).
69. F. Birch, *Phys. Rev.* 71, 809 (1947).
70. F. Birch, *J. Geophys. Res.* 57, 227 (1952).
71. M. T. Yin and M. L. Cohen, *Phys. Rev.* B 26, 5668 (1982).
72. R. Biswas, R. M. Martin, R. J. Needs, and O. H. Nielsen, *Phys. Rev.* B 30, 3210 (1984).
73. B. G. Pfrommer, M. Cote, S. G. Louie, and M. L. Cohen, *Phys. Rev.* B 56, 6662 (1997).
74. R. Biswas, R. M. Martin, R. J. Needs, and O. H. Nielsen, *Phys. Rev.* B 35, 9559 (1987).
75. J. D. Joannopoulos and M. L. Cohen, *Phys. Rev.* B 7, 2644 (1973).
76. V. I. Trefilov and Y. V. Milman, *Sov. Phys. Dokl.* 8, 1240 (1964).
77. T. Suzuki and T. Ohmura, *Philos. Mag.* A 74, 1073 (1996).
78. J. J. Gilman, *J. Appl. Phys.* 46, 5110 (1975).
79. O. N. Grigor'ev, V. I. Trefilov, and A. M. Shatokhin, *Soviet Powder Metallurgy and Metal Ceramics* (Poroshkovaya Metallurgiya), 1028 (1983).
80. K. Gaál-Nagy, A. Bauer, M. Schmitt, K. Karch, P. Pavone, and D. Strauch, *Phys. Stat. Sol.* (b) 211, 275 (1999).
81. J. J. Gilman, *J. Appl. Phys.* 39, 6086 (1968).
82. T. Figielski, *Phys. Stat. Sol.* (a) 4, 773 (1971).
83. M. J. Hill and D. J. Rowcliffe, *J. Mater. Sci.* 9, 1569 (1974).
84. J. Castaing, P. Veyssiere, L. P. Kubin, and J. Rabier, *Philos. Mag.* A 44, 1407 (1981).

85. A. P. Gerk and D. Tabor, *Nature* 271, 732 (1978).
86. M. C. Gupta and A. L. Ruoff, *J. Appl. Phys.* 51, 1072 (1980).
87. J. J. Gilman, *Philos. Mag.* B 67, 207 (1993).
88. Y. Q. Wu and Y. B. Xu, *J. Mater. Res.* 14, 682 (1999).
89. Y. Q. Wu, X. Y. Yang, and Y. B. Xu, *Acta Mater.* 47, 2431 (1999).
90. O. Shimomura, S. Minomura, N. Sakai, K. Asaumi, K. Tamura, J. Fukushima, and H. Endo, *Philos. Mag.* 29, 547 (1974).
91. A. B. Mann, D. van Heerden, J. B. Pethica, and T. P. Weihs, *J. Mater. Res.* 15, 1754 (2000).
92. J. E. Bradby, J. S. Williams, J. Wong-Leung, M. V. Swain, and P. Munroe, *Appl. Phys. Lett.* 77, 3749 (2000).
93. R. A. Cowley, in "The Raman Effect" (A. Anderson, Ed.), Vol. 2, p. 95, Marcel Dekker, Inc., New York, 1973.
94. J. H. Parker, D. W. Feldman, and M. Ashkin, *Phys. Rev.* 155, 712 (1967).
95. X. S. Zhao, Y. R. Ge, J. Schroeder, and P. D. Persans, *Appl. Phys. Lett.* 65, 2033 (1994).
96. R. J. Nemanich, S. A. Solin, and R. M. Martin, *Phys. Rev.* B 23, 6348 (1981).
97. I. H. Campbell and P. M. Fauchet, *Solid State Commun.* 58, 739 (1986).
98. S. H. Chen, *Phys. Rev.* 163, 532 (1967).
99. K. J. Chang and M. L. Cohen, *Phys. Rev.* B 31, 7819 (1985).
100. S. P. Lewis and M. L. Cohen, *Phys. Rev.* B 48, 3646 (1993).
101. H. Olijnyk, *Phys. Rev. Lett.* 68, 2232 (1992).
102. R. J. Needs and R. M. Martin, *Phys. Rev. B* 30, 5390 (1984).
103. D. Knight and W. B. White, *J. Mater. Res.* 4, 385 (1989).
104. D. Weaire, M. F. Thorpe, R. Alben, and S. Goldstein, in "Eleventh International Conference on the Physics of Semiconductors," p. 470, Polish Scientific, Warsaw, 1972.
105. R. Alben, M. F. Thorpe, D. Weaire, and S. Goldstein, unpublished; cited in Reference [56].
106. G. Weill, J. L. Mansot, G. Sagon, C. Carlone, and J. M. Besson, *Semicond. Sci. Tech.* 4, 280 (1989).
107. R. J. Kobliska, S. A. Solin, M. Selders, R. K. Chang, R. Alben, M. F. Thorpe, and D. Weaire, *Phys. Rev. Lett.* 29, 725 (1972).
108. M. Hanfland and K. Syassen, *High Pressure Res.* 3, 242 (1990).
109. H. Olijnyk and A. P. Jephcoat, *Phys. Stat. Sol.* (b) 211, 413 (1999).
110. K. Winer and F. Wooten, *Phys. Stat. Sol.* (b) 136, 519 (1986).
111. M. H. Brodsky, in "Light Scattering in Solids I" (M. Cardona, Ed.), Vol. 8, p. 205, Springer, Berlin, 1983.
112. J. E. Smith, Jr., M. H. Brodsky, B. L. Crowder, M. I. Nathan, and A. Pinczuk, *Phys. Rev. Lett.* 26, 642 (1971).
113. N. Maley, D. Beeman, and J. S. Lannin, *Phys. Rev.* B 38, 10611 (1988).
114. M. Marinov and N. Zotov, *Phys. Rev.* B 55, 2939 (1997).
115. J. S. Lannin, in "Semiconductors and Semimetals" (R. K. Willardson and A. C. Beer, Eds.), Vol. 21, p. 159, Academic Press, New York, 1984.
116. G. Lucazeau and L. Abello, *J. Mater. Res.* 12, 2262 (1997).
117. M. T. Yin, *Phys. Rev.* B 30, 1773 (1984).

*118.* Y. Gogotsi and V. Domnich, to be published.

*119.* A. Kailer, Y. G. Gogotsi, and K. G. Nickel, in "International Conference Micro-Materials '97" (B. Michel and T. Winkler, Eds.), Springer, Berlin, 1997.

*120.* J. B. Pethica, R. Hutchings, and W. C. Oliver, *Philos. Mag.* A 48, 593 (1983).

*121.* S. V. Hainsworth, A. J. Whitehead, and T. F. Page, in "Plastic Deformation of Ceramics" (R. C. Bradt, C. A. Brookes, and J. L. Routbort, Eds.), p. 173, Plenum Press, New York, 1995.

*122.* J. S. Williams, Y. Chen, J. Wong-Leung, A. Kerr, and M. V. Swain, *J. Mater. Res.* 14, 2338 (1999).

*123.* V. Domnich, Y. Gogotsi, and S. Dub, *Appl. Phys. Lett.* 76, 2214 (2000).

*124.* M. Bowden and D. Gardiner, *Appl. Spectrosc.* 51, 1405 (1997).

*125.* Y. Gogotsi and M. A. Gardner, *The Americas Microscopy and Analysis*, March, 2000.

*126.* C. S. Menoni, J. Z. Hu, and I. L. Spain, *Phys. Rev.* B 34, 362 (1986).

*127.* S. B. Qadri, E. F. Skelton, and A. W. Webb, *J. Appl. Phys.* 54, 3609 (1983).

*128.* F. P. Bundy and J. S. Kasper, *Science* 139, 340 (1963).

*129.* R. J. Nelmes, M. I. McMahon, N. G. Wright, D. R. Allan, and J. S. Loveday, *Phys. Rev.* B 48, 9883 (1993).

*130.* C. H. Bates, F. Dachille, and R. Roy, *Science* 147, 860 (1965).

*131.* S. Minomura, *J. Phys. (Paris) Colloq.* 42, C4 (1981).

*132.* A. G. Lyapin, V. V. Brazhkin, S. V. Popova, and A. V. Sapelkin, *Phys. Stat. Sol.* (b) 198, 481 (1996).

*133.* Y. Gogotsi, M. S. Rosenberg, A. Kailer, and K. G. Nickel, in "Tribology Issues and Opportunities in MEMS" (B. Bhushan, Ed.), p. 431, Kluwer, 1998.

*134.* A. Kailer, K. G. Nickel, and Y. G. Gogotsi, *J. Raman Spectrosc.* 30, 939 (1999).

*135.* G. M. Pharr, W. C. Oliver, and D. R. Clarke, *J. Elec. Mater.* 19, 881 (1990).

*136.* J. R. Chelikowsky, *Phys. Rev.* B 34, 5295 (1986).

*137.* M. Bukowinski, *Nature* 398, 372 (1999).

*138.* A. L. Ruoff and T. Li, *Ann. Rev. Mater. Sci.* 25, 249 (1991), and references therein.

*139.* S. Froyen and M. L. Cohen, *Phys. Rev.* B 28, 3258 (1983).

*140.* S.-H. Wei and H. Krakauer, *Phys. Rev. Lett.* 55, 1200 (1985).

*141.* S. B. Zhang and M. L. Cohen, *Phys. Rev.* B 35, 7604 (1989).

*142.* Z. W. Lu, D. Singh, and H. Krakauer, *Phys. Rev.* B 39, 10154 (1989).

*143.* J. Crain, R. O. Piltz, G. J. Ackland, S. J. Clark, M. C. Payne, V. Milman, J. S. Lin, P. D. Hatton, and Y. H. Nam, *Phys. Rev.* B 50, 8389 (1994).

*144.* R. J. Nelmes, M. I. McMahon, N. G. Wright, and D. R. Allan, *J. Phys. Chem. Solids* 56, 545 (1995).

*145.* R. J. Nelmes, M. I. McMahon, N. G. Wright, D. R. Allan, H. Liu, and J. S. Loveday, *J. Phys. Chem. Solids* 56, 539 (1995).

*146.* M. I. McMahon and R. J. Nelmes, *Phys. Stat. Sol.* (b) 198, 389 (1996).

*147.* R. J. Nelmes and M. I. McMahon, *Phys. Rev. Lett.* 79, 3668 (1997).

*148.* V. Ozoliņš and A. Zunger, *Phys. Rev. Lett.* 82, 767 (1999).

*149.* A. Mujica, R. J. Needs, and A. Muñoz, *Phys. Rev.* B 52, 8881 (1995).

*150.* A. Mujica, R. J. Needs, and A. Muñoz, *Phys. Stat. Solidi* (b) 198, 461 (1996).

*151.* A. Mujica and R. J. Needs, *Phys. Rev.* B 55, 9659 (1997).

*152.* T. Huang and A. L. Ruoff, in "Proc. of 9th AIPAPT Conf." (C. Homan, R. K. Maccone, and E. Whaley, Eds.), Vol. 22, p. 37, North-Holland, New York, 1984.

153. M. I. McMahon and R. J. Nelmes, *Phys. Rev.* B 78, 3697 (1997).

154. M. I. McMahon, R. J. Nelmes, D. R. Allan, S. A. Belmonte, and T. Bovornratanaraks, *Phys. Rev. Lett.* 80, 5564 (1998).

155. A. A. Kelsey, G. J. Ackland, and S. J. Clark, *Phys. Rev.* B 57, R2029 (1998).

156. A. Mujica, A. Muñoz, and R. J. Needs, *Phys. Rev.* B 57, 1344 (1998).

157. S. T. Weir, Y. K. Vohra, C. A. Vanderborgh, and A. L. Ruoff, *Phys. Rev.* B 39, 1280 (1989).

158. A. Mujica and R. J. Needs, *J. Phys.: Condens. Matter* 8, L237 (1996).

159. E. Le Bourhis and G. Patriarche, *Philos. Mag. Lett.* 79, 805 (1999).

160. J. M. Besson, J. P. Itie, A. Polian, G. Weill, J. L. Masot, and J. Gonzalez, *Phys. Rev.* B 44, 4214 (1991).

161. R. W. Stimets, J. Waldman, J. Lin, T. S. Chang, R. J. Temkin, and G. A. N. Connell, in "Amorphous and Liquid Semiconductors" (J. Stuke and W. Brenig, Eds.), Vol. 2, p. 1239, Taylor & Francis, London, 1974.

162. A. Kailer, "Locale Schädigung von Oberflächen und Phasenumwandlungen in harten, spröden Materialien verursacht durch mechanischen Kontakt." Eberhard-Karls Univ., Tübingen, 1998.

163. T. Soma, *Phys. Stat. Solidi* (b) 99, 701 (1980).

164. U. D. Venkateswaran, L. J. Cui, B. A. Weinstein, and F. A. Chambers, *Phys. Rev.* B 43, 1875 (1991).

165. R. J. Nelmes, M. I. McMahon, P. D. Hatton, J. Crain, and R. O. Piltz, *Phys. Rev.* B 47, 35 (1993).

166. M. D. Banus and M. C. Lavine, *J. Appl. Phys.* 16, 495 (1969).

167. R. J. Nelmes and M. I. McMahon, *Phys. Rev. Lett.* 77, 663 (1996).

168. M. Mezouar, J. M. Besson, G. Syfosse, J. P. Itie, D. Häusermann, and M. Hanfland, *Phys. Stat. Sol.* (b) 198, 403 (1996).

169. P. Feltham and R. Banerjee, *J. Mater. Sci.* 27, 1626 (1992).

170. A. Garcia and M. L. Cohen, *Phys. Rev.* B 47, 4215 (1993).

171. L. G. Berry and R. M. Thompson, *Geol. Soc. Amer. Mem.* 1, 26 (1962).

172. F. P. Bundy and J. S. Kasper, *J. Chem. Phys.* 46, 3437 (1967).

173. P. P. Edwards, T. V. Ramakrishnan, and C. N. R. Rao, in "Metal-Insulator Transitions Revisited" (P. P. Edwards and C. N. Rao, Eds.), p. xv, Taylor & Francis, London, 1995.

174. A. L. Ruoff, H. Luo, and Y. K. Vohra, *J. Appl. Phys.* 69, 6413 (1991).

175. M. P. Surh, S. G. Louie, and M. L. Cohen, *Phys. Rev.* B 45, 8239 (1995).

176. F. Gygi and A. Baldereschi, *Phys. Rev. Lett.* 62, 2160 (1989).

177. A. L. Ruoff and H. Luo, *J. Appl. Phys.* 70, 2066 (1991).

178. O. H. Nielsen, *Phys. Rev.* B 34, 5808 (1986).

179. P. E. Van Camp, V. E. Van Doren, and J. T. Devreese, *Solid State Commun.* 84, 731 (1992).

180. M. T. Yin and M. L. Cohen, *Phys. Rev. Lett.* 50, 2006 (1983).

181. S. J. Clark, G. J. Ackland, and J. Crain, *Phys. Rev.* B 52, 15035 (1995).

182. A. K. McMahan, *Phys. Rev.* B 30, 5935 (1984).

183. S. Scandolo, G. L. Chiarotti, and E. Tosatti, *Phys. Stat. Sol.* (b) 198, 447 (1996).

184. L. F. Vereshchagin, E. H. Yakovlev, G. N. Stepanov, and B. V. Vinogradov, *JETP Lett.* 16, 270 (1972).

185. L. F. Vereshchagin, E. H. Yakovlev, B. V. Vinogradov, V. P. Sakum, and G. N. Stepanov, *JETP Lett.* 17, 301 (1973).
186. L. F. Vereshchagin, E. H. Yakovlev, K. K. Bibaev, and B. V. Vinogradov, *JETP Lett.* 16, 169 (1972).
187. A. L. Ruoff, *High Pressure Res.* 8, 639 (1992).
188. Y. K. Vohra, T. S. McCauley, and S. S. Vagarali, in "Advances in New Diamond Science and Technology" (S. Saito, N. Fujimori, O. Fukugawa, M. Kamo, K. Kobashi, and M. Yoshikawa, Eds.), p. 391, MYU, Tokyo, 1994.
189. I. J. McHolm, "Ceramic Hardness." Plenum, New York, 1990.
190. Y. V. Milman, in "Materials Science of Carbides, Nitrides and Borides" (Y. G. Gogotsi and R. A. Andrievski, Eds.), p. 223, Kluwer, Dordrecht, 1999.
191. Y. V. Milman, *Problems of Strength* 1, 52 (1990).
192. H. Li and R. C. Bradt, *Diam. Relat. Mater.* 1, 1161 (1992).
193. N. V. Novikov and S. N. Dub, *J. Hard Mater.* 2, 3 (1991).
194. V. N. Bakul, M. G. Loshak, and V. I. Mal'nev, *Sint. Almazy* 6, 16 (1973).
195. C. A. Brookes, *Nature* 228, 660 (1970).
196. Y. G. Gogotsi, A. Kailer, and K. G. Nickel, *J. Appl. Phys.* 84, 1299 (1998).
197. H. K. Mao and R. J. Hemley, *Nature* 351, 721 (1991).
198. I. Pócsik, M. Hundhausen, M. Koós, and L. Ley, *J. Non-Cryst. Solids* 227–230, 1083 (1998).
199. Y. G. Gogotsi, A. Kailer, and K. G. Nickel, *Nature* 401, 663 (1999).
200. N. Wada, P. J. Gaczi, and S. A. Solin, *J. Non-Cryst. Solids* 35/36, 543 (1980).
201. S. A. Solin and R. J. Kobliska, in "Amorphous and Liquid Semiconductors" (J. Stuke and W. Brenig, Eds.), Vol. 2, p. 1252, Taylor & Francis, London, 1974.
202. K. E. Spear, A. W. Phelps, and W. B. White, *J. Mater. Res.* 5, 2277 (1990).
203. S. Prawer, K. W. Nugent, and D. N. Jamieson, *Diam. Relat. Mater.* 7, 106 (1998).
204. D. Beeman, J. Silverman, R. Lynds, and M. R. Anderson, *Phys. Rev.* B 30, 870 (1984).
205. C. Z. Wang and K. M. Ho, *Phys. Rev. Lett.* 71, 1184 (1993).
206. J. L. Hoard and R. E. Hughes, in "The Chemistry of Boron and Its Compounds" (E. L. Muetterties, Ed.), p. 26, Wiley, New York, 1967.
207. T. B. Massalski, "Binary Alloy Phase Diagrams", American Society for Metals, Metals Park, OH, 1986.
208. D. Emin, *Phys. Today* 20, 55 (1987).
209. R. J. Nelmes, J. S. Loveday, R. M. Wilson, W. G. Marshall, J. M. Besson, S. Klotz, G. Hamel, T. L. Aselage, and S. Hull, *Phys. Rev. Lett.* 74, 2268 (1995).
210. R. Lazzari, N. Vast, J. M. Besson, S. Baroni, and A. Dal Corso, *Phys. Rev. Lett.* 83, 3230 (1999).
211. P. S. Kisly, M. A. Kuzenkova, N. I. Bodnaruk, and B. L. Grabchuk, "Karbid Bora." Naukova Dumka, Kiev, 1988.
212. F. Thevenot, in "Properties of Ceramics" (G. de With, R. A. Terpstra, and R. Metselaar, Eds.), Vol. 2, p. 2.1, Elsevier Appl. Sci., London and New York, 1989.
213. T. F. Page, L. Riester, and S. V. Hainsworth, *Mat. Res. Soc. Symp. Proc.* 522, 113 (1998).
214. S. J. Bull, T. F. Page, and E. H. Yoffe, *Philos. Mag. Lett.* 59, 281 (1989).

215. U. Kuhlmann and H. Werheit, *J. Alloy. Compd.* 205, 87 (1994).
216. T. L. Aselage, D. R. Tallant, and D. Emin, *Phys. Rev.* B 56, 3122 (1997).
217. K. Shirai and S. Emura, *J. Phys.: Condens. Matter* 8, 10919 (1996).
218. A. Addamiano, in "Silicon Carbide – 1973" (R. C. Marshall, J. J. W. Faust, and C. E. Ryan, Eds.), University of South Carolina Press, Columbia, 1974.
219. K. J. Chang and M. L. Cohen, *Phys. Rev.* B 35, 8196 (1987).
220. K. Karch, F. Bechstedt, P. Pavone, and D. Strauch, *Phys. Rev.* B 53, 13400 (1996).
221. M. Yoshida, A. Onodera, M. Ueno, K. Takemura, and O. Shimomura, *Phys. Rev.* B 48, 10587 (1993).
222. W. H. Gust, A. C. Halt, and E. B. Royce, *J. Appl. Phys.* 44, 550 (1973).
223. T. Sekine and T. Kobayashi, *Phys. Rev.* B 55, 8034 (1997).
224. J. Lankford, *J. Mater. Sci.* 18, 1666 (1983).
225. X.-J. Ning, N. Huvey, and P. Pirouz, *J. Am. Ceram. Soc.* 80, 1645 (1997).
226. K. Karch, P. Pavone, W. Windl, O. Shütt, and D. Strauch, *Phys. Rev.* B 50, 17054 (1994).
227. R. B. Sosman, "The Phases of Silica." Rutgers Univ. Press, New Brunswick, NJ, 1965.
228. R. J. Hemley, in "High-Pressure Research in Mineral Physics" (M. H. Manghnani and Y. Syono, Eds.), p. 347, American Geophysical Union, Washington, DC, 1987.
229. K. J. Kingma, R. J. Hemley, H. K. Mao, and D. R. Veblen, *Phys. Rev.* B 70, 3927 (1993).
230. K. J. Kingma, H. K. Mao, and R. J. Hemley, *High Press. Res.* 14, 363 (1996).
231. J. Tse, D. D. Klug, Y. Le Page, and M. Bernaskoni, *Phys. Rev.* B 56, 10878 (1997).
232. N. N. Ovsyuk and A. V. Goryainov, *Phys. Rev.* B 60, 14481 (1999).
233. R. M. Wentzcovitch, C. da Silva, J. R. Chelikowsky, and N. Binggeli, *Phys. Rev. Lett.* 80, 2149 (1998).
234. D. M. Teter, R. J. Hemley, G. Kresse, and J. Hafner, *Phys. Rev. Lett.* 80, 2145 (1998).
235. M. Grimsditch, *Phys. Rev. Lett.* 52, 2379 (1984).
236. P. McMillan, B. Piriou, and R. Couty, *J. Chem. Phys.* 81, 4234 (1984).
237. F. L. Galeener, A. J. Leadbetter, and M. W. Stringfellow, *Phys. Rev.* B 27, 1052 (1983).
238. R. J. Bell, N. F. Bird, and P. Dean, *J. Phys.* C 1, 299 (1968).
239. F. L. Galeener, *Phys. Rev.* B 19, 4292 (1979).
240. S. K. Sharma, J. F. Mammone, and M. F. Nikol, *Nature* 292, 140 (1981).
241. F. L. Galeener, *Solid State Commun.* 44, 1037 (1982).
242. C. R. Kurkjian, G. W. Kammlott, and M. M. Chaudhri, *J. Am. Ceram. Soc.* 78, 737 (1995).
243. C. C. Ferguson, G. E. Lloyd, and R. J. Knipe, *Can. J. Earth Sci.* 24, 544 (1987).
244. E. H. Kisi and C. J. Howard, in "Zirconia Engineering Ceramics: Old Challenges, New Ideas" (E. Kisi, Ed.), p. 1, Trans. Tech. Publications, Zurich, Switzerland, 1998.
245. H. G. Scott, *J. Mater. Sci.* 10, 1527 (1975).
246. R. H. J. Hannink, P. M. Kelly, and B. C. Muddle, *J. Am. Ceram. Soc.* 83, 461 (2000).
247. O. Ohtaka, T. Yamanaka, S. Kume, N. Hara, H. Asano, and F. Izumi, *Proc. Japan. Acad.* Ser. B 66, 193 (1990).
248. R. C. Garvie, R. H. J. Hannink, and R. T. Pascoe, *Nature* 258, 703 (1975).

249. G. Behrens, G. W. Dransmann, and A. H. Heuer, *J. Am. Ceram. Soc.* 76, 1025 (1993).
250. G. Behrens and A. H. Heuer, *J. Am. Ceram. Soc.* 79, 895 (1996).
251. M. Suganuma, *J. Am. Ceram. Soc.* 82, 3113 (1999).
252. E. Anastassakis, B. Papanicolaou, and I. M. Asher, *J. Phys. Chem. Solids* 36, 667 (1975).
253. A. P. Mirgorodsky, M. B. Smirnov, and P. E. Quintard, *Phys. Rev. B* 55, 19 (1997).
254. C. M. Phillippi and K. S. Mazdiyasni, *J. Am. Ceram. Soc.* 54, 254 (1971).
255. C. G. Kontoyannis and M. Orkoula, *J. Mater. Sci.* 29, 5316 (1994).
256. D. J. Kim, H. J. Jung, and I. S. Yang, *J. Am. Ceram. Soc.* 76, 2106 (1993).
257. G. Morell, R. S. Katiyar, D. Torres, S. E. Paje, and J. Llopis, *J. Appl. Phys.* 81, 2830 (1997).
258. A. Celli, T. Gutema, P. Bracali, R. Gropetti, L. Esposito, and A. Tucci, *Am. Ceram. Soc. Bull.* August, 1999.
259. B. Lawn and R. Wilshaw, *J. Mater. Sci.* 10, 1049 (1975).
260. G. M. Hamilton and L. E. Goodman, *J. Appl. Mech.* 33, 371 (1966).
261. B. R. Lawn, *Proc. Roy. Soc. Lond.* A279, 307 (1967).
262. P. N. Blake and R. O. Scattergood, *J. Am. Ceram. Soc.* 73, 949 (1990).
263. W. S. Blackley and R. O. Scattergood, *J. Eng. Ind.* 116, 263 (1994).
264. K. Minowa and K. Sumino, *Phys. Rev. Lett.* 69, 320 (1992).
265. B. N. Tanikella, A. H. Somasekhar, A. T. Sowers, R. J. Nemanich, and R. O. Scattergood, *Appl. Phys. Lett.* 69, 2870 (1996).
266. Y. Gogotsi, G. Zhou, S. S. Ku, and S. Cetincunt, *Semicond. Sci. Tech.* 16, 345 (2001).
267. P. S. Pizani, F. Lanciotti, Jr., R. G. Jasinevicius, J. G. Duduch, and A. J. V. Porto, *J. Appl. Phys.* 87, 1280 (2000).
268. S. Webster, D. N. Batchelder, and D. A. Smith, *Appl. Phys. Lett.* 72, 1478 (1998).
269. R. Komanduri, N. Chandrasekaran, and L. M. Raff, *Wear* 219, 84 (1998).
270. D. C. Thompson and J. A. Patten, 1998 ASPE Spring Topical Meeting on Silicon Machining, ASPE, Carmel-by-the-Sea, CA, 1998.
271. T. G. Bifano, T. A. Dow, and R. O. Scattergood, *J. Eng. Ind.* 113, 184 (1991).
272. K. E. Puttick, C. Jeynes, M. Rudman, A. E. Gee, and C. L. Chao, *Semicond. Sci. Tech.* 7, 255 (1992).
273. J. Verhey, U. Bismayer, B. Guttler, and H. Lundt, *Semicond. Sci. Tech.* 9, 404 (1994).
274. R. R. Kunz, H. R. Clark, P. M. Nitishin, M. Rothschild, and B. S. Ahern, *J. Mater. Res.* 11, 1228 (1996).
275. Y. Gogotsi, C. Baek, and F. Kirscht, *Semicond. Sci. Tech.* 14, 936 (1999).
276. L. Q. Chen, X. Zhang, T. Y. Zhang, H. Y. Lin, and S. Lee, *J. Mater. Res.* 15, 1441 (2000).
277. Y. Gogotsi and V. Domnich, in "NSF Design & Manufacturing Research Conference," Vancouver, Canada, 2000 (CD-ROM).

# Index

## A

Absorbed hydrocarbon polymers, 123
Acceleration, 7
Actinolite, 164
Adatoms, 19
Adsorbate layers, in STM, 81
Adventitious carbon (AC), 121, 135
   general, 123
   sputtering effect, 123
Aggregation, 306–308
Ag/Si(111), 56
Al($2p$) chemistry, 148
Al—Ti—N systems, ESCA or XPS studies, 146
Alloy surfaces, in STM, 73
Al oxidation, binding energies, 133
Alumina
   binding energy, 131
   bulk or thick forms, 131
   interpretation of different forms, 131
Alumina films, 131
Aluminosilicates, 159
   ESCA early studies, 157
Aluminum, 131
   mixed structures, 160
   single type structures, 158
Amorphization, 356
Amphiboles, 162
Angle of incidence, 18
Angular profiles, 35
Annealing time, 381
Anthophyllite, 164
Applied areas, solid surfaces in, 269
ARGUS, 15
Atomic force microscopy (AFM), 1–58, 58–89, 89–101
Atomic structure, determination of parameters, MVV SEFS, 256–259
Auger capture, 15

Auger deexcitation, 15
Auger electron spectra, 51
Auger electron spectroscopy (AES), 24, 40, 56, 113, 141
Au/Si(111), 57
Autodetachment, 15
Autoionization, 15
Average contact pressures (ACP), 392
   during indentation, 392
   in boron carbide, 410
Azimuthal angle, 19

## B

B—N—Ti system, 141
   ESCA or XPS results, 141
Beam collimator, 7
Benzophenone, 303
Berkovich indentations, in boron carbide, 409
Berkovich nanoindentations, 384
Binding energy (BE), 114
   determinations, brief summary, 135
   initial vs final state, 115
   interpretation of, 137
   referencing, 135, 136
   representative Ti—B—N deposition system, 144
   Ti + BN, 146
Bond breaking, 2
Boron carbide, 408
   high pressure behavior, 408
Botany, ESCA studies, 177
Bragg equation, 27
Bragg peaks, 30
Bulk oxides, ESCA analysis, 136

447

ISBN 0-12-475985-8